ALLE ZEIT WACH
1842

Springer Series in Synergetics

Editor: Hermann Haken

Synergetics, an interdisciplinary field of research, is concerned with the cooperation of individual parts of a system that produces macroscopic spatial, temporal or functional structures. It deals with deterministic as well as stochastic processes.

Volume 1 **Synergetics** An Introduction 3rd Edition
By H. Haken

Volume 2 **Synergetics** A Workshop Editor: H. Haken

Volume 3 **Synergetics** Far from Equilibrium Editors: A. Pacault and C. Vidal

Volume 4 **Structural Stability in Physics** Editors: W. Güttinger and H. Eikemeier

Volume 5 **Pattern Formation by Dynamic Systems and Pattern Recognition**
Editor: H. Haken

Volume 6 **Dynamics of Synergetic Systems** Editor: H. Haken

Volume 7 **Problems of Biological Physics** By L. A. Blumenfeld

Volume 8 **Stochastic Nonlinear Systems** in Physics, Chemistry, and Biology
Editors: L. Arnold and R. Lefever

Volume 9 **Numerical Methods in the Study of Critical Phenomena**
Editors: J. Della Dora, J. Demongeot, and B. Lacolle

Volume 10 **The Kinetic Theory of Electromagnetic Processes** By Yu. L. Klimontovich

Volume 11 **Chaos and Order in Nature** Editor: H. Haken

Volume 12 **Nonlinear Phenomena in Chemical Dynamics** Editors: C. Vidal and A. Pacault

Volume 13 **Handbook of Stochastic Methods** for Physics, Chemistry and the Natural Sciences
By C. W. Gardiner

Volume 14 **Concepts and Models of a Quantitative Sociology**
The Dynamics of Interacting Populations By W. Weidlich and G. Haag

Volume 15 **Noise-Induced Transitions** Theory and Applications in Physics, Chemistry, and Biology By W. Horsthemke and R. Lefever

Volume 16 **Physics of Bioenergetic Processes** By L. A. Blumenfeld

Volume 17 **Evolution of Order and Chaos** in Physics, Chemistry, and Biology
Editor: H. Haken

Volume 18 **The Fokker-Planck Equation** By H. Risken

Volume 19 **Chemical Oscillations, Waves, and Turbulence** By Y. Kuramoto

Volume 20 **Advanced Synergetics** By H. Haken

Volume 21 **Stochastic Phenomena and Chaotic Behaviour in Complex Systems**
Editor: P. Schuster

Volume 22 **Synergetics – From Microscopic to Macroscopic Order** Editor: E. Frehland

Volume 23 **Synergetics of the Brain** Editors: E. Başar, H. Flohr, H. Haken, and A. J. Mandell

Volume 24 **Chaos and Statistical Methods** Editor: Y. Kuramoto

Chaos and Statistical Methods

Proceedings of the Sixth Kyoto Summer Institute,
Kyoto, Japan
September 12 – 15, 1983

Editor: Y. Kuramoto

With 123 Figures

Springer-Verlag
Berlin Heidelberg New York Tokyo 1984

Professor Dr. Yoshiki Kuramoto
Research Institute for Fundamental Physics, Yukawa Hall, Kyoto University
Kyoto 606, Japan

Series Editor:
Professor Dr. Dr. h. c. Hermann Haken
Institut für Theoretische Physik der Universität Stuttgart, Pfaffenwaldring 57/IV,
D-7000 Stuttgart 80, Fed. Rep. of Germany

ISBN 3-540-13156-6 Springer-Verlag Berlin Heidelberg New York Tokyo
ISBN 0-387-13156-6 Springer-Verlag New York Heidelberg Berlin Tokyo

Offset printing: Beltz Offsetdruck, Hemsbach. Bookbinding: J. Schäffer OHG, Grünstadt
2153/3130-543210

Preface

The 6th Kyoto Summer Institute devoted to "Chaos and Statistical Mechanics" was held from September 12 to 15, 1983, at the Research Institute for Mathematical Sciences, Kyoto University, and at Hotel Kuniso. The meeting was aimed at clarifying various aspects of chaotic systems appearing in different scientific disciplines, critically examining related mathematical methods developed so far, thus preparing for possible breakthroughs, among others, for the opening of a new period of statistical mechanics of deterministic systems. The number of participants was 135, of which 24 were from abroad. We believe that the well-prepared lecture of each speaker and lively discussions among many participants from various research fields led the meeting to a successful conclusion.

The 6th KSI was organized by the Research Institute for Fundamental Physics. A number of young chaos researchers in Japan also participated actively in the organization. We were also in close contact with the organizer of the IUTAM Symposium on "Turbulence and Chaotic Phenomena in Fluids" (Kyoto Kaikan Conference Hall, Kyoto, September 5-10 1983).

This volume contains most of the lectures presented at the 6th KSI. We are very grateful to all the authors for their efforts in preparing such excellent manuscripts.

The 6th KSI was supported by the Ministry of Education, Science and Culture and the Yamada Science Foundation. The organizing committee acknowledges gratefully their generous financial support. Finally, thanks are due to Dr. M. Toya and Miss T. Sumide for their invaluable assistance.

Kyoto, January, 1984 *Yoshiki Kuramoto*

Opening Address

Ladies and gentlemen,

It is a great pleasure and privilege for me to present the opening address at this 6th Kyoto Summer Institute on Chaos and Statistical Mechanics. On behalf of the Research Institute for Fundamental Physics, Kyoto University, I am delighted to welcome all of you to this international meeting. And I would like to thank all participants, especially those who have traveled far distances to contribute to the meeting. I wish everyone a pleasant stay in Kyoto.

The series of Kyoto Summer Institute, called in short "KSI", started in 1978, and ever since it has been held every year. The topics of KSI in the past varied from year to year, and they ranged from particle physics to condensed matter physics. They were all carefully selected topics so that they could cover most important and active research areas of fundamental physics. This year we selected a topic on chaos. I believe this timely subject is particularly suited to the spirit of KSI. In fact, everyone knows the marvelous progress achieved recently in this area both in theory and experiment. Moreover, the notion of chaos is becoming far more familiar than in the past in various fields of science. This seems to be true not only for such fields as fluid dynamics and statistical mechanics,but also for quite different fields such as condensed matter physics, plasma physics, nuclear physics, chemical reactions and even some branches of biology.

There exists certainly added significance to this 6th KSI, which I would like to mention briefly. As is well known, the Research Institute for Fundamental Physics was founded by the late Dr. Hideki Yukawa, which was exactly 30 years ago. In this connection, we regard the present 6th KSI as one of the most important activities in the celebration program of the 30th anniversary of its foundation. It is gratifying that the 6th KSI has taken up an attractive field of highly interdisciplinary nature, because our institute has, since its foundation, always encouraged and contributed itself to such research fields that looked still premature but might possess great potentiality for future development.

Finally, I hope that this meeting will lead to better understanding of chaos and statistical mechanics through lively discussions among scientists from different parts of the world and from different scientific disciplines.

Before closing my address, I should like to thank the Ministry of Education, Science and Culture and the Yamada Science Foundation for their financial support , and the Research Institute for Mathematical Sciences for kindly providing us with such a nice lecture hall.

With these few words, I now wish to declare the 6th KSI open. Thank you.

Ziro Marki

Director of the
Research Institute
for Fundamental Physics

Contents

Part IV One-Dimensional Mappings

Part V Bifurcations and Normal Forms

Part VI Soliton Systems

Part VII Fluid Dynamics

Part I

General Concepts

Coarse Graining Revisited – The Case of Macroscopic Chaos†

K. Tomita

Department of Physics, Faculty of Science, University of Kyoto
Kyoto 606, Japan

Abstract

The concept of 'coarse graining' in statistical physics is reexamined in relation to the appearance of macroscopic chaos in dynamical systems. Two aspects are pointed out which seem to deserve due consideration.
(i) Concerning the formal aspect, the onset of chaos should be modelled after 'intrinsic coarsening', due essentially to the limitation in self-evaluation. This implies that the 'kinematic' many-body aspect is in fact irrelevant.
(ii) Concerning the physical aspect, the proper description of the macroscopic chaos should be bilateral, i.e.,covering the deterministic aspect as well as the stochastic. This leads naturally to the concept of 'coherent randomness'.

1. Introduction

When a phenomenological (macroscopic) law of motion is derived from the microscopic dynamics, the intuitive term 'coarse graining' is often used to describe the process of information contraction through kinematic projection. The concept is based on that of molecular chaos, and is usually associated with [1] :

i) Many degrees of freedom N on the dynamic level, which are to be reduced to a few degrees of freedom $n(<< N)$ on the phenomenological level, and

ii) Average over the residual degrees of freedom $N-n$, which are to be hidden or contracted by projection.

When chaos appears on the macroscopic level, there are aspects which proceed parallel with the molecular chaos. The concept of 'coarse graining', however, should receive a careful reexamination. The reason is the following.

(i) The macroscopic chaos appears in a system with few degrees of freedom, and

(ii) Introduction of a simplistic average often entails difficulty, e.g., appearance of ordered or coherent motions in a turbulent phase.

In this situation a new point of view is proposed , in which the 'coarseness' in observation is not necessarily identified with the average over residual degrees of freedom.

In conventional statistical mechanics 'coarse graining' appears controllable in principle, for it is a result of kinematic projection or folding of the orbit. In contrast, one here is concerned with an uncontrollable onset of coarsening which is due to nonlinear dynamic folding of the orbits. A new concept 'intrinsic coarsening' is proposed,which is associated with the limited reproducibility due to self-recurrence, rather than the many-body structure of the system to be observed. In order to facilitate the extension of our view analogies are pointed out with situations outside physics, which include the problem of computability on machine and that of provability of the consistency of a formal arithmetic system.

In §2 several examples are discussed which involve 'intrinsic coarsening', i.e., (1) Onset of turbulence (system of differential equations), (2) Ecological system (System of difference equations), (3) Computer System and (4) The formal arithmetic system, and common features shared by these examples are pointed out.

In the following sections the necessity of bilateral description of 'chaos' is pointed out, covering 'regular' as well as 'random' aspects of the situation.

2. Intrinsic Coarsening and Examples

In physics one is concerned with the outcome of observation. Single measurement at one point in space-time, however, may not directly be associated with any systematic prediction. Therefore, a non-trivial observation in physics is bound to be a two-point concept, which is associated with two gate measurements corresponding to preparation and detection of the system. It should then be remembered that a physical preparation can never be precise owing to an uncontrollable noise input from the outside. It is important, therefore, whether this unavoidable noise input affects the later course of the system appreciably or not.

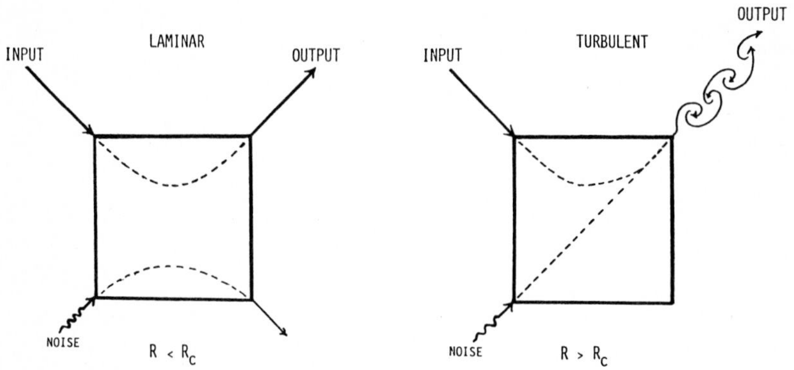

FIG. 1

In a dissipative system there is a tendency for the effect of noise to be damped. Suppose the systematic input is not so large the output is also systematic and reflects the real input character, for the effect of noise is damped out or at least remains at the same order. This enables us to predict the future behavior of the system, given the initial condition. When the systematic input is large enough, however, the effect of noise starts to be amplified, so that the output character becomes very much different from that of the systematic input. Accordingly, an accurate prediction of the future behaviour is impossible, even if the initial condition is known. It is proposed to call this kind of situation 'Intrisic Coarsening' or 'Chaos'.

In what follows several known examples are described with the purpose of finding common features among different kinds of phenomena.

FIG. 2

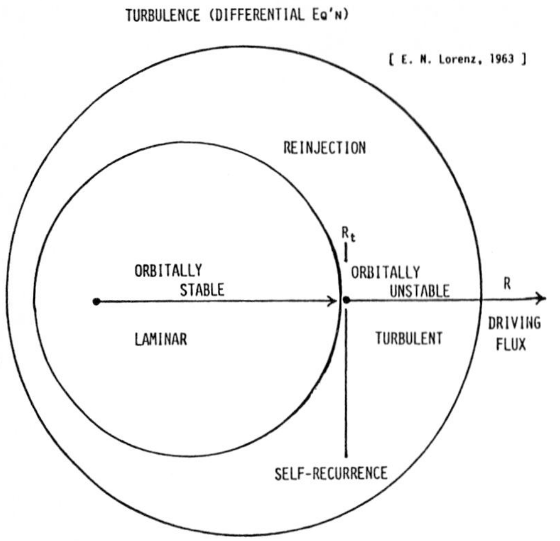

2.1 Onset of Turbulence (System of differential equations)

Let us consider the character of solution of a system of differential equations, keeping the appearance of fluid turbulence in mind. The condition for the onset of turbulence has been discussed in many ways. A simple and abstract way of argument is as follows. When the level of driving flux, or pumping from outside, is increased, at a certain level it may counterbalance the effect of internal damping (the critical point), and beyond that level a new situation arises concerning certain hyperbolic fixed point in a projected phase space. At the critical point the outgoing stream along the unstable manifold of the fixed point begins to flow back along the stable manifold of the same point. However, as the orbit cannot be closed, it flows out again, and repeats an eddy-like reinjection with no exact periodicity. The resulting locus is distributed around itself and the orbital stability which characterizes the laminar phase is not assured any more. The sensitive dependence on the initial condition is considered characteristic of this turbulent phase (cf. Lorenz model [2]). The above formal considerations may apply not only to fluid turbulence, but also to other systems, provided it is described by a set of differential equations, including the noted example of chemical reaction.

2.2 Ecological System (System of Difference Equations)

In order to facilitate the discussion of the solution of differential equations, Poincaré introduced a discrete mapping or a difference equation, which describes the behaviour of the cross-sections of the continuous orbit with a plane having a constant phase. Also there are many problems in ecology which should be described by a difference equation from the outset. A noted example is the logistic equation

$$x_{n+1} = Ax_n(1 - x_n), \qquad (2.1)$$

for which the appearance of solutions with complicated behaviour was first investigated by R. May with respect to the irregular trans-generation behaviour of insect

FIG. 3

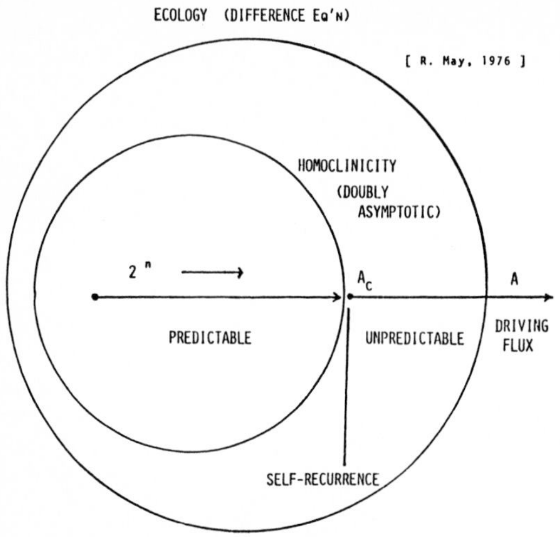

population. [3] The result is summarized as follows. For lower levels of driving force, which is parametrized by A in (2.1), the solution exhibits simple and predictable behaviour, i.e. fixed points or closed orbits. When the force is increased to a certain level, however, a cascade of period doubling appears in a typical case, and the sequence converges at $A = A_C \equiv 3.57\cdots$. Beyond this level the orbit comes back to a certain interval repeatedly, but never to be closed.

In this situation the behaviour is unpredictable, in the sense that the result can never be reproduced even if the same initial condition is adopted physically, for the accuracy is bound to be finite. This is the phenomenon called sensitive dependence on the initial condition.

The above example is of one dimension. When a two-dimensional mapping is considered, however, a closer analogy to the case of differential equations becomes clear. When the driving flux is increased to reach a critical level, a new situation arises in the flow around a hyperbolic point. Namely, there appears a contact between the unstable and the stable manifolds of the same fixed point, i.e. the homoclinic condition. The fixed point is then called doubly asymptotic, which is a different way of describing a self-recurrence. On the appearance of self-recurrence the orbital character of solution becomes extremely complex, and the invariant manifold is often called 'strange attractor'. The behaviour is again unpredictable physically and the situation is similar to the case of turbulence. One may in fact discuss the fluid turbulence in a discrete representation by taking Poincaré section.

2.3 Computer System

In order to obtain insight into the crux of the problem a computer system is invoked in this subsection in its relation to computability. The rôle played by a computer may be described as follows.

i) The computer is expected to 'process' the input data, i.e. precisely to evaluate its content of information, and code the result in the form of algorithms.

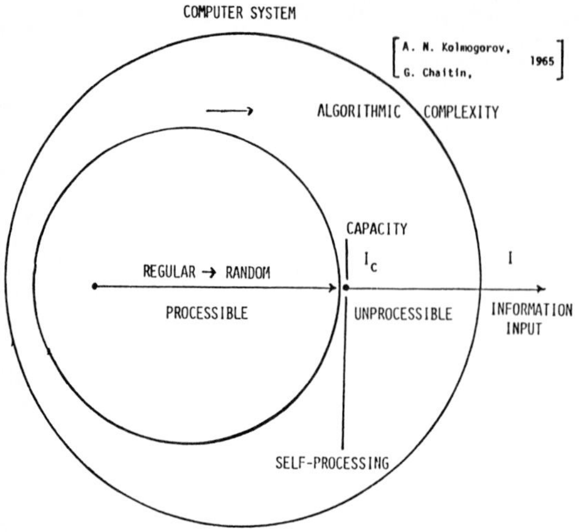

FIG. 4

ii) By way of the algorithm thus obtained it may 'reproduce' the input as an output whenever it is required.

The uniqueness in the expression of information content is secured if one concentrates on the shortest possible programme. The number of bits I contained in the minimal programme is called 'algorithmic complexity' (Kolmogorov 1965 [4], Chaitin 1965 [5]), which provides a measure of the information content, and as such it may be used to classify the input data. Accordingly, this quantity may be taken as an analogue to the driving flux in the foregoing sections.

One may further define a relative complexity by

$$K(I_i) \equiv I/I_i < 1 \qquad \text{(Kolmogorov complexity)} \tag{2.2}$$

with respect to the input number of digits I_i. The smaller the value of $K(I_i)$, the greater the amount of order in the input. For $K(I_i) \sim 1$, however, it is difficult to sort out the regularity. Accordingly, it was proposed by Kolmogorov to define an input as 'random' when $K(I_i) = 1$, i.e. one cannot replace the input by any shorter alternatives. It is known that the relative frequency of occurrence of random signal increases exponentially with I_i. Note that the concept of 'randomness' is inherent to the signal itself, something like irreducibility.

Concerning the above procedure of evaluation, however, one should be aware that the capacity of the machine is finite. It is defined by the maximum number of digits, or bits, I_c, which may be processed by the computer. Given an input data I_i, the computer, with capacity I_c, is asked to evaluate its algorithmic complexity I.

This can be done only if $I < I_i < I_c$. When $I \ll I_c$, there is little problem in the precise evaluation. When $I \lesssim I_i \lesssim I_c$, the input data should be termed highly 'random', and the evaluation of I is not very easy. When I_i reaches the level I_c and beyond, however, a precise evaluation of I becomes impossible, i.e. the input data becomes unprocessible with this computer. Put in other words the marginal situation is reached when the computer is required to 'self-process', and beyond that point precise reproduction of the input as an output cannot be expected any more. This situation has close similarity to that in fluid turbulence, and it may be associated with the term 'chaos'. In contrast, one should remember that the 'randomness' defined above is a concept which belongs to the processible range.

2.4 Formal Arithmetic System

It was G. Chaitin [5] who related the problem of computability, discussed in the last section, to Gödel's incompleteness theorem in metamathematics [6][7]. The latter theorem established that — "It is not possible to prove the consistency of arithmetic system inside the system itself at least through finite steps."

Following the analogy to computer system, one has theorems in mathematics corresponding to input messages. Corresponding to the processing of the input message, one has here the process of proof starting from fundamental axioms. Corresponding to the possibility of processing, therefore, there appears the possibility of proof.

Gödel devised a method of mapping through which every statement on number, which appears in the process of the proof of a theorem, can be replaced uniquely by a single number, i.e. a product of powers of prime numbers, which now is called Gödel number. By this method he could co-ordinate to every proof a single (Gödel) number in a unique manner. Corresponding algorithmic complexity in the previous section, one here finds the Gödel number of the minimal proof. One is now asked whether one may find the Gödel number for a closed proof.

Pursuing the analogy, then, it is naturally expected that there will appear a limited capacity for proof in the arithmetic system, too. In this case it is described as follows.

Let us consider a statement: "It is impossible to prove the 'statement' which is represented by Gödel number G." Following Gödel, one may find a new Gödel number, say G', for the whole "statement".

Replacing G (for the smaller 'statement') by G', one may proceed to find G" (for the larger "statement"), and so forth. This recursive progression has a

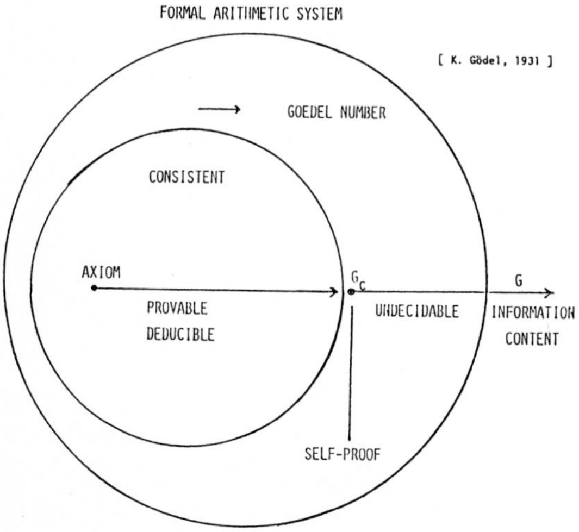

FIG. 5

unique singular situation, namely the fixed point G_C of the recursion corresponding to 'self-proof'. It is possible to proceed infinitely close to G_C in the provable domain; however, G_C itself is strictly outside this domain. —— This is the fact proved by K. Gödel. Namely, the statement represented by G_C, i.e. 'self-proof', is undecidable inside the formal arithmetic system at least through finite steps. It is clear that the last statement corresponds to the fact that self-processing, i.e. $I = I_C$, belongs to the domain which is unprocessible in the previous section.

2.5 Common Feature of Various Examples

In the foregoing subsections a number of different systems have been described, which reveal lack of predictability, processibility and provability, in the circumstances in which self-recurrence, self-processing and self-proof become unavoidable, respectively. One may summarize this kind of phenomenon by using a common language as much as possible as follows.

These systems are all required to process a proper information input (with information content I), and reproduce it as an information output as accurately as possible. However, the information processor has its own capacity I_C, i.e. the largest possible information content which can be processed, due to inherent finiteness of the system. Accordingly, at least the following two situations are to be distinguished.

(i) $I < I_C$

In this case there is little problem in processing the information input and to reproduce it as an output information. —— This implies that, even if a macroscopic noise is accompanied by the systematic input, the system may detect and suppress the effect of noise before it affects the output information. —— Accordingly, given the output information, one may retrace back to the information input, if one wishes.

(ii) $I \geq I_C$

In this case it is impossible to reproduce the input information as an information output at least in a precise manner. This implies that when a macroscopic noise is accompanied by the systematic input, its effect is amplified in consecutive steps, and the output may have a different systematic pattern from that involved in the systematic input. It follows naturally that even if one is given the output, one may not hope to infer the input at least in an accurate manner.

The change of situation from (i) to (ii) at $I = I_C$ we propose to term "*Intrinsic Coarsening*", which is inherent to the system and outside the artificial control. The new concept is not unrelated to the conventional 'coarse graining', but it is restricted neither by 'many-body nature' nor by 'average', therefore it may be taken as an extension of our view point to cover a wider range of phenomena.

A few remarks may be due here. Examples have been taken up in the order from concrete physical to abstract mathematical in character; however, the last theorem in metamathematics was in fact the earliest to be put forth. The point to be stressed is that the recognition of the common character may be a good help in understanding concrete problems.

In contrast to the case of mathematics, one is bound in physics to compare two different systems, i.e. experimental and theoretical findings, put in parallel. In this case the above-mentioned isomorphism plays by far the more significant rôle.

Finally a related proposal: In view of the above definition of 'randomness' in the completely processible range, it is proposed that the term 'chaos' should be reserved for the case of intrinsic coarsening, i.e. the unprocessible range, strictly. This will help us to avoid possible conceptual confusion.

3. Need of Bilateral Description

In the previous section several different systems were considered individually and the existence of formal analogy was pointed out among their structures, which was termed 'intrinsic coarsening'. In physics, however, one is bound to compare two different systems, i.e. experimental observation and theoretical estimation, put in parallel, rather than analyse a single system. These are the minimum necessary for the physical understanding of nature.

Looked from this view point, remarkable facts are revealed concerning the first two examples discussed in the previous section (§2.1 and §2.2). Namely, the intrinsic coarsening pointed out in these examples refers to the experimental side only, and not to the theoretical estimation based on the determininstic equation of motion.

Let us re-examine the onset of turbulence in particular. On the experimental side the physical predictability is clearly lost through the transition from laminar to turbulent flow at the critical value of driving flux, which is an example of intrinsic coarsening.

On the theoretical side, however, an ill-predictive behaviour of the differential equation (or of the corresponding computing mechanism) appears at a many orders of magnitude larger value of the driving flux. This implies that from a theoretical point of view both sides of the transition under consideration belong to the processible range, which is the natural merit of the theory. (cf. Fig. 6)

However, this does not necessarily mean that the turbulent state is represented by a simple theoretical solution. On the contrary, the theoretical solution itself is complicated to such an extent that it might well be described as 'random'. At the transition point a discontinuous increase appears in the 'complexity' of the solution. In this sense the turbulence has a random complexion on the theoretical side, and its treatment is far from easy.

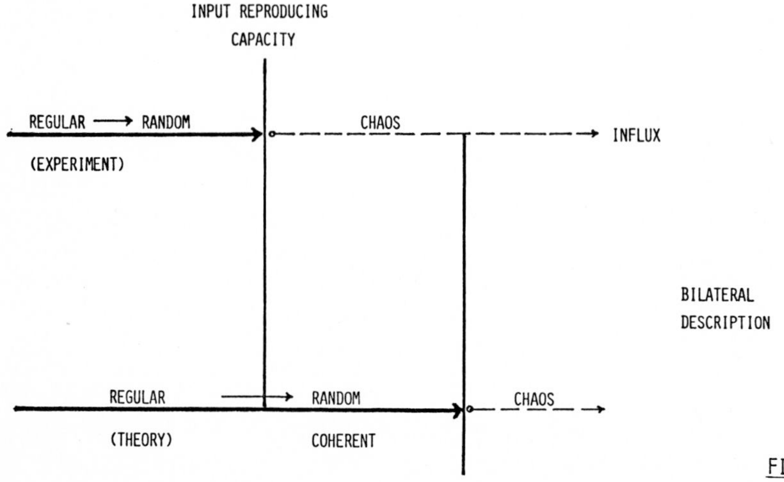

FIG. 6

Facing this kind of complex situation, statistical theory proposes to replace the dynamics by a statistical average based on an ensemble. This idea had an amazing success in the derivation of phenomenological law in the neighbourhood of equilibrium from the molecular dynamics. When the macroscopic chaos far from equilibrium is under consideration, however, this strategy is too narrow and definitely unsatisfactory, for the macroscopic fluctuation can be as large as the mean motion and ordered motions are known to be appreciable in the turbulent range.

It is proposed, therefore, that one should carefully retain deterministic aspects as well as stochastic in the treatment of macroscopic chaos, i.e. the intrinsic coarsening. One of the important recognitions derived from this bilateral description is that 'random' phenomena can be 'coherent', and this recognition throws a significant light on the interpretation of puzzles existing hitherto. In the next section examples are given which indicate 'coherent randomness'.

4. Coherent Randomness

In order to illustrate what is meant by '*coherent randomness*', simple examples are described in this section. Let us consider two interacting chaotic systems, using the logistic chaos described in §2.2. The first is a case of unilateral chaotic modulation, i.e. chaotic rate modulation, and the second is a bilateral coupling of two chaotic systems.

(A) Chaos modulated by chaos [8]

This example is an analogue of random frequency modulation in the circuit theory, and is defined by

$$x_{n+1} = 4y_n x_n(1 - x_n) \quad , \tag{4.1a}$$

$$y_{n+1} = By_n(1 - y_n) \quad . \tag{4.1b}$$

It is known that (4.1a) exhibits chaotic behaviour for a fixed value of y_n not smaller than 0.9. It is also clear that for $B \lesssim 4$ (4.1b) exhibits the logistic chaos. Here the latter is used as a chaotic modulator, with a value of B close

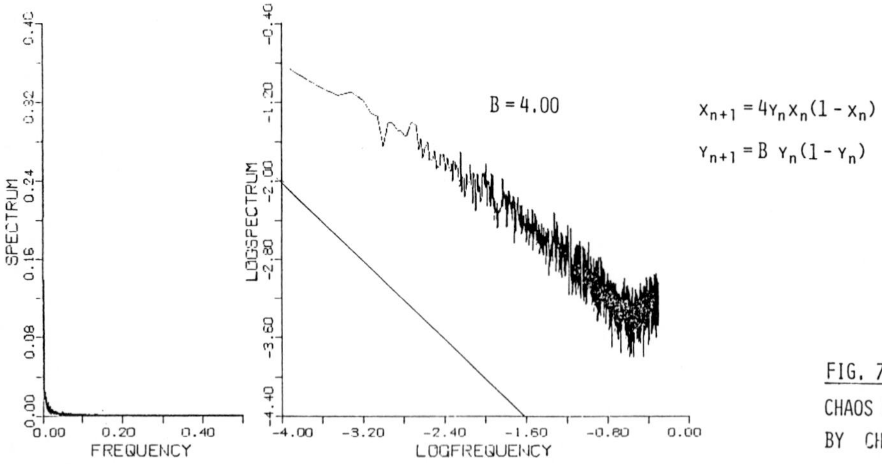

FIG. 7
CHAOS MODULATED BY CHAOS

to 4, to modulate the rate constant of the former, i.e. (4.1a). The behaviour of the total system was analysed by using: (a) Transfer function of x_n, (b) Correlation spectrum of x_n and (c) The two Liapunov exponents of the total system, i.e. λ_1 and λ_2.

(a) exhibits a clear intermittent character, (b) indicates a nearly 1/f noise spectrum (See Fig.7), and the second Liapunov exponent λ_2 from (c), corresponding to x degree of freedom, yields a small negative value, the absolute value of which behaves as

$$|\lambda_2| \propto (4 - B)^{\nu} \quad , \qquad (4.2)$$

where $\nu \sim 3/10$.

Three different kinds of aspects indicated above join to suggest that the degree of chaos has been reduced to the minimum through the chaotic rate modulation. Clearly this would never have been observed, if the behaviour of one or both systems were replaced by any statistical distribution from the outset. Some other examples are reported of noise-induced order. [9]

(B) Coupled chaos [10]

In this case two equivalent logistic systems are connected through a linear bilateral coupling, i.e.

$$x_{n+1} = Ax_n(1 - x_n) + D(y_n - x_n) \quad , \qquad (4.3a)$$

$$y_{n+1} = Ay_n(1 - y_n) + D(x_n - y_n) \quad , \qquad (4.3b)$$

where $0 < x_i,\ y_i < 1$, $A > 0$ and $D > 0$ stand for the coefficients of the rate of the bare system and the coupling, respectively.

For D = 0 chaos is known to appear for $A^*_\infty (\equiv 3.57) < A < 4.0$ for each one-dimensional system. The behaviour of the total system for $D \neq 0$ has been investigated sys-

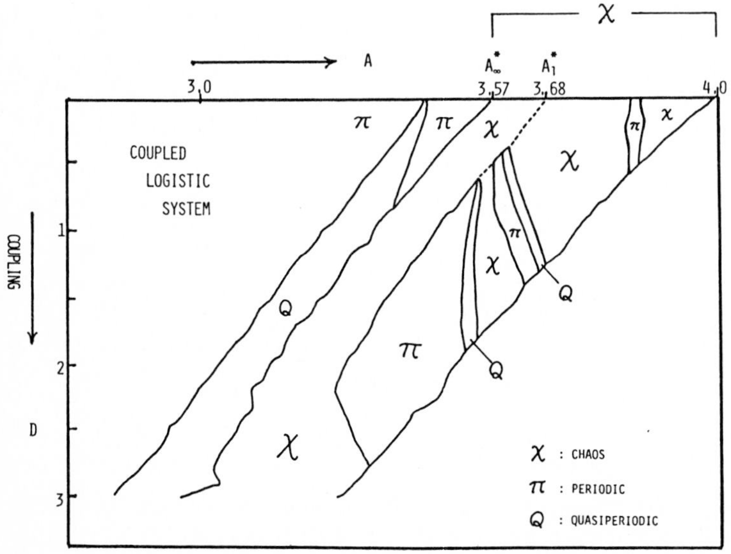

FIG. 8

tematically by calculating the Liapunov exponents λ_1 and λ_2, and the results are shown in Fig. 8 as a phase diagram in A - D plane.

The right-hand extreme region of this diagram is unphysical in that the orbits tend to infinity asymptotically. It should be noted, however, that the position of the boundary of this unphysical region changes roughly in proportion to D, and so does that of the other boundary of chaotic region (indicated by χ). In other words the region of chaos is shifted towards smaller values of A, keeping its overall span, roughly in proportion to D. For small enough values of D this is the only change ; however, when D is further increased there appear various changes inside the region of original chaos as is shown in the phase diagram. (Fig. 8)

It is clear that the region of chaos corresponding to $3.68 < A < 4.0$ (at $D = 0$) is affected appreciably by the increase of D beyond 0.5, and in a wide range chaos is replaced by periodic (indicated by π) or quasiperiodic (indicated by Q) modes of motion. In the region corresponding to $3.57 < A < 3.68$ (at $D = 0$) no such change in modes of motion is recognized; however, the value of the Liapunov numbers decreases monotonically with the increase of D. The tendency in both regions may be summarized as an increased suppression or diminution of chaos with the increase in coupling D. A similar tendency is reported in the coupled chaos described by differential equations. [11]

It would be very difficult to expect this kind of suppression of chaos, if the uncoupled original systems were replaced by statistical distributions before the introduction of coupling.

Both of the above findings with simple models, (A) and (B), indicate clearly that the original chaoses are in fact 'coherently random', so that the randomness may be subdued or suppressed by an appropriate modulation or coupling. This provides the ground for the need of bilateral description of chaos, which is often replaced simply by a statistical distribution.

5. Discussion and Summary

In the previous section evidence has been provided for the coexistence of 'randomness' and 'coherence' by using simple models. Here several further indications are listed towards the same direction. ——

(A) Very often the behaviour of a system is investigated by changing the single control parameter only. In this case, however, it is usual that one observes a complicated alternation of 'regular' and 'random' phases in a small range of the parameter. The noted window structure is one such example. Suppose one concentrates on just one variety of the two phases, then a 'fractal dimension' is needed in general. In this way there are cases in which one has to say that 'chaos emerges in a chaotic way', provided one sticks to one-dimensional cross-sections of the phase diagram. [12]

The behaviour on the two-dimensional cross-section is, however, much easier to understand than that on one-dimensional cross-sections. The phase diagram of the coupled logistic system is one such example, on which one may easily locate regions of regular and random modes of motion, respectively.

(B) The intermittency phenomenon in phase space, which is closely related to 1/f spectrum, clearly indicates bilateral character, in the sense that it is a chaos closest to the regular periodicity, something between purely regular and purely random modes of motion. The situation is associated with long time tail and self-similarity, which strongly reminds us of the critical regime in equilibrium phase transition. However, in the case of turbulence it is highly probable that this kind of situation persists up to large values of driving flux. This, then, implies that bilateral character is essential not only in the neighbourhood of the transition but also in the whole range of turbulence.

(C) The bilateral character of chaos may be thought to provide a conceptual basis for the variational principle which is often invoked in deriving a regular equation of motion. The idea is that the sample functional appearing in variation is not just an artifact, but may be representative of the random counterpart of the modes of motion, whether it is explicit or latent, out of which regularity is sought.

Finally one may summarize the content of the paper in the following way. — The concept of "coarse graining" in statistical physics is re-examined in relation to the appearance of macroscopic chaos in dynamical systems, and two aspects are pointed out which seem to deserve due consideration.

(i) Concerning the formal aspect, the onset of chaos should be modelled after "intrinsic coarsening", due essentially to the limitation in self-evaluation. This implies that the 'kinematic' many-body aspect is really irrelevant.

(ii) Concerning the physical aspect, the proper description of the macroscopic chaos should be bilateral, i.e. covering the coherent aspect as well as the stochastic. This leads naturally to the concept of 'coherent randomness'.

Either of these points suggests the necessity for reconsideration and extension of the conventional view on coarse graining, and the new extended view indicated by "intrinsic coarsening" seems to have applications to wider range of subjects.

References

1. See, e.g., R. Kubo, M. Toda, N. Hashitsume, Statistical Physics (Springer Series in Solid-State Sciences vol.30-31, 1983)
2. E.N. Lorenz, J. Atmos. Sci. 20 (1963) 130
3. R. May, J. Theor. Biol. 51 (1975) 511; Nature 261 (1976) 459
4. A.N. Kolmogorov, Tri podhoda k opredeleniju ponjatija "Kolicestvo informacii". Problemy peredaci informacii 1 (1965) 3
5. G. Chaitin, J. Assoc. for Computing Machinary 21 (1974) 403; Sci. American (1975) May, p.47
6. K. Gödel, Monatschefte für Mathematik und Physik. 38 (1931) 173
7. E. Nagel and J.R. Newman: Gödel's Proof (Newyork University press, 1959)
8. H. Sakaguchi and K. Tomita, to be published
9. K. Matsumoto and I. Tsuda, J. Stat Phys. 31 (1983) 87
10. Y.S. Lee and K. Tomita, to be published
11. H. Fujisaka and T. Yamada, Progr. Theor. Phys. 69 (1983) 82
12. D. Rand, S. Ostlund, J. Sethna and E.D. Siggia, Phys. Rev. Lett. 49 (1982) 132

† The content of this paper was first reported on Jan. 7, 1983 at a Symposium on "Investigation and Control of Turbulent Phenomena" in Kyoto.

Gibbs Variational Principle and Fredholm Theory for One-Dimensional Maps

Y. Takahashi

Department of Mathematics, College of Arts and Sciences, University of Tokyo
Komaba, Meguro, Tokyo 153, Japan

0. Introduction

In the present paper the following formal power series associated with a 1-dim map F plays an important role:

(1) $D_\beta(z) = \exp[-\sum z^n Q_n(\beta)/n], \quad Q_n(\beta) = \sum_{F^n t = t} |(F^n)'(t)|^{-\beta}$

where $F^n t = F(F(\ldots(F(t))\ldots))$ (n times).

The case $\beta=0$ corresponds to the dynamical properties on the topological level. $1/D_0(z)$ is called Artin-Mazur zeta function and is a generating function of the number of periodic points. The topological entropy of F coincides with the value at $\beta=0$ of the quantity

(2) $P_\beta = \limsup Q_n(\beta)$. (Roughly to say, $D_\beta(\exp(-P_\beta))=0$.)

The case $\beta=1$ is the most important. $D_1(z)$ carries much information on the chaotic/non-chaotic behaviour of F which is observed by numerical experiments. For example ,

(a) classification of chaos: observable and non-observable [according to the sign of P_1]
(b) number of islands(or bands) [the multiplicity of the zero z=1]
(c) power spectrum, such as 1/f [the behaviour near z=1] .

Furthermore, there is a structure which is statistical mechanical. Let us denote the metrical(or Kolmogorv-Sinai) entropy of an F-invariant probability Borel measure μ by $h_\mu(F)$. Then,

(3) $P_\beta = \max_\mu [\, h_\mu(F) - \beta \langle \log|F'| \rangle_\mu \,]$ (variational principle).

All these results follow, heuristically to say, from the suitable identification of $D_1(z)$ with the Fredholm determinant $\det(I-zL)$ of the so-called Perron-Frobenius operator L defined by

(4) $(Lu)(t) = \sum_{Fs=t} u(s)/|F'(s)|$ on $L^1(dt)$.

The PF operator plays the role of the transfer operator in equlibrium statistical mechanics and the "partition function" $Q_n(1)$ is the formal trace of L^n.

The meaning of the case $\beta\neq 0,1$ is still obscure from the mathematical point of view but a thermodynamical formalism can be given[2]. Indeed, the equilibrium measure at $\beta=\infty$ can be interpreted as the ground state. Moreover the value of β where $P_\beta=0$ is the Hausdorff dimension of the support of the asymptotic measure (whose definition is found in 2), at least when F is piecewise linear (not necessarily continuous).

In sections 1-4 the general theory is stated and in the final section 5 several interesting simple examples are discussed.

1. Entropy and Variational Principle: Large Deviation Formalism

Let us begin with a look at the mathematical mechanism underlying the Gibbs VP which describes the relationship between thermodynamical limit, entropy and mean energy. For simplicity, let us consider a 1-dim. spin system. Let $U(s_0;s_j,j\neq 0)$ be the potential energy of spin s_0 at site 0 when the spins s_j are located at other sites $j\neq 0$ and assume the shift (or translation) invariance. Then the partition function of the periodic system of size n is given as

$$(5) \qquad Z_n(\beta)=\sum_{s_0}\cdots\sum_{s_{n-1}}\exp[-\beta\sum_{i=0}^{n-1}U(s_i;s_{i+j},j\neq 0)].$$

Up to a normalizing constant, the summation in (5) can be regarded as the integral with respect to the uniform product measure, say m. Then the thermodynamical limit can be expressed (up to an additive constant) as

$$(6) \qquad \begin{aligned} P_\beta(U) &= \lim (1/n)\log \langle\exp[-\beta\sum_{i=0}^{n-1}U(T^i x)]\rangle_m \qquad x=(s_j),\ T^i x=(s_{j+i}) \\ &= \lim (1/n)\log\langle\exp[-n\beta\textstyle\int \ell_{n,x}(dy)U(y)]\rangle_m \end{aligned}$$

where $\ell_{n,x}$ is the measure with mass 1/n at each Dirac measure $\delta_{T^i x}$ $i=0,1,\ldots,n-1$, on the spin configuration space, say X. Now let us introduce the following functional of a state (=probability Borel measure) μ on X:

$$(7) \qquad q(\mu)=\lim_G[\lim_n (1/n)\log m(x;\ \ell_{n,x}\varepsilon G)]$$

where the "lim" in G is taken along open subsets of X which shrink to μ. Then it is proved that (up to the additive constant -log 2)

$$(8) \qquad \begin{aligned} q(\mu) &= \text{the mean energy per site of the state } \mu \\ &= \text{the metrical entropy } h_\mu(T) \text{ of the dynamical system } (T,\mu). \end{aligned}$$

This is the simplest case of the large deviation theory of Donsker and Varadhan[4], which is given for Markov processes. It can be generalized to abstract dynamical system[7]. The following theorem is a special case . Note that the measure m used as reference measure to define P(U) and $q(\mu)$ is not assumed to be T invariant, which means that one can treat not only equilibrium system but also stationary non-equilibrium systems.

Theorem 1. Let A be a finite set, $X=\{x=(s_j);\ s_j\varepsilon A,\ j=0,1,\ldots\}$ and m a probability Borel measure on X which is non-singular with respect to the shift $T:x=(s_j)\to Tx=(s_{j+1})$. Assume that the Jacobian j_m is continuous on X. Then,

(9) $P_\beta(U) = \max_\mu [q(\mu) - \beta\langle U\rangle_\mu]$ and

(10) $q(\mu) = h_\mu(T) + \langle \log j_m \rangle_\mu$.

Remark. In the case stated above, j_m is constant with value 1/2. In the case of a Markov chain, $j_m = j_m(x)$ is, by definition, a function of s_0 and s_1 and (10) remains valid when j_m is replaced by the transition probability. In the following we shall see that m corresponds to the Lebesgue measure on the interval and j_m is the reciprocal of the Jacobian of the map F.

2. Asymptotic Measures and Observable Chaos

Now let us consider a piecewise smooth map F of a bounded closed interval J. There are many mathematical results on the topological properties of F as a dynamical system, e.g., Sarkovskii and other orderings , 'Period Three Implies Chaos' etc. But there is a gap between those results and experimental results. For example, in the case where a periodic point is superstable, the numerical experiment shows that "every" orbit converges to the superstable periodic orbit while period three implies chaos. The initial data put in computer or set in real experiments should not be considered as arbitrarily taken but should be understood to be restricted to almost all data (with respect to the Lebesgue measure). Thus the topological study of dynamical systems is not sufficient for understanding chaos and we are led to the ergodic theory or the study of "ensembles".

Definition 1.([7,8]) A probability measure μ on J is called an <u>asymptotic measure</u> of F on an open subset J' of J if, on the set J'

(11) $\lim_{n\to\infty} (1/n) \sum_{i=0}^{n-1} u(F^i t) = \langle u\rangle_\mu$ a.a. t for every continuous u.

The asymptotic measures are always F invariant but they are not necessarily absolutely continuous. In the superstable case the asymptotic measure is unique and is the uniform measure on the superstable periodic orbit.

Definition 2.([1,7,8]) A map F is said to show <u>observable chaos</u> on an open subset J' if there exists the asymptotic measure μ on J' and $h_\mu(F)$ is positive, or equivalently to say, if there exists a weakly mixing asymptotic measure.

A map F is said to show <u>topological chaos</u> when a Bernoulli trial (coin-tossing, or 2-shift) can be embedded into the orbit space of F^i for some i. It is also equivalent to the positivity of the topological entropy of F or to the existence of an F-invariant probability measure with h (F) positive. Hence one may say that there is a <u>non-observable (or latent) chaos</u> when F shows topological chaos but it does not show observable chaos on any open subsets. Such a case corresponds to the phenomenon called a window by R.May.

<u>Theorem 2</u>.([1,7,8]) The quantity P_1 is positive if and only if there is a strictly stable periodic orbit.

If one ignores the critical cases where there is a stable but not strictly stable periodic orbit, then one obtains the following classification of chaos:

- top. non-chaotic .... $P_0=0$ and $P_1>0$
- top. chaotic
  - non-observable ..... $P_0>0$ and $P_1>0$
  - observable ..... $P_0>0$ but $P_1\leq 0$.

It is conjectured in [1] that $P_1>0$ implies the existence of an attractor which is a Cantor set. When the map F is piecewise linear, it is true and, besides it, a finer classification of the case $P_1=0$ is also possible as is stated in Theorem 4 below. The ignored cases include the case which may be interpreted as tricritical points in the route of intermittent chaos.

3. Fredholm Determinant of the Perron-Frobenius Operator

The analysis of asymptotic measures is reduced to the study of the PF operator L defined by (9) since the dynamics of density functions u under the map F is described by L as is seen from the relation*

$$(12) \qquad \int_J u(t)v(Ft)dt = \int_J (Lu)(t)v(t)dt \qquad (u\varepsilon L^1,\ v\varepsilon C)$$

where $L^1=L^1(J,dt)$ is the space of all integrable functions on J and $C=C(J)$ is the space of all continuous functions on J.

The ergodic-theoretical properties can be rephrased in terms of L, e.g., concerning absolutely continuous invariant measures,

(13) $u^*\varepsilon L^1$, $Lu^*=u^*$ $\leftrightarrow$ u^* is the density of an F-invariant measure μ^*

(14) $\lim L^n u=\text{const.}u^*$ for each u $\leftrightarrow$ (F,μ^*) is mixing

(15) $\lim (u+\ldots+L^n u)/n=\text{const.}u^*$ for each u $\leftrightarrow$ (F,μ^*) is weakly mixing.

Thus the problem is reduced to the spectral analysis of the positivity preserving operator L.

Nevertheless, L has a strange property in nature: every complex number with modulus less than 1 is an eigenvalue of L with infinite multiplicity**. ([5])

It is known under some natural assumptions that L is quasi-compact if it is restricted to the space BV=BV(J) of functions with bounded total variation. Note that the spectrum of L on the original space L^1 may be different from that of L on BV.

Example. Let F have a strictly stable periodic orbit, say $O=\{t_i\}$, $Ft_i=t_{i+1 \bmod p}$. Then, $c=\Pi|F'(t_i)|<1$. Consider the function u such that u=0 outside O and

$$u(t_i)= c^{-i/p}|F'(t_i)\ldots F'(t_p)| \qquad i=1,\ldots,p.$$

---

*) It holds if $F'\neq 0$ a.e. (Otherwise, the dynamics is trivial or it has a trivial subdynamics.)

**) It holds if F(J)=J and F is not one-to-one.

Then, $Lu \geq \lambda u$ with $\lambda = c^{-1/p} > 1$. Since L preserves positivity, the spectrum on BV has intersection with the interval $[\lambda, \infty)$. On the other hand, the operator norm of L on L^1 is exactly 1 and so the spectrum there is contained in the closed unit disc. By the way, the function u is 0 a.e. and so it is trivial in L^1. But it indicates that the uniform measure on O, a finite set, is F invariant.

The identification of the formal power series $D_1(z)$ with the Fredholm determinant det(I-zL) will be done through the following result.

Theorem 3. Let F be a piecewise linear continuous map of a bounded closed interval J onto itself, i.e.,

$$(16) \qquad F(t) = [(t_i - t)F(t_{i-1}) + (t - t_{i-1})F(t_i)]/(t_i - t_{i-1}) \quad \text{if } t_{i-1} \leq t \leq t_i.$$

Assume (a) No perodic orbit is contained in $\{t_i\}$

(b) There is no non-empty open subinterval J' such that $F^n(J') = J'$ for some n.

Then, for a complex number z in the domain of convergence of $D_1(z)$, the relation $D_1(z)=0$ is equivalent, including its multiplicity, to the fact that 1/z belongs to the spectrum of L restricted to the subspace A of BV(J) spanned by the indicator functions of the intervals with end points in $\{F^n t_i\}$*

Remark. If the condition (a) or (b) is violated, then the multiplicity may change. The proof of Theorem 3 is given in [5] where the number of t_i's is assumed to be three for the sake of the simplicity of the proof, but the number can be infinite.

Conjecture: If a continuous map F of J onto itself is piecewise C^2 and satisfies ess inf $|F'| > 0$ and ess sup $|F''| < \infty$, then the theorem above remains valid under a suitable choice of the subspace A of BV(J).

Now let us recall the definition of the Fredholm determinant of a compact operator of trace class. An operator L is called trace class if its spectrum is discrete and consists of eigenvalues λ_i with finite multiplicity m_i and $\Sigma\, m_i |\lambda_i|$ is finite. Then the Fredholm determinant D(z)=det(I-zL) is defined as

$$(17) \qquad D(z) = \Pi(1 - z\lambda_i)^{m_i} = \exp[-\, \Sigma\ (1/n)\,\mathrm{trace}\ L^n],$$

and, hence, D(z)=0 is equivalent to the fact that 1/z belongs to the spectrum of L (including the multiplicity). Hence the following definition is quite natural.

Definition 3. The formal power series $D_1(z)$ defined by (1) with $\beta=1$ is called the Fredholm determinant of the Perron-Frobenius operator L associated with the map F or, simply, the Fredholm determinant of the map F.

---

*) The choice among four types of intervals [t,t'],[t,t'),etc.,depends on the property of F, the precise statement of which requires an argument on the structure of orbit spaces and so I omit it here (cf. [6]).

A similar definition is possible for $\beta \neq 1$ using the operator L_β, which is not defined here but whose form may be self-evident from the form of $Q_n(\beta)$.

As Theorem 3 suggests it, if there is an F-invariant absolutely continuous measure, then $D_1(1)=0$ and the converse is also true when F is piecewise linear. But the measure may actually not be a probability measure.

Theorem 4. Let F be piecewise linear as in Theorem 3. Assume that $D_1(1)=0$. Then the following statements are valid:

(i) If $D_1'(1-0)$ is infinite, then there exists an F-invariant absolutely continuous measure with infinite total mass.

(ii) If $D_1'(1-0)$ is finite and there are exactly p zeros of $D_1(z)$ on the unit circle $|z|=1$, then there exists an F-invariant probability measure which is absolutely continuous and is supported by exactly p open subintervals. Furthermore, the iterate F^p is mixing on each subinterval.

(iii) If $D_1'(1-0)$ is finite and z=1 is the zero of $D_1(z)$ with multiplicity m, then there are m absolutely continuous invariant probability measures.

The proof can be given by the method in [5] and it will be almost clear from the computation given below in simple examples, which also serve as the examples of (i) and (ii).

4. A Formula for the Metrical Entropy and the Variational Principle

In order to clarify the connection between the spectral analysis of the PF operator L and the variational principle (3), let us state a key lemma in the proof of Theorem 1 on the shifts. It is a formula for the metrical entropy in terms of the transfer operators.

Lemma 1.([7,8]) Under the same notations as in Theorem 1, put

(18) $$(Mu)(x) = \Sigma_{Ty=x}\, u(y).$$

Then the following identity holds for each shift invariant probability Borel measure μ on X:

(19) $$h_\mu(T) = \inf\{\langle \log(Mu/u)\rangle_\mu\ ;\ u>0,\ \text{continuous}\}.$$

This formula may be regarded as the definition of the metrical entropy for the shifts. On the other hand, the largest lower bound on the right-hand side of (19) may regarded as the "maximal eigenvalue" or the principal eigenvalue of M on the support of μ.

Corollary. Let U be a continuous function on X and put

(20) $$(L_U u)(x) = \Sigma_{Ty=x}\, u(y).\exp[-U(y)].$$

Then, for each T invariant μ on X,

(21) $$h_\mu(T) - \langle U\rangle_\mu = \inf\{\langle \log(L_U v/v)\rangle_\mu\ ;\ v>0,\ \text{continuous}\}.$$

The proof is immediate: put $v=u\exp U$, then, $L_U v=Mu$.

Historically, the VP such as Theorem 1 is proved for simple cases by a constructive method using the solutions of the eigenvalue problems

$$(22) \qquad L_U u=\lambda u, \qquad u>0,\ \lambda>0$$

and their dual eigenvalue problems (cf. D.Ruelle[9] or R.Bowen's Springer Lecture Notes in Math. vol.470,1975).

Now it is natural to expect a similar formula in terms of the PF operator L. But there lies a mathematical difficulty due to the non-expansiveness and the structural instability. Nevertheless, the following formula is proved by using the results on the detailed structure of the orbit space of F as symbolic dynamics. ([6,7])

Lemma 2. For an ergodic F-invariant probability Borel measure μ,

$$(23) \qquad h_\mu(F)-\langle\log|F'|\rangle_\mu = \inf\{\langle\log(Lu/u)\rangle_\mu \ ;\ u>0,\ \text{continuous}\}.$$

The additional assumption here, the ergodicity of μ, is necessary since the functional $q(\mu)$ may lose the affine property* when the under-lying dynamical system is not structurally stable. The relation (23) can be extended to the case $\beta\neq 1$.

From Lemma 2 and the structure of orbit space of F one can prove the following:

Theorem 5. Let F be a piecewise C^1 map** of a bounded closed interval J and assume that ess inf $|F'|>0$. Then the variational principle given as (3) is valid.

5. Examples of the Unit Interval

In the following (and everywhere in the proofs of theorems and lemmas which are already stated), the notion of symbolic dynamics or shifts is used essentially but I omit the detailed argument, a summary of which is given in [8] and the detail will be found in [6].

A. Tent map:

$$Ft = \begin{cases} t/b & \text{on } I_0=[0,b], & b>0 \\ (1-t)/a & \text{on } I_1=[b,1], & a=1-b>0. \end{cases}$$

In this case, $F(I_0)=F(I_1)=J=[0,1]$ and F is one-to-one on each I_i. This implies the following: if one makes an observation of F orbits that $F^n t\ I_0$ or not, then the totality of two symbol sequences is obtained. Usually, this fact states that F is realized by 2-shift. Consequently one obtains

---

*) $q(t\mu+(1-t)\mu')= tq(\mu)+(1-t)Q(\mu')$, $0\leq t\leq 1$.
**) F' exists except at division points t_i where $F'(t_i-0)$ and $F'(t_i+0)$ exist.

$D_0(z)= 1-2z,$ $\qquad P_0= \log 2$

$D_1(z)= 1-(a+b)z= 1-z,$ $\qquad P_1=0,$ $\qquad D_1(1)=0,$ $\quad D_1'(1)= -1$ (finite).

These facts agree with the well-known properties:

the topological entropy= log 2

there is a mixing absolutely continuous F-invariant probability measure (Lebesgue measure itself)

B. <u>β transformations</u>: $Ft= \beta t \mod 1, \ \beta>1.$

Take the integer N such that $N-1<\beta\leq N$ and put $I_i=[i/\beta,(i+1)/\beta)$ for $i=0,1,\ldots,N-2,$ and $I_{N-1}=[(N-1)/\beta,1)$. Then, for $i\neq N-1$, $F(I_i)=J=[0,1)$. Next put $I_i^1=I_{N-1}\cap F^{-1}I_i$ if $i<\beta F(1-0)=\beta[\beta-(N-1)]$. Then, $F^2(I_i^1)=J$. By a similar procedure one can construct subintervals $I_i^0=I_i$, I_i^1, $I_i^2,\ldots$ such that $F^{n+1}(I_i^n)=J$ and which form a partition of J. In symbolic dynamic cases, I call such a system $\{I_i^n\}$ the free orbit basis[6]. Let a_n be the number of words of length n+1 in the free orbit basis, i.e., of the intervals I_i^n , n being fixed. Then, generally,

$$D_0(z)= 1- \Sigma\, a_n z^{n+1}, \quad a_n\geq 0. \tag{24}$$

In this case, the construction shows that $D_0(1/\beta)=0$ and so $P_0=\log \beta$ is the topological entropy. Since F' is constant,

$$D_1(z)= 1- \Sigma\, a_n(z/\beta)^{n+1}, \quad P_1=0, \quad D_1(1)=0, \quad D_1'(1) \text{ finite.}$$

This implies that F shows observable chaos. In this case, the invariant density is $u^*(t)=$ const. $1[t\ F^n(1-0)]/\ ^{n+1}$.

C. <u>Unimodal Linear Maps</u>: $Ft= (a+b-1+t)/b$ on $I_0=[0,1-a]$

$\qquad\qquad\qquad\qquad\quad (1-t)/a$ on $I_1=[1-a,1]$

where $0<a<1$, $b>0$ and $a+b\geq 1$. The case a+b=1 is the tent map. The family UML was studied in detail by S.Ito, H.Nakada and S.Tanaka in Tokyo J.Math. vol.2(1979). It is shown, in my terminology, that a free orbit basis exists and so F shows observable chaos except for the following two cases ($p\geq 3$):

(a) $ab^{p-1}<1,\quad a(a+b)b^{p-2}\geq 1,\quad a(1+b+\ldots+b^{p-2})<1<a(1+b+\ldots+b^{p-1})$

(b) $ab^{p-1}>1,\quad a(1+b+\ldots b^{p-2})<1<a(1+b+\ldots+b^{p-1})$.

(Precisely said, there is one more exceptional case where $ab^{p-1}=1$, which is ignored here since that case is pathological.)

In either case, there are p subintervals which are cyclically mapped to each other by F. The series $D_\beta(z)$ is expressed as a product $d_{\beta,1}(z^p)d_{\beta,2}(z)$, where $d_{\beta,i}(z)$ are of the form (24).

In case (a), $d_{1,1}(1)=0$, $P_1=0$ and p islands(bands) appear.

In case (b), $d_{1,1}(1)\neq 0$, $P_1>0$ and F shows window phenomena, i.e., non-observable topological chaos.

D. <u>Countable Markov Maps which show Intermittent Chaos</u>

Let $c_0=1>c_1>\ldots>\lim c_n=0$. The map F on J=[0,1] is defined by the conditions that it is continuous, that it is linear on each subinterval $I_i=[c_i,c_{i-1}]$ and that $F(c_i)=c_{i-1}$ $(i\geq 1)$ and $F(c_0)=0$. Then,

$$F(I_i)=I_{i-1} \quad (i\geq 1) \qquad \text{and} \qquad F(I_0)=J.$$

Consequently, $\{I_i\}$ plays the role of the free orbit basis. It then follows immediately that

$$D_0(z)= 1- \Sigma\, b_n z^{n+1}, \quad P_0= \log 2, \qquad \text{with} \quad b_n=|I_n|=c_n-c_{n+1}$$

$$D_1(z)= 1- \Sigma\, b_n z^{n+1}, \quad P_1=0.$$

Hence the invariant density exists and is a probability density or not according to

$$-D_1'(1-0)= \Sigma\,(n+1)b_n <\infty \quad \text{or} \quad = \infty.$$

For instance, take $b_n=\text{const.}(n+1)^{-\beta}$. Since $\Sigma\, b_n=1$, $\beta>1$. Then a probability density exists or does not according to $\beta>2$ or $2\geq\beta>1$. In the case $2\geq\beta>1$, the asymptotic measure is degenerated to the Dirac measure δ_0 at the origin.

The time correlation function can also be described by L. Since F is piecewise linear, the spectrum of L on the space of piecewise linear functions with division points c_i (which are sufficient for the time correlation function) is essentially reduced to the reciprocals of zeros of $D_1(z)$. Especially, the asymptotics of the time correlation or the singularity at f=0 of the power spectrum p(f) is completely determined by them. The resultant singularity as $f\to 0$ is

$$p(f)\propto f^{-\alpha} \quad \text{with} \quad \alpha= \begin{cases} 0 & \text{if } \beta>3 \\ 3-\beta & \text{if } 3\geq\beta>2 \\ \beta-1 & \text{if } 2\geq\beta>1. \end{cases}$$

In particular, $p(f)\propto 1/f$ at $\beta=2$.(Precisely said, a log correction is necessary for b_n to obtain the exact asymptotics.)

References

1. Y.Oono and Y.Takahashi, Chaos, external noise and Fredholm theory, Prog.Theor.Phys.63(1980) 1804-1807
2. Y.Takahashi and Y.Oono, Towards Statistical Mechanics of Chaos, preprint
3. S.Ito, H.Nakada and S.Tanaka, Unimodal linear transformations I,II, Tokyo J.Math.2(1979) 221-259
4. M.D.Donsker and S.R.S.Varadhan, Asymptotic evaluation of certain Markov process expectations I,III, Comm.Pure Appl.Math.28(1975) 1-45, 29(1976) 389-461
5. Y.Takahashi, Fredholm determinant of unimodal linear maps, Sci.Papers Coll.Gen.Educ.Univ.Tokyo,31(1981) 61-87
6. Y.Takahashi, Shifts with orbit basis and realization of 1-dim. maps, to appear in Osaka J.Math.20-3(1983)
7. Y.Takahashi, Entropy functional(free energy) of dynamical systems and their random perturbations, to appear in Proc. Taniguchi Symp. on Stochastic Analysis, Katata and Kyoto, 1982
8. Y.Takahashi, An ergodic theoretical approach to the chaotic behaviours of dynamical systems, to appear in Publ.RIMS Kyoto Univ. and also references therein
9. D.Ruelle, Thermodynamical Formalism, Encyclopedia in Math.5, Addison-Wesley,1979

Truncated Development of Chaotic Attractors in a Map when the Jacobian is not Small

T. Short and J. A. Yorke[1]

Department of Mathematics and Institute for Physical Science and Technology
University of Maryland
College Park, MD 20742 USA

Abstract

Numerical studies of a map that models the rotation of a periodically kicked rotor with friction reveal chaotic attractors. At parameter values that precede the existence of a horseshoe a chaotic attractor is found. The growth and destruction of this attractor are studied in relation to the early stages of the formation of a horseshoe in the region. A pattern, also seen in a Hénon map, is that of a "crisis" which occurs as the parameter increases[2]. The chaotic attractor has 2^n pieces upon collision with an orbit of period 3×2^n. This crisis marks the destruction of the attractor. The value of n depends on the Jacobian of the map.

In dynamical systems a horseshoe (as described by Smale) demonstrates the existence of chaotic (irregular aperiodic) orbits. The existence of a horseshoe, however, does not imply the existence of a chaotic attractor. If a horseshoe is shown to exist in a system there can be no attractor within the horseshoe, since almost every trajectory in a horseshoe must eventually leave. (At parameter values for which a horseshoe is still in its formative stages, chaotic attractors are sometimes found.) In this range of parameter values, some results are known for dynamical system maps that are area decreasing (the Jacobian of the map has absolute value less than 1). As the parameters are changed to form a horseshoe in such a map, a period-doubling cascade of attracting periodic point must occur. See [1] for the detailed hypotheses. Usually in numerical studies such a cascade is seen to develop into a chaotic attractor. It is our intent to study numerically how chaotic attractors evolve, and then to study the crisis which destroys the attractor.

To study the parallel development of horseshoes and associated chaotic attractors, we chose one of the simplest physical models that yields an area decreasing map with chaotic dynamics. The map models a pulsed rotor with friction (see appendix). The equations are:

$$x_{n+1} = (x_n + y_n) \bmod 2\pi$$

$$y_{n+1} = Ay_n - G\cos(x_n + y_n),\quad x \in [0,2\pi] \text{ and } y \in \mathbb{R} \quad . \qquad \text{(R)}$$

[1] This research was partially supported by the Air Force Office of Scientific Research under grant AFOSR-81-0217 and by the National Science Foundation under grant MCS 81-17967.

[2] As a parameter in a map is increased the sudden disappearance of a chaotic attractor due to its coincident collision with an unstable periodic orbit is called a "crisis", and in particular a "boundary crisis". See [3].

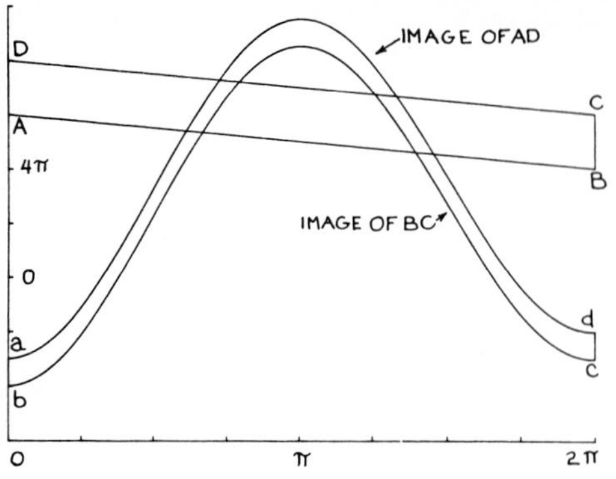

Fig. 1 An example of a horseshoe for the system (R) with $A = 1/2$ and $G = 6\pi$. Under the map (R) the rectangle ABCD has stretched and bent its image to a "horseshoe" shape under the map. Points a, b, c, and d denote images of A, B, C, and D. For the image of the rectangle to be horseshoe, the images ad and bc of the sides of the rectangle must both cross the rectangle twice

For this map all numerically observed bifurcations are regular rather than inverted. (That is, when an attracting periodic orbit undergoes a period-doubling bifurcation, the new orbit with the doubled period is attracting.) The system is also regular in that, as the parameters G and A are varied, the only way attracting periodic orbits are seen to become unstable is through period-doubling bifurcations. The unstable orbit, from this bifurcation, persists as the parameter G is increased further and A remains fixed.

Holmes proved the existence of fully developed horseshoes in this map for sufficiently large values of the parameter G. He studied this system as an approximation of a ball bouncing on a vibrating table [2]. In this paper we study (R) numerically, fixing A, (the dissipation[3] constant), and using a range of values for G, (the magnitude of the driving force). Thus the early stages of the transition to chaos and their relation to the development of horseshoes are explored numerically in this map. (When the image of a curvilinear rectangle is stretched, bent over and mapped back across itself under one iteration of some map, as in Fig. 1, the result is called a (topological) horseshoe for the map.)

If the parameter values of a map are set below the range at which the rectangle yields a horseshoe image, the search for a horseshoe may yield a rectangle which is mapped entirely into itself. (The term "rectangle" is used to denote any figure with four curvilinear sides.) At this stage we say the map has a proto-horseshoe. A proto-horseshoe by its contractive nature contains an attractor, possibly a single attracting point. A horseshoe develops from this proto-horseshoe in map (R) as G is increased. A study of the development from proto-horseshoe to horseshoe has been carried out for the Hénon map in [3].

In searching for either a proto-horseshoe or horseshoe it is possible to make an optimal choice of the initial (topological) rectangle. This optimal choice will insure that the image of the rectangle is either: a proto-horseshoe (or horseshoe), or that its image is not a proto-horseshoe (or horseshoe) and there is no slightly different choice of an initial curvilinear rectangle that yields a proto-horseshoe (or horseshoe) image. (Some higher iterate of the map might produce a horseshoe in this vicinity, but we restrict our attention to the first iterate.)

---

[3]The Jacobian determinant of (R) is A.

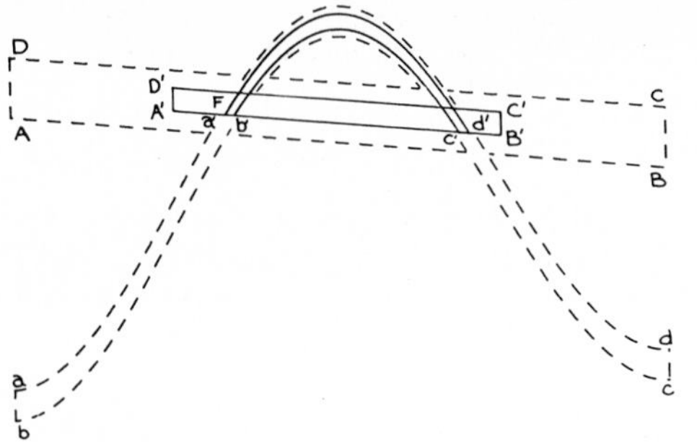

Fig. 2

An example of an optimal rectangle (A'B'C'D') which produces a "reduced" horseshoe (a'b'c'd'). The rectangle ABCD has horseshoe image abcd (----). (This is from Fig. 1.) When the smaller curvilinear rectangle A'B'C'D' is chosen so the sides A'B' and C'D' correspond to the stable manifold of the saddle point (F), then side C'D' is mapped to c'd' and side A'B' is mapped to a'b'. Thus the feet of the horseshoe image (a'b'c'd') of the rectangle (A'B'C'D') are mapped to the lower boundary of the rectangle. The reduced horseshoe is a trimmed version of the more general horseshoe (abcd). The image is taken under map (R) with $A = 1/2$, $G = 6\pi$

The key is to choose the rectangle so that the feet of its proto-horseshoe (or horseshoe) image rests on the lower boundary of the rectangle, as in Fig. 2 (solid line) and Fig. 3, rather than extending beyond (as does the image of the solid line rectangle of Fig. 1.) In Fig. 2, side A'B' is mapped to a'b' (which lies on A'B', so D'C', is another piece of the same stable manifold. The proto-horseshoe (or horseshoe) that results is called "reduced", since it is the image of a (curvilinear) rectangle which can be created by reducing the sides of a non-optimal rectangle to coincide with sections of the stable manifold of a nearby fixed point.

Figure 3 shows a reduced proto-horseshoe and its relation to both the stable and unstable manifold of its saddle point (F). In this picture the unstable manifold winds into a periodic orbit of period one (a stable fixed point) interior to the proto-horseshoe. This fixed point bifurcates from the saddle point (F) when G is increased past 3π. When $G \leqslant 3\pi$ no proto-horseshoe image is possible in the region of the graph shown in Fig. 3. To form a proto-horseshoe, it is necessary to have both a saddle point and a second periodic point (initially an attractor). By con-

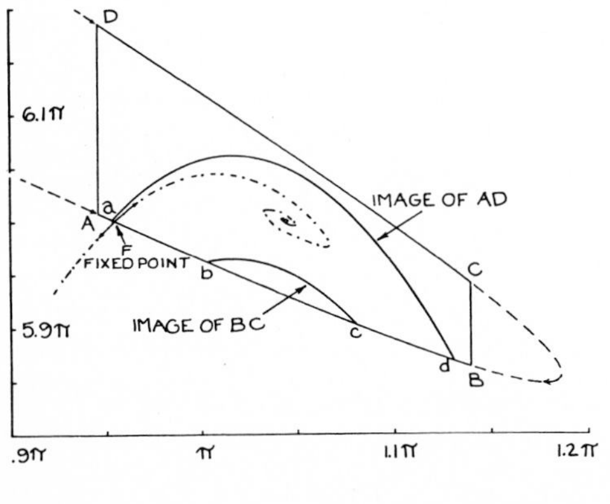

Fig. 3 An example of a reduced proto-horseshoe (abcd) for (R) in relation to its stable manifold (----), unstable manifold (-.-.-) and its initial (curvi-linear rectangle ABCD, with Jacobian parameter A=1/2, and G=3.03π

struction, the saddle point always lies on the boundary of the optimal rectangle and thus the other fixed point lies in the interior of the rectangle's image.

Starting at a value of G just greater than 3π, the proto-horseshoe develops as G is increased. Since the optimal choice of (curvilinear) rectangle is always used, it depends solely on the stable manifold of the saddle point). It is sufficient to keep track of the manifolds of the saddle point (F). If the unstable manifold of (F) were to intersect the portions of the stable manifold of (F) that form the top (AB) and bottom (CD) sides of the optimal rectangle ABCD in Fig. 3, no choice of rectangle in the region would produce a proto-horseshoe image under the map.

Figure 4 shows just such an intersecting alignment of manifolds ($G = 3.093\pi$). In Fig. 4 fingers of the unstable manifold of (F) extend out of the region bounded by the stable manifold of (F). These are points that would be mapped from the interior of an optimal (curvilinear) rectangle. The image of the optimal (curvilinear) rectangle (which has two boundaries chosen to lie on the stable manifold) would not lie entirely in the rectangle and therefore could not, by definition, be a proto-horseshoe. In map (R) the parameter value $G = 3.09\pi$ marks a transition at which the stable and unstable manifolds are tangent. For an interval of intermediate values of G the rectangle is neither a proto-horseshoe nor a horseshoe.

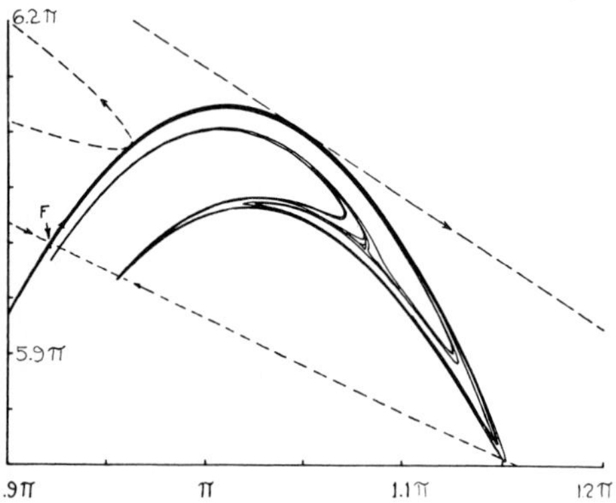

Fig. 4

At $A = 1/2$, $G = 3.095\pi$ no choice of initial topological rectangle in the area pictured will produce a proto-horseshoe image under map (R). The intersection of the stable manifold (----) of (F) with the unstable manifold (——) of (F) means that points interior to the optimal choice of a (curvilinear) rectangle will be mapped outside the rectangle. Higher iterations of the map may produce a horseshoe image, but under the first image of map (R) neither a proto-horseshoe nor horseshoe is formed in this region. A periodic attractor (not shown) lies inside the region bounded by the dashed lines.

For values of the parameter G in this interval just beyond the transition value it is possible to find a horseshoe for higher iterates of the map, but under the first iteration of map (R), the image of the (curvilinear) rectangle is at an intermediate stage. In this "intermediate" interval of values of G, interior to the image of the optimal rectangle, there can exist a chaotic attractor. Under one iteration of the map, the image of the optimal (curvilinear) rectangle will not be a horseshoe until G is increased to the point where the bottom arch of the image clears the top boundary of the optimal (curvilinear) rectangle. Even at the intermediate stage the map does not yet satisfy the hyperbolicity properties required by Smale.

The development of attractors in map (R) is related to the pattern of development of the chaotic attractors in the Hénon map:

$$x_{n+1} = 1 - Gx_n^2 - Ay_n, \quad y_{n+1} = x_n . \tag{H}$$

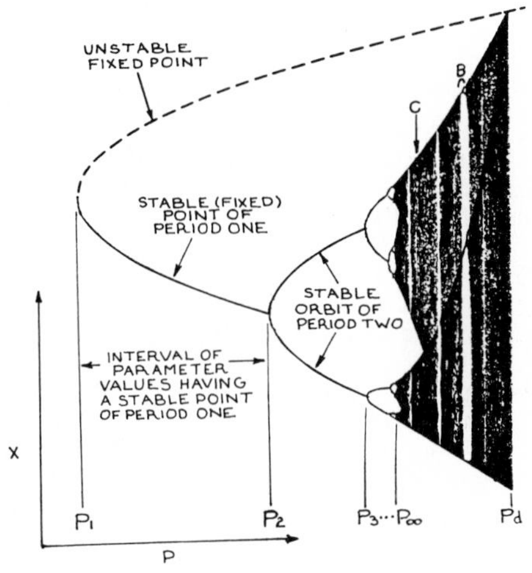

Fig. 5 The standard evolution for small Jacobian. A cascade of period-doubling bifurcations culminates at the limiting parameter value P_∞, beyond which bands of chaos result. A vertical cross section taken at parameter value = C would show two dark bands corresponding to a two-piece chaotic attractor. At P_d the attractor collides with a periodic point. Beyond P_d the attractor no longer exists. This bifurcation example is taken from the Hénon map, the Jacobian parameter A is -0.3. The Hénon map has two variables (x and y), but only x is plotted against the parameter on the nonlinear term, P, here. The narrow band B shows the period 3 attractor, a feature which is missing when A = +0.3. Thus for a Jacobian of ÷0.3 the development of the chaotic attractor is truncated in this regard

In both maps there is a single parameter that determines the Jacobian A, and parameter on the nonlinear term (G). For the Hénon map (with A = - .3), the attractor develops from a periodic orbit of period one (a fixed point). As pictured in Fig. 5, the period of this orbit doubles at intervals that are successively shorter until there is a critical parameter, such that, for values of the parameter beyond this value the orbit is chaotic. For these parameter values beyond the critical value, the chaotic attractor consists of several pieces which join together as the parameter increases, to form eventually a one-piece attractor. (For simplicity we are ignoring the narrow bands that indicate brief intervals during which the attractor becomes periodic.) The sudden destruction of the attractor in the Hénon map (H) occurs at the transition parameter value that marks the beginning of the intermediate stage of the image of the optimal rectangle. In contrast the destruction of the attractor in map (R) occurs at values of G significantly larger than the transition value (which marks the tangency of the stable and unstable manifolds of the saddle (fixed) point).

The picture shown in Fig. 5 is representative of the development of a chaotic attractor when the Jacobian is small in absolute value, but for a larger Jacobian this picture is truncated. In both the Hénon map and the rotor map (R) there is seen a special saddle periodic orbit whose collision with the chaotic attractor, as the parameter on the nonlinear term is increased, marks the destruction of the attractor. The stable manifold of this orbit delineates the boundary of the chaotic

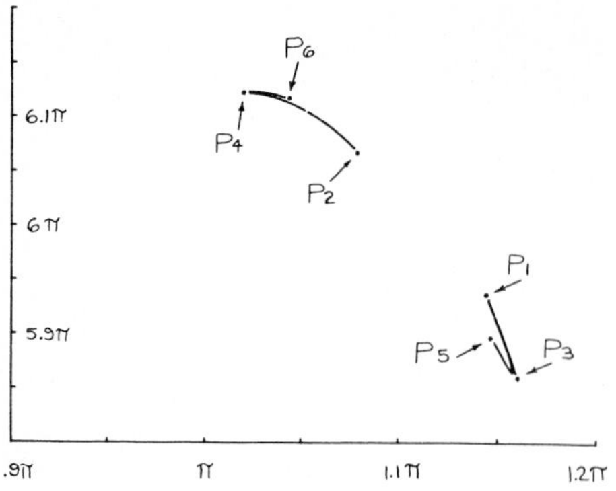

Fig. 6 The attractor of map (R) very near destruction, A=1/2, G=3.1986π. The two-piece chaotic attractor is shown relative to the unstable orbit of period six (arrows). As the map is iterated the point cycles P1→P2...→P6→P1. The distance between the points of the orbit and the attractor has been exaggerated slightly for clarity. The actual gap is too small to be seen

attractor's basin of attraction. If the chaotic attractor is two (Figs. 6 and 7) or more pieces when it is destroyed, the period of this colliding orbit is three times the number of pieces. For both maps the number of pieces the chaotic attractor is composed of on destruction is determined by the size of the Jacobian.

In the Hénon map (H), with a Jacobian $A = -0.3$, the attractor is a single piece[4] when it collides with the unstable orbit of period one from which it originally bifurcated. In map (R) at $A = 1/2$, (which yields a Jacobian of 1/2) the attracting orbit goes through a cascade of bifurcations similar to that portion of Fig. 5 for the values of $P \leqslant C$. The attractor does not continue to expand to form a one-piece attractor, but is destroyed while it consists of two pieces. Each of these pieces collides with three points of a period 6 orbit. The destruction of the two-piece attractor occurs at a value well into the "intermediate" range of parameter G (values which produce an intermediate stage image of the optimal curvilinear rectangle). The larger the Jacobian A (assuming $|A| < 1$) the more pieces the attractor is seen to have when it is destroyed. In map (R), $A = .9$ for example, there is a period 24 orbit which destroys a chaotic attractor of 8 pieces as G is increased. Each piece of the attractor collides with three of the 24 points of the orbit. The collision that destroys an attractor does not affect the periodic orbit which is far more robust!

We have not examined exhaustively the behavior of map (R) for all Jacobian values, and we do not know what additional features might have been missed,but one common pattern seems to have emerged. As our last numerical example, we chose the Hénon map with Jacobian $(0.3)^{1/64} \approx 0.98$. Here we saw the boundary crisis when there were 64 pieces to the attractor. The attractor collided with an orbit of period 192.

The authors would like to thank D. Dichman for valuable comments.

---

[4]One might wonder why Hénon chose a Jacobian value of $A = -0.3$ for his main example. In his example his choice of the value of the parameter on the nonlinear term seems reasonable; 1.4 is just before the crisis that destroys the attractor. Yet at this Jacobian near the crisis, the attractor is quite "thin", that is the dimension is small (less than 1.3), indicating (as we see in this paper) that if the Jacobian had been much larger, Hénon would not have gotten a one-piece attractor for any value of the parameter on the nonlinear term. This is presumably why he chose -0.3 as a Jacobian value.

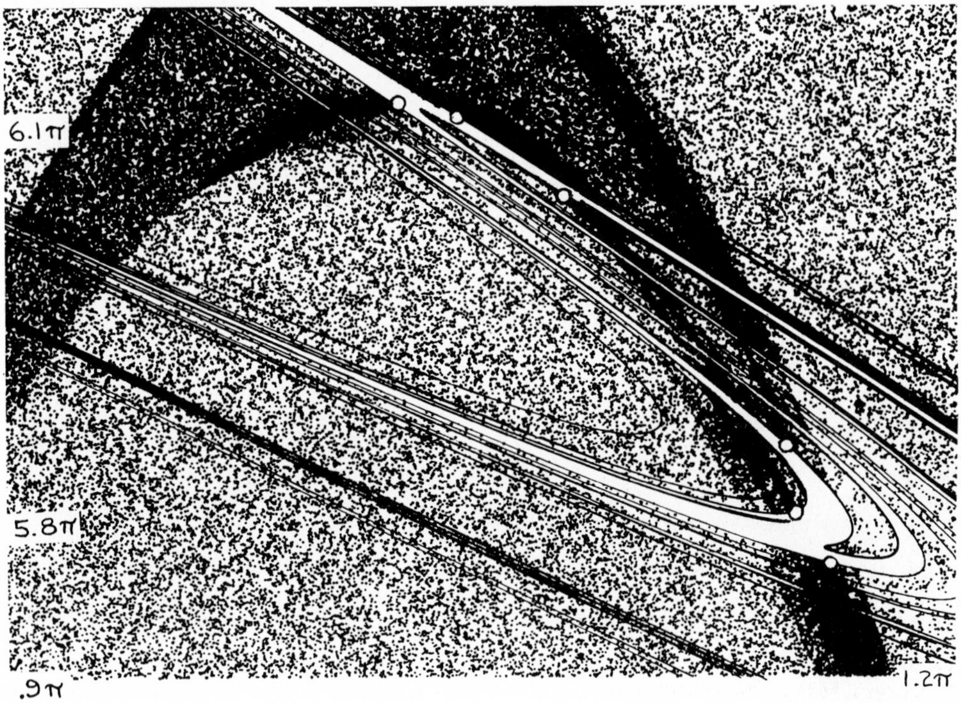

Fig. 7

The basin or attraction for the two-piece chaotic attractor in map (R) $A = 1/2$, $G = 3.1986\pi$. The figure is formed by choosing initial points randomly and iterating. Points whose trajectories eventually leave the region and do not return have their trajectories plotted, forming the speckled region. Points in the white region are drawn into the two-piece chaotic attractor (not shown). The open dots indicate the location of the period six orbit whose stable manifold (continuous line) forms the boundary of the region of attraction

Appendix

The Map Associated with a Pulsed Rotor with Friction

An example of a damped system rotating around a fixed pivot point driven by a periodic force: a one unit long rigid rod, rotating with friction, that is sharply kicked at regular times; $(t = 0,1,\ldots)$. In such a system, (we ignore gravity) the angular acceleration is given by:

$$\ddot{\theta}(t) = -p\,[\sum_n \delta(t-n)\cos(\theta(t))] - \alpha\dot{\theta}(t) \qquad \text{(A1)}$$

where P is the magnitude of the kick (uniformly applied in a single direction) and α is the frictional constant (see figure below).

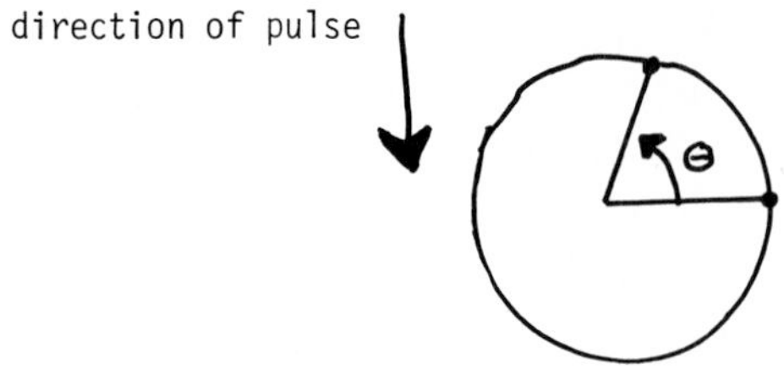

Integrating over the unit time interval (between n and n+1) to solve the differential equation (A1) we obtain:

$$x_{n+1} = (x_n + y_n) \bmod 2\pi$$

$$x_{n+1} = ay_n - G \cos (x_n + y_n) \qquad \text{(R)}$$

where:

x_n = angle at time n (taken mod 2π)

y_n = distance traveled between times n and n+1 (this is not taken mod 2π).

$a = e^{-\alpha}$; $G = P\left(\frac{1-e^{-\alpha}}{\alpha}\right)$.

HOLMES [2] uses this system to model a ball bouncing on a vibrating table. To arrive at this system he makes an approximating assumption that the time of flight of the ball can be estimated purely from the speed of the ball leaving the table, ignoring the phase of oscillation of the table. The system (R) is a variant of the standard map of CHIRIKOV [3]. The map (R) differs from the standard map in that dissipation is added (for CHIRIKOV $A = 1$). A more general derivation of a map representing a perturbed nonlinear oscillator is available in ZASLOVSKY [5]; although similar to map (R) in some respects, his variables are defined differently. In his paper [5] ZASLOVSKY established some formal conditions for the existence of a chaotic attractor in map (R). ZASLOVSKY and RACHKO elaborate in [6].

References

1. J. A. Yorke and K. T. Alligood: "Cascades of Period-Doubling Bifurcations: A Prerequisite for Horseshoes", Bull. Amer. Math. Soc. (1983), in press. A more detailed paper is in preparation
2. P. J. Holmes: "The Dynamics of Repeated Impacts with a Sinusoidally Vibrating Table", Journal of Sound and Vibration 84(2), pp. 173-189 (1983)
3. C. Grebogi, E. Ott and J. A. Yorke: "Crises, Sudden Changes in Chaotic Attractors and Transient Chaos", Physica D 7, pp. 181-201 (1983)
4. B. V. Chirikov: "A Universal Instability of Many-Dimensional Oscillator Systems", Physics Reports 52, 265-379 (1979)
5. G. M. Zaslovsky: "The Simplest Case of a Strange Attractor", Physics Letters 69A(3), pp. 145-147 (1978)
6. G. M. Zaslovsky and Kh.-R. Ya. Rachko: "Singularities of the Transition to a Turbulent Motion", Soviet Physics JETP 49(6), 1039-1055 (1979)

Part II

Fractals in Dynamical and Stochastic Systems

On the Dynamics of Iterated Maps VIII: The Map $z \to \lambda(z+1/z)$, from Linear to Planar Chaos, and the Measurement of Chaos

Benoit B. Mandelbrot
IBM Thomas J. Watson Research Center
Yorktown Heights, New York 10598 USA

0. Introduction

While the terms "chaos" and "order in chaos" prove extremely valuable, they elude definition and it remains important to single out instances when the progress to planar chaos can be followed in detailed and objective fashion. This paper proposes to show that an excellent such example is provided by the iterates of the map $z \to g(z) = \lambda(z+1/z)$, when z and λ are both complex. This map is touched upon in [1], but only on page 465, which was added in second printing. Therefore, the present paper is self-contained.

The map g(z) was singled out because of its valuable properties. A) Within a broad domain of λs, there are two distinct limit cycles, symmetric of each other with respect to $z = 0$. B) Suitable changes in λ cause both cycles to bifurcate simultaneously into $n>2$ times larger cycles. C) The chaos which prevails for certain λ extends over the whole z plane. These features A), B) and C) all fail to hold for the complex map $z \to f^*(z) = z^2-\mu$ (e.g.,[1], [2], [3], [4]). Indeed, for every μ, one of the limit cycles of f^* reduces to the point at infinity; this point never bifurcates; and chaos, when it occurs, consists in motion over a small subset of the z plane.

1. Summary. Relativity of the Notion of Chaos

The bulk of this paper consists in explanations for a series of figures that illustrate, for diverse λ, the shape of the Julia set $\mathcal{F}^*$, that is, of boundary of the open domains of attraction of the stable limit points and cycles. Different sequences of figures follow different "scenarios" of variation in λ, and yield maps that transform gradually from linear chaos and planar order, to either questionable or unquestioned planar chaos.

In order to put these illustrations in perspective, the paper includes comparisons with the polynomial maps. To begin with, the special map $z \to z^2-2$ restricted to the real interval $[-2,2]$ is called thoroughly chaotic. However, the very same map generalized to the complex plane should be called almost completely orderly, since all z_0 that are not in the real interval $[-2,2]$ iterate to ∞. As is well known [5], there are many other μ's for which the maps $z \rightarrow z^2 - \mu$ are chaotic on a suitable real interval. But the very same maps are *least* chaotic in the plane, in the sense that the domain of exceptional z_0 that fail to iterate to ∞ is smaller for a chaotic μ_0 than for any of the nonchaotic μ that can be found arbitrarily close to μ_0.

Thus, there is a clear need for an objective measure of the progress towards chaos. An obvious candidate for measuring orderliness is the fractal dimension D of the Julia set $\mathcal{J}^*$. This paper finds that D is indeed appropriate for some scenarios, but raises very interesting complications for other scenarios, when $\mathcal{J}^*$ involves more than one shape, hence more than one dimension.

The best known scenario, pioneered by J. Myrberg and very well explored in many contexts [5], proceeds from linear to planar chaos by an infinite series of finite bifurcations of arbitrary order. When this scenario is applied to g(z) (Section 5), $\mathcal{J}^*$ remains a fractal curve whose D grows from 1 in to 2, hence its codimension 2-D is indeed an acceptable measure of orderliness.

In one alternative scenario, which can be credited to C. L. Siegel, D also varies steadily from 1 to 2, but intuition tells us that the limit is very incompletely chaotic in the plane. The key of this paradox is that the corresponding $\mathcal{J}^*$ involves two different shapes, hence two distinct dimensions.

In the third scenario to be examined, planar chaos is approached without bifurcation, and D tends to 2.

2. When $|\lambda| > 1$ or λ are real, Iteration is Orderly Except on $\mathcal{J}^*$

For $\lambda = 0$, all points other than 0 and ∞ move in one step to 0, henceforth the motion is indeterminate. For $|\lambda| > 1$, there is an attractive fixed point at ∞, which contradicts our requirement A).

For real $\lambda > 0$, the map g(z) preserves the sign of Re(z), and for real &la<0, the iterated map $g_2(z)$ preserves Re(z). More generally the Julia set is the imaginary axis for all real $\lambda \neq 0$.

3. Non-real λ's that Satisfy $|\lambda| < 1$

For these λ, the Julia set $\mathcal{J}^*$ is either the whole complex plane or a fractal curve. In the latter case, $\mathcal{J}^*$ has the following properties.

$\mathcal{J}^*$ is (obviously) symmetric with respect to $z = 0$, and is self-inverse with respect to the circle $|z| = 1$.

$\mathcal{J}^*$ includes $z = 0$ and is unbounded. This is obvious when the fixed points $z = \pm\sqrt{\lambda}/\sqrt{1-\lambda}$ are stable: if z_0 iterates to one of the fixed points, $-z_0$

iterates to the other fixed point, hence the circle of radius mod(z_0) must intersect $\mathcal{F}^*$. (The origin $z = 0$ must be added because $\mathcal{F}^*$ is a closed set.)

$\mathcal{F}^*$ is asymptotically self-similar for $z \to \infty$. Indeed, if $|z| \gg 1$ and z_0 iterates into some cycle, $\lambda(z+1/z) \sim \lambda z$ iterates into the same cycle. Since $\mathcal{F}^*$ is self-inverse, it also follows that $\mathcal{F}^*$ is asymptotically self-similar for $z \to 0$. When $\mathcal{F}^*$ is topologically a line, it winds around a logarithmic spiral for $z \to \infty$ and for $z \to 0$. These spirals wind in the same direction, but *are not* the continuation of each other, because scale invariance fails near $|z| = 1$.

We wish to start with a real λ for which $\mathcal{F}^*$ is a straight line, and then to change λ and follow $\mathcal{F}^*$ as it changes from a straight line to an increasingly wiggly curve. This requires drawing the semi-open variant of the $\mathcal{M}$ set ("Mandelbrot set") as defined in MANDELBROT [2].

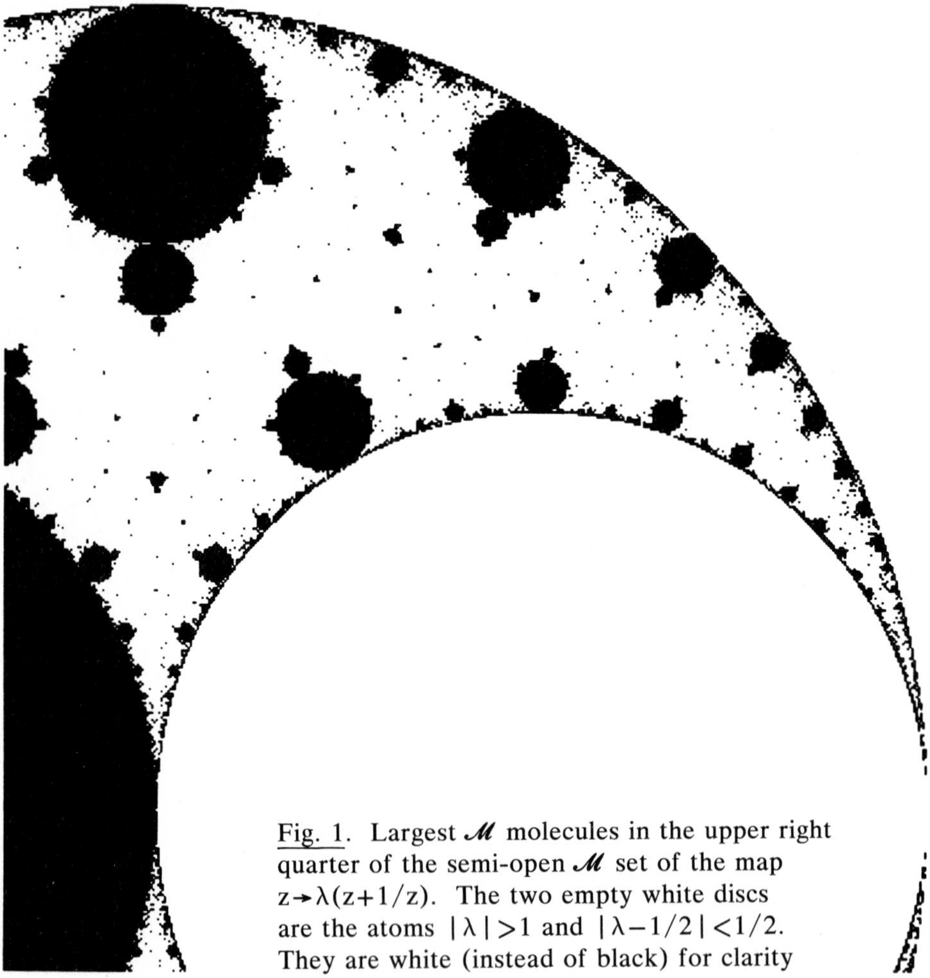

Fig. 1. Largest $\mathcal{M}$ molecules in the upper right quarter of the semi-open $\mathcal{M}$ set of the map $z \to \lambda(z+1/z)$. The two empty white discs are the atoms $|\lambda| > 1$ and $|\lambda - 1/2| < 1/2$. They are white (instead of black) for clarity

The semi-open $\mathcal{M}$ set is the maximal set of λ's, such that the iteration of the map has a finite limit cycle. Its closure of is the ordinary $\mathcal{M}$ set. The semi-open $\mathcal{M}$ set of $z \to \lambda(z+1/z)$ is shown on Fig. 1 (reproducing Plate x of [1], second and later printings. Note that p. 465, which explains Fig. x of [1] omits to say it is the semi-open $\mathcal{M}$ set, and replaces z by iz.)

Inspection shows that the semi-open $\mathcal{M}$ set is made of $\mathcal{M}$ molecules, each made of $\mathcal{M}$ atoms, both shapes being the same in the case of the map $\lambda(z+1/z)$ and in the deeply studied case of the maps $z^2-\mu$ [2, 3, 4].

The present study is concerned with three different scenarios that start from the extreme order represented by λ's in the real interval]0,1[—hence a Julia set identified with the imaginary axis—and end in planar chaos. We focus on the $\mathcal{M}$ molecule that includes the disc-shaped $\mathcal{M}$ atom $|\lambda-1/2|<1/2$. It is easy to see that this atom collects all λ's for which the iteration of f(z) has 2 limit points, $z = \pm\sqrt{\lambda}/\sqrt{1-\lambda}$.

4. From a Flat Sea to a Great Wave: the Computer's Homage to Katsushika Hokusai (1760-1849) (Fig. 2)

For all λ in the disc $|\lambda-1/2|<1/2$, the Julia $\mathcal{F}^*$ is topologically a straight line that winds for $z \to 0$ or $z \to \infty$ around logarithmic spirals symmetric of each other with respect to 0. The spirals are both nicest and most educational when they are neither too loose not too tight. Let us therefore scatter a few parameter values, well within the $\mathcal{M}$ atom $|\lambda-1/2|<1/2$, between $\lambda = 1/2$ and the neighborhood of 1/2+i/2. To deemphasize the non-spiral complications near $|z| = 1$, the window (portion of the complex plane that is shown) is 200 units wide, and the $\mathcal{F}^*$ sets are rotated to become easier to compare. The $\mathcal{F}^*$ sets show as the boundaries between black "water" and white "air", which are the domains of attraction of two limit points. As intended, the first part of Fig. 2 evokes a completely flat black sea, hence planar order. And the figures that follow counter-clockwise evoke increasingly threatening black waves.

In parallel, the fractal dimension D of $\mathcal{F}^*$ increases. In this context, D tells how many decimals of z_0, in the counting base b, are needed to know whether z_0 is attracted to $\sqrt{\lambda}/\sqrt{1-\lambda}$, or to $-\sqrt{\lambda}/\sqrt{1-\lambda}$. To establish this fact, draw on our window a collection of boxes of relative side $r_1 = 1/b$. Roughly b^D of these boxes intersect $\mathcal{F}^*$. Write $\beta = b^{D-2}$, and choose $= x_0+iy_0$ at random in the box. With the probability $1-\beta$, the first b decimals of x_0 and y_0 suffice to determine where z_0 is attracted. More generally, the first k b decimals of x_0 and y_0 suffice with the probability $(1-\beta)\beta^{k-1} \propto b^{k(D-2)} \propto r_k{}^{2-D}$. On the average, the number of b decimals needed to determine the limit is $1/(1-\beta)$. When 2-D is small, the expected number of base e "decimals" needed to determine the limit is $\log_e b/(1-\beta) \sim 1/(2\text{-}D)$.

Fig. 2. Examined counterclockwise from here, Julia $\mathcal{F}^*$ sets of $z \rightarrow \lambda(z+1/z)$ for several selected values of λ. "From a Flat Sea to a Great Wave". Homage to Katsushika Hokusai

5. First Path Beyond the Great Wave. The Myrberg Scenario of Bifurcations

Figure 3 represents the $\mathcal{F}^*$ sets for two values of λ. In the top graph, λ lies past a bifurcation into 4, close to (but short of) a second bifurcation into 3. In the bottom graph, λ is reached by two successive bifurcations into 4, followed by a bifurcation into 3. Thus, the first λ lies off the center in an $\mathcal{M}$ atom off the atom $|\lambda-1/2|<1/2$. And the second λ lies near the nucleus in a small $\mathcal{M}$ atom off a small $\mathcal{M}$ atom off a small $\mathcal{M}$ atom attached to the $\mathcal{M}$ atom $|\lambda-1/2|<1/2$. (A fourth λ is seen,p. viii of [1], second printing.)

The first bifurcation forms "white water" through the breakdown of connected water and connected air into larger drops, some of them quite large. The bifurcations that follow break these drops into smaller ones, without end. It is clear that one watches a gradual progression towards the ultimate replacement of separate black water and white air by something that is neither water nor air. One cannot help evoking the critical temperature of physics.

The fractal dimension D of $\mathcal{F}^*$ tends toward 2 as planar chaos is approached, and the factor β tends to 1.

6. Second Path Beyond the Great Wave. A Scenario of Spiraling Towards Chaos (Fig. 4)

Now select λ to be within the atom $|\lambda-1/2|<1/2$ but extremely close to $\lambda = 1$. The $\mathcal{F}^*$ set is illustrated by Fig. 4. It is clear that as $\lambda\to 1$, $D\to 2$, hence $\beta\to 1$, and that *chaos is approached without bifurcation.* The facts are perhaps easier to visualize in terms of the parameter $\mu = 1/\lambda$ and the variable $u = 1/z$. This change of variable does not change $\mathcal{F}^*$. For $|\mu|<1$, there is one limit point at $u = 0$. As μ crosses 1, this limit point bifurcates into two limit points that coexist in a chaotic situation.

7. Third Path Beyond the Great Wave. The Siegel Scenario

Figure 5 represents the $\mathcal{F}^*$ set for a value of λ within the atom $|\lambda-1/2|<1/2$, but very close to a point on boundary, namely $\lambda_S = 1/2+(1/2)\exp(2\pi i\gamma)$, where γ is the irrational number (<1) whose continued fraction expansion is (4,1,1,1,... ad infinitum).

This path is interesting because its limit is hard to label as chaotic or non-chaotic. One can understand the difficulty and the opportunity by viewing the boundary between colors on Fig. 5, first, as an approximate $\mathcal{F}^*$ set for $\lambda \equiv \lambda_S$ and, next, as an approximate $\mathcal{F}^*$ set for λ just short of λ_S on the way from $\lambda = 1/2$. The same figure can serve two purposes because the differences between the corresponding exact figures are erased by the inevitable limitations of actual computation. Furthermore, the roughly circular white

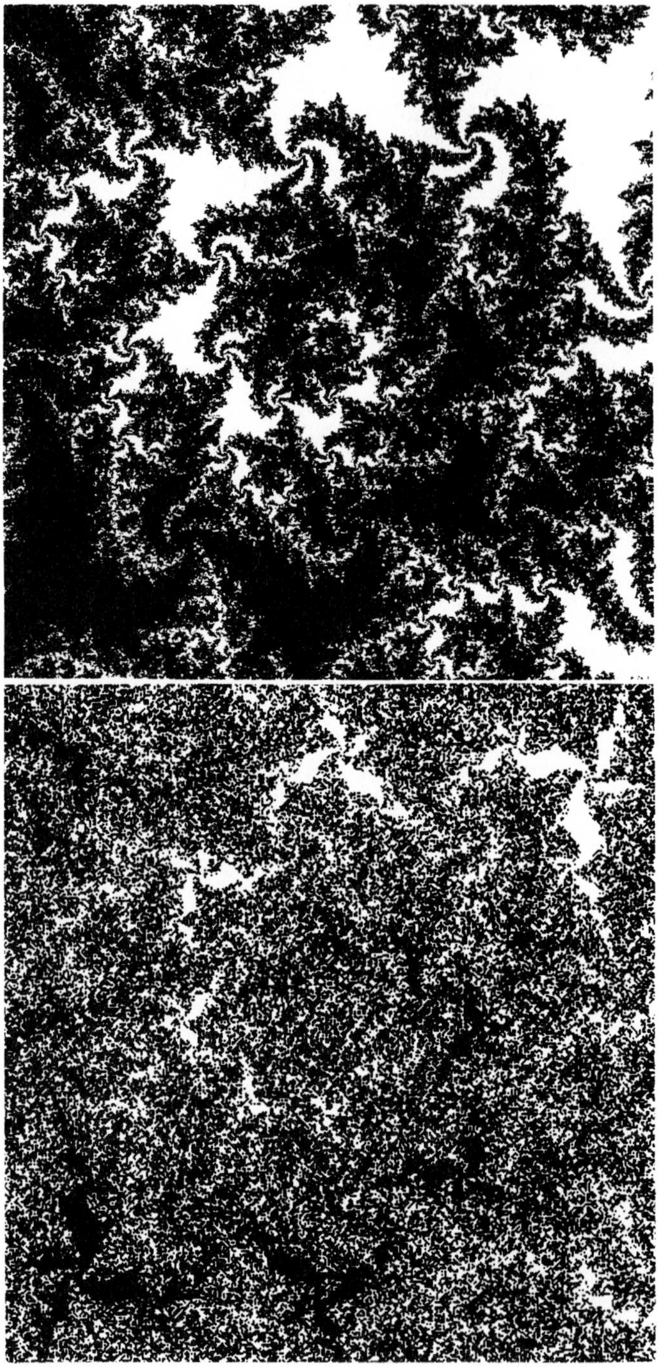

Fig. 3. Julia sets for two λ that yield near totally chaotic maps g(z), along the Myrberg scenario of repeated bifurcations

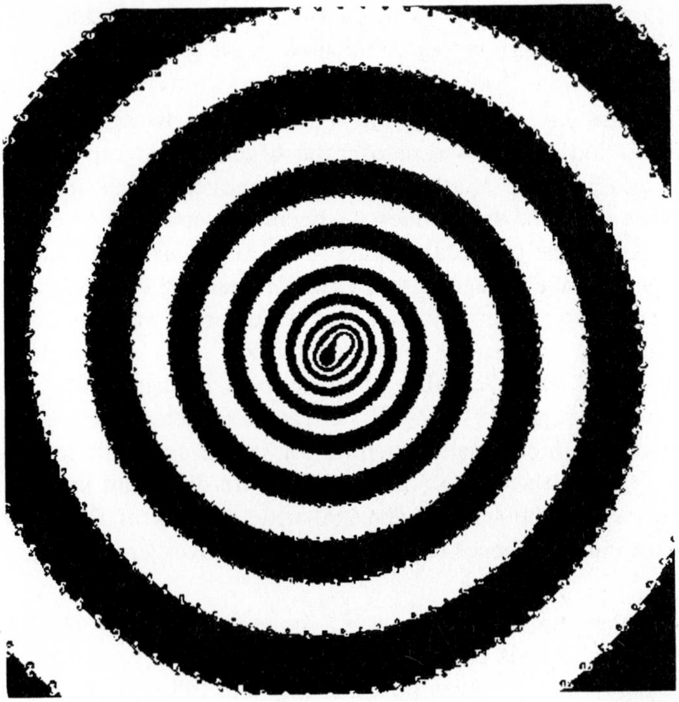

Fig. 4. Julia set for λ that yields a near totally chaotic map and is attained by yet another scenario. Topologically, this curve is a straight line

Fig. 5. Julia Set for λ that yields a questionably chaotic map g(z), and is attained by the Siegel scenario

spots do not contribute to $\mathcal{F}^*$, because they too are computation artifacts: the values of z_0 that had failed to converge to either limit point after 3000.

The behavior for $\lambda \equiv \lambda_S$ is known from a theory due to SIEGEL [6]. Focus on the "cracks" that seem to separate the black wave into roughly circular black discs, and imagine that these cracks converge and join. It follows that water—and air also, by symmetry—becomes separated into discs attached to each other by single punctual bonds. Two of the discs include the points $\pm\sqrt{\lambda}/\sqrt{1-\lambda}$ and are called Siegel discs; let the remaining discs be called Siegel pre-discs. At each inter-disc bond, air and water cross each other but over most of the plane they are clearly separated by $\mathcal{F}^*$. Not unexpectedly, the fractal dimension of $\mathcal{F}^*$ takes a value of D_S that is unquestionably less than 2. One needs on the average $\sim 1/(2-D_S)$ decimals to determine whether a point z_0 is black or white. Incidentally, there is no limit point or limit cycle, but the iterates of the z_0's in the white (black) Siegel pre-discs end up in the white (black) Siegel disc. On the scale of the 200-wide window of Fig. 3, the Siegel discs are so small that the Siegel regime looks like convergence.

Next, in order to achieve an idea of how $\mathcal{F}^*$ looks for λ just short of λ_S, it is necessary to know that Siegel discs are created when a curve $\mathcal{F}^*$ that is topologically a line folds up and becomes domain or plane filling, as described and illustrated in [4, Paper VII]. When λ is just short of λ_S, the cracks invoked in the preceding paragraph have not converged and joined. Instead, the interior of each of the black discs is partly split by many (here, 157) very narrow "fjords", that penetrate deep into the white domains, without quite meeting, but coming close to meeting near the center of a spurious white spot. In symmetric fashion, one must visualize black fjords thrusting into the white domain.

Since the boundaries of both white and black fjords are part of $\mathcal{F}^*$, the curve $\mathcal{F}^*$ is very close to filling the whole plane. Its dimension being arbitrarily close to 2 tempts us to conclude that the corresponding map g(z) is completely chaotic. But it is not. In fact, the Siegel scenario reveals an important and subtle point: We need a close look at the factor β. For very tiny values of the cell side r_k, we find $\beta \propto r_k^{2-D}$. However, as long as $r_k > \xi$, with ξ a function of 2-D, we find $\beta \equiv 1$. Thus, the smallness of 2-D expresses that *every* cell of side $>\xi$ will be intersected by $\mathcal{F}^*$. But this does not say anything about the relative proportions of black and white in the cells $>\xi$. In the present case, the fjords are so narrow that a cell $>\xi$ is mostly black or mostly white, depending on whether it is in a domain that Fig. 5 shows as solid black or solid white. The expected number of decimals of base b depends upon whether one wants to know the color of z_0 precisely, or with high probability. Absolute precision requires $1/[1-b^{D-2}]$ decimals, high probability requires only $1/[1-b^{D_S-2}]$.

In other words, the overall appearance that computer limitations give to Fig. 5 is not misleading at all. In fact, it helps reveal a basic truth. When λ is very near λ_S, the shape of $\mathcal{F}^*$ is ruled by *two* distinct dimensions: its own and that of the $\mathcal{F}^*$ corresponding to the nearest Siegel value of λ.

We must agree that near-complete planar chaos should require that all small cells a) intersect $\mathcal{F}^*$ and b) be about half black and half white. Under these conditions, a nearly space-filling $\mathcal{F}^*$ set of dimension nearly 2 is *not* sufficient for complete chaos. The presentation of further results on this topic must be postponed to a later occasion.

Acknowledgement

The illustrations were prepared by James A. Given, using computer programs by Alan Norton.

References

1. B. B. Mandelbrot: The Fractal Geometry of Nature (W. H. Freeman, New York 1982).

2. B. B. Mandelbrot: Fractal aspects of the iteration of $z \to \lambda z(1-z)$ for complex λ and z. Non Linear Dynamics, Ed. R. H. G. Helleman. Annals of the New York Academy of Sciences, **357**, 249-259 (1980).

3. B. B. Mandelbrot: On the quadratic mapping $z \to z^2 - \mu$ for complex μ and z: the fractal structure of its $\mathcal{M}$-set and scaling. Physica **7D**, 224-239 (1983); also in Order in Chaos, Ed. D. Campbell (North Holland, Amsterdam).

4. B. B. Mandelbrot: On the dynamics of iterated maps III: The individual molecules of the $\mathcal{M}$ set, self-similarity properties, the N^{-2} rule, and the N^{-2} conjecture. IV: The notion of "normalized radical" $\mathcal{R}$ of the $\mathcal{M}$ set, and the fractal dimension of the boundary of $\mathcal{R}$. V: Conjecture that the boundary of the $\mathcal{M}$ set has a fractal dimension equal to 2. VI: Conjecture that certain Julia sets include smooth components. VII: Domain-filling ("Peano") sequences of fractal Julia sets, and an intuitive rationale for the Siegel discs. Chaos, Fractals and Dynamical Systems, Ed. P. Fischer and W. Smith. (Marcel Dekker, New York 1984).

5. P. Collet and J. P. Eckmann Iterated Maps on the Interval as Dynamical Systems (Birkhauser, Boston 1980).

6. C. L. Siegel: Iteration of analytic functions. Annals of Mathematics **43**, 607-612 (1942).

Self-Similar Natural Boundaries of Non-Integrable Dynamical Systems in the Complex t Plane

H. Yoshida

Department of Astronomy, University of Tokyo
Tokyo 113, Japan

Self-similar structure of the set of singularities (*natural boundary*) in the complex t plane is explained by the expansion of solution about singularities.

1. Introduction

Recently much attention has been paid to the relation between the integrability (non-integrability) of dynamical systems and the distribution of singularities of solution in the complex t plane. Among others, CHANG et al.[1),2)] found numerically in the Hénon-Heiles system that the set of singularities have a self-similar structure and form natural boundaries of fractal dimension ($D>1$) roughly parallel to the real t axis. The purpose of this note is to clarify the following two points; i.e. (i) Why do the set of singularities have the self-similar structure ? and (ii) What determines the parameters of self-similar structure such as its fractal dimension ?

In an algebraically integrable system, the solution has multi-periodicity in the complex direction and no natural boundaries appear. Thus, the appearance of natural boundaries is one characteristic in non-integrable systems. The details will be published elsewhere.

2. Numerical Procedure - Automatic Detection for Singularities

The model problem is the 2-degrees of freedom Hamiltonian system

$$H = \frac{1}{2}(p_1^2+p_2^2)+q_1^2 q_2 + \frac{\varepsilon}{3}q_2^3 \qquad (1)$$

with canonical variables q_i, p_i and a constant parameter ε. The initial condition for $t=0$ is fixed as $q_1=5, q_2=p_1=p_2=0$. First, the Hamiltonian equation is integrated numerically along the real-t axis from $t=0$ to $t=t_1(>0)$. From there we search the integration path and integrate the equation numerically in the complex t plane so that the value of $|q_2(t)|$ is constant along the path. By this way we obtain a contour $|q_2(t)|=const.$ At each point on this contour, CHANG's three-term test[3)] is employed to detect the position of the nearest singularity and its leading order. This test, which works well when and only when there is only one singularity on the convergence circle of the Taylor series, gives a consistent distribution of singularities.

Fig.1 shows the adopted contour of $|q_2(t)|$ and detected singularities for $\varepsilon=-1$. The Taylor expansion of solution at $t=0$ with our initial condition proves that the distribution of singularities has four lines $Re(t)=0, Im(t)=0, Re(t)=Im(t), Re(t)=-Im(t)$ as its axes of symmetry.

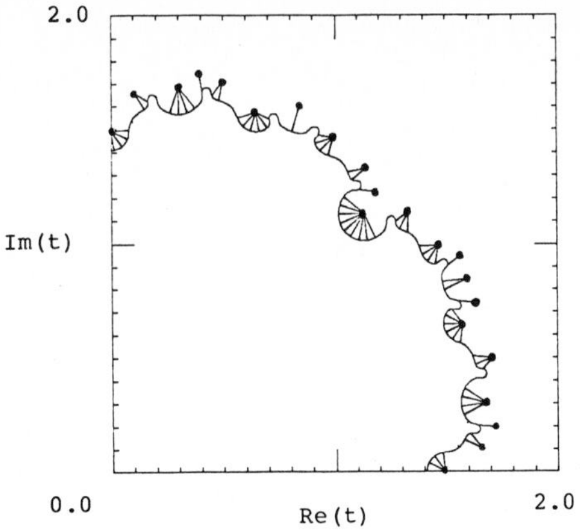

Fig.1 : A contour $|q_2(t)| = const.$ and detected singularities

The set of singularities has locally the same self-similar structure as obtained by CHANG et al. If we adopt an outer contour to detect singularities, we find more singularities and deeper structure among them. Further numerical computations show that although the global symmetric structure strictly depends on the adopted initial condition at $t=0$, the local self-similar structure does not depend on the initial condition.

3. Reason for Self-Similar Structure - Imaginary Kowalevski's Exponent

Suppose the solution of Hamiltonian system (1) has a singularitiy at a point $t=t_*$ in the complex t plane. Then in the neighborhood of t_*, $q_i(t)$ has, in general, the following series expansion [4),5)],

$$q_i = t^{-2}\{c_i + P_i(I_{\rho_1}t^{\rho_1}, I_{\rho_2}t^{\rho_2}, I_6 t^6)\}, \qquad (t \equiv t - t_*) \tag{2}$$

where $I_{\rho_1}, I_{\rho_2}, I_6$ are three arbitrary constants and $P_i(x,y,z)$ is a Taylor series in its arguments without constant term, which converges for sufficiently small $|x|, |y|, |z|$. Constants c_i and exponents ρ_1, ρ_2 (Kowalevski's exponents) are determined by direct substitution of (2) into the Hamiltonian equation. According to two choices of c_1, c_2, i.e.

$$(i) \quad c_1 = \pm\sqrt{9(2-\varepsilon)}, \; c_2 = 3, \quad \text{and} \quad (ii) \quad c_1 = 0, c_2 = -6/\varepsilon, \tag{3}$$

the exponents are defined as the roots of the quadratic equations

$$(i) \quad \rho^2 - 5\rho + 6(2-\varepsilon) = 0, \quad \text{and} \quad (ii) \quad \rho^2 - 5\rho + 6(1-2/\varepsilon) = 0, \tag{4}$$

respectively. YOSHIDA [5)] showed that the appearance of at least one irrational or imaginary Kowalevski's exponent means the non-existence of another algebraic (and analytic) first integral independent of Hamiltonian. In the range $-48 < \varepsilon < 0$, the exponents in both cases become imaginary (complex conjugate pair), and therefore the Hamiltonian system (1) is non-integrable. In this case it can be proved that H *(Hamiltonian)* is proportional to the arbitrary constant I_6, and since $H=0$ for our initial condition, it follows that $I_6=0$. Thus, the expansion of solution reduces to

$$q_i = t^{-2}\{c_i + P_i(I_\rho t^\rho, I_{\rho^*} t^{\rho^*})\}. \tag{5}$$

where ρ^* is complex conjugate of ρ.

For simplicity, first consider the case of single imaginary exponent

$$q_i = t^{-2}\{c_i + P_i(I_\rho t^\rho)\}, \tag{6}$$

obtained by putting $I_{\rho^*} = 0$ in (5). On the convergence circle of the Taylor series $P_i(x)$ in (6), there is at least one singularity, say, at $x = x_0$. Put t_0 as a solution of $x_0 = I_\rho t^\rho$. This means that at $t=t_0$ the function $q_i(t)$ in (6) becomes singular. Then we find that at a sequence of points $t_n (n = 1,2,\ldots -1,-2,\ldots)$ defined by $t_n = s^n t_0$ with $s = exp(2\pi i/\rho)$, the solution becomes singular since $(t_n)^\rho = exp(2\pi i n)(t_0)^\rho = (t_0)^\rho$ (Fig.2). Therefore, around a singularity with imaginary Kowalevski's exponent, other singularities are located on a logarithmic spiral or *convergent spiral* so that the set of singularities form a sequence of similar triangles (*fundamental triangle*) defined by a complex number s. If we put $s = re^{i\theta}$, then r and θ are expressed as

$$r = exp\{2\pi\rho_I/|\rho|^2\}, \qquad \theta = 2\pi\rho_R/|\rho|^2, \tag{7}$$

where $\rho = \rho_R + i\rho_I$. For the type (i) singularity, we have by direct computation

$$r_1 = exp\{-\pi\sqrt{23-24\varepsilon}\ /6(2-\varepsilon)\}, \qquad \theta_1 = 5\pi\ /\ 6(2-\varepsilon). \tag{8}$$

For the type (ii) singularity, the expansion coefficients of $P_i(x)$ in (6) vanish alternately. So, the *effective* exponent becomes 2ρ and the fundamental triangle should be defined by $s = exp(\pi i/\rho)$. Thus,

$$r_2 = exp\{-\pi\sqrt{-1-48/\varepsilon}\ /12(1-2/\varepsilon)\}, \qquad \theta_2 = 5\pi\ /\ 12(1-2/\varepsilon). \tag{9}$$

Here, we remark that $\theta_1 + \theta_2 = 75^\circ$ independent of ε.

In our problem, the expansion of solution (5) has two complex conjugate exponents. We can show that in the domain of complex t plane such that $|I_\rho t^\rho| \gg |I_{\rho^*} t^{\rho^*}|$, the distribution of singularities of (5) can be well approximated by that of (6). On the contrary, in the domain such that $|I_\rho t^\rho| \ll |I_{\rho^*} t^{\rho^*}|$, we can use (6) with the substitution $\rho \to \rho^*$. Hence, around a singularity with two complex conjugate exponents, there are two convergent *semi*-spirals ; one is clockwise and another anti-clockwise.

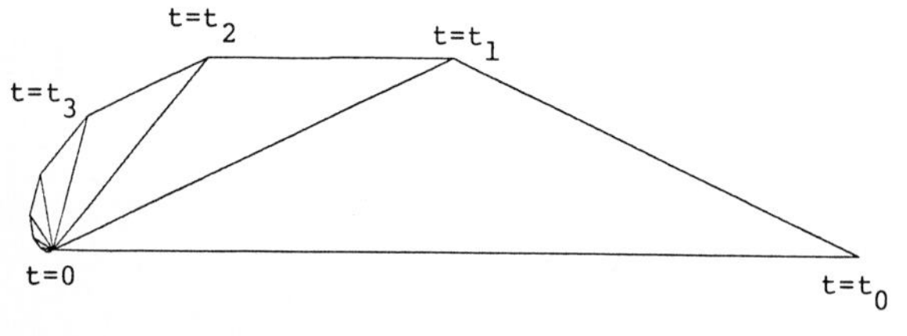

Fig.2 : Distribution of singularities around $t = 0$ ($t = t^*$)

Fig.3 shows the construction of two fundamental triangles around type (i) and type (ii) singularities for the case $\varepsilon = -1$. In Fig.3, we find by formula (8) and (9) that $\theta_1 = 50^\circ$, $\theta_2 = 25^\circ$, $r_1 = r_2^2 = 0.302$. Therefore, both triangle ABD and triangle DBC approximately become isosceles triangles with base angle 25° . Continuing the process of Fig.3 *ad infinitum* we have Fig.4, where the set of singularities form a natural boundary of some fractal dimension. The full natural boundary for the case $\varepsilon = -1$ is shown in

Fig.5(a) . Other cases of $\varepsilon = -8$ and $\varepsilon = -0.1$ are shown in Fig.5(b) and Fig.5(c). These figures well coincide with the set of singularities obtained by the numerical procedure of the preceding section.

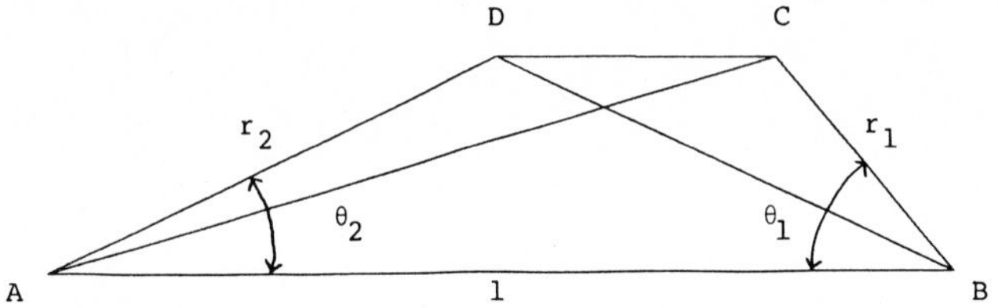

Fig.3 : Two fundamental triangles and definition of r_1, θ_1, r_2, θ_2

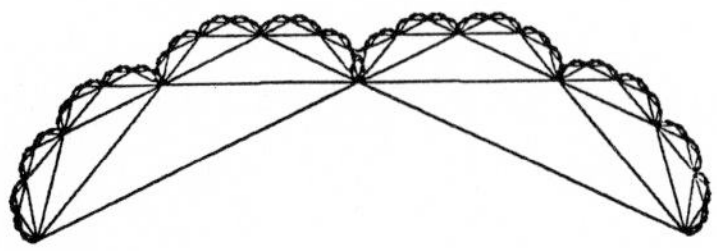

Fig.4 : Geometrical construction of natural boundary ($\varepsilon = -1$)

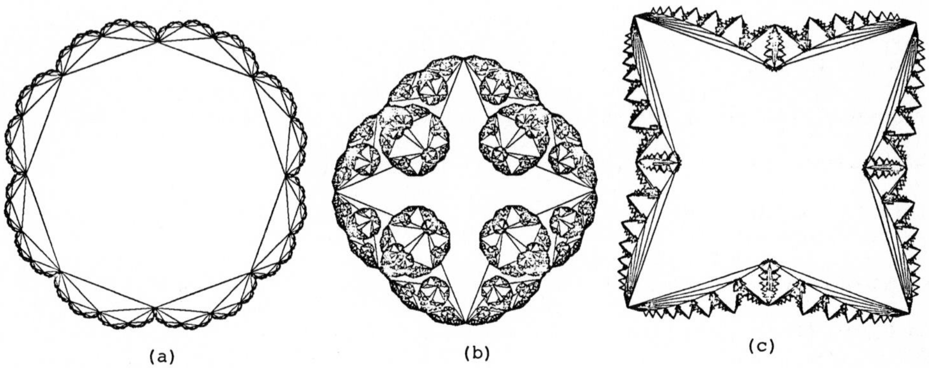

Fig.5 : Global natural boundaries. (a) $\varepsilon = -1$, (b) $\varepsilon = -8$, (c) $\varepsilon = -0.1$

4. Concluding Remarks

In this note, it was shown that the self-similar structure of natural boundaries of non-integrable systems can be well explained by the rigorous expansion of solution about singularities. More precisely, the set of Kowalevski's exponents solely determine the fundamental triangles and precise self-similar structure. Investigations on (i) Hamiltonian system with non-homogeneous potential, and (ii) Hamiltonian system with three (or more) degrees of freedom, still remain.

References

1) Y.F.Chang, M.Tabor, J.Weiss and G.Corliss : *Phys. Lett.* A85, 211 (1981).
2) Y.F.Chang, M.Tabor and J.Weiss : *J. Math. Phys.* 23, 531 (1982).
3) Y.F.Chang and G.Corliss : *J. Inst. Math. Applics.* 25, 349 (1980).
4) H.Yoshida : *Kôkyûroku RIMS, Kyoto University*, 472, 143 (1982). *in Japanese.*
5) H.Yoshida : "Necessary condition for the existence of algebraic first integrals, Part I and Part II", *Celestial Mech.* in press.

Topological Phase Transitions

M. Widom

Department of Physics, Harvard University, Cambridge, MA 02138, USA

S.J. Shenker

Department of Physics, Cornell University, Ithaca, NY 14850, USA

Abstract

We use methods of statistical mechanics to study the Julia set of the mapping $f(z)=z^2+p$. For most values of p these methods allow extremely accurate determination of the fractal dimension and escape rate. At special values of p the Julia set undergoes a change in topology. We study the value $p=1/4$ in detail and find singularities which we interpret as a phase transition.

1. Introduction

Julia sets are unstable invariant sets of complex analytic mappings. They possess chaotic dynamical behavior, a fractal dimension, and sometimes sensitive dependence on parameters. Despite these remarkable properties, which are shared by strange attractors, the theory of Julia sets is well established in an elegant and extensive mathematical literature.

In this article we consider the fractal dimension and escape rate of Julia sets. We utilize theorems of RUELLE [1] and BOWEN [2] in numerical measurements of these quantities. In this respect this article is a refinement of an earlier paper [3] which discussed the Julia set of

$$f(z) = z^2 + p \tag{1.1}$$

in the limits of large and small p. The statistical mechanical description of Julia sets [4,5] allows extremely accurate measurements of the fractal dimension and escape rate when p is small.

The present article is also an extension of [3] because we explore a new range of values of the parameter p. In particular, we study phenomena on and near the boundary of the Mandelbrot set [6]. The Mandelbrot set, M, is the set of all values of p for which Julia set of (1.1) is connected. This is the set of parameters for which the critical point does not iterate to infinity [7].

Consider a subset of M consisting of the set of parameter values for which f has a stable fixed point. The Julia set in this case is the boundary of the basin of attraction and is a Jordan curve. The critical point lies in the basin of attraction. In order to determine the parameter values which constitute this subset we compute the derivatives of the mapping at the fixed points

$$\lambda = 1 - \sqrt{1-4p} \tag{1.2}$$

$$\mu = 1 + \sqrt{1-4p} \quad . \tag{1.3}$$

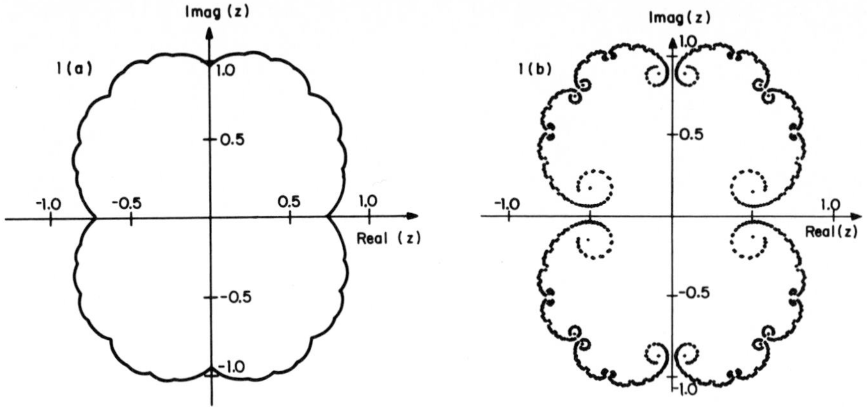

Figure 1. The Julia set of the mapping (1.1) with (a) $p = 0.20$, and (b) $p = 0.26$

The mapping (1.1) has a stable fixed point for any parameter value inside the cardioid

$$C = \{p : |\lambda(p)| < 1\} \quad . \tag{1.4}$$

Imagine varying the parameter p in such a way that

$$\theta = \arg(\lambda) \tag{1.5}$$

is held fixed while $|\lambda|$ increases. In this article we concentrate on the value $\theta = 0$. As p increases from 0 to 1/4, λ increases from 0 to 1. The value $p = 1/4$ lies on the boundary of C. For $p > 1/4$ the critical point iterates to infinity, and the Julia set is completely disconnected. Thus the value $p = 1/4$ lies on the boundary of M and as the parameter crosses this boundary the Julia set undergoes a transition of its topology (see Figure 1).

2. Small p

Many authors have noted an analogy between a statistical mechanical system on a one-dimensional lattice and a mapping [5,8-11]. In this analogy a bond on the lattice corresponds to an iteration of f, the state space at each lattice site corresponds to the domain and range of f, and the interaction between lattice sites can be represented by a transfer operator with kernel

$$T_D(z',z) = \delta^{(2)}(z' - f(z))\left|\frac{df}{dz}\right|^{2-D} \tag{2.1}$$

where $\delta^{(2)}$ is the two-dimensional delta function, and D is a parameter which will play the role of inverse temperature.

The partition function on a lattice with N sites and periodic boundary conditions is

$$Z_N = \int dz T_D^N(z,z) = \sum_{\text{Fix } f^N} \frac{\left|\frac{df^N}{dz}\right|^{2-D}}{\left|1 - \frac{df^N}{dz}\right|^2} \qquad (2.2)$$

We choose to restrict the integration in (2.2) so that the sum includes only unstable cycles. RUELLE [4] has shown that the spectrum of T is discrete and has a real nondegenerate largest eigenvalue, provided that p is not on the boundary of M. Thus

$$Z_N(p,D) = \sum_{m=0}^{\infty} \lambda_m^N \qquad (2.3)$$

where $\lambda_0 > 0$ and $\lambda_0 > |\lambda_1| \geqslant |\lambda_2| \geqslant \cdots$. In addition RUELLE [1] showed that the λ_m are real analytic functions of D and p.

This theorem possesses enormous utility in numerical computations of the fractal dimension and escape rate for small p. The fractal dimension is defined by [1,2]

$$\lambda_0(D_F) = 1 \qquad (2.4)$$

and the D-dimensional escape rate, R, is defined by [3]

$$R = \log \lambda_0(D) \quad . \qquad (2.5)$$

Our numerical approach is to compute the partition function Z_N for several values of N and fit this to a truncated sum of exponentials as in (2.3). Remarkably we find that the eigenvalues λ_m are all real for $0 \leqslant p < 1/4$. Thus we can fit L eigenvalues exactly to L values of Z_N. We have carried out this procedure for lattices of up to $N = 12$ sites. In Table 1 we show the value of D for which $\lambda_0(D) = 1 \pm 10^{-13}$ as computed in an $L = 1,2,3,4$ eigenvalue fit.

Table 1. Convergence of D_F when $p = 0.025$

L	D_F
1	1.00021
2	1.00023457
3	1.0002345919
4	1.00023459189

3. Phase Transition

Notice that the derivative at the unstable fixed point, μ, equals 1 when $p = 1/4$. This means that while thermodynamic averages may still exist, the partition function Z_N becomes infinite. By choosing a more complicated transfer operator than (2.1) we can define a new partition function

$$\hat{Z}_N = \sum_{\text{Fix } f^N} \left| \frac{df^N}{dz} \right|^{-D} \quad . \tag{3.1}$$

When p is not on the boundary of M the cycles are all unstable and the sum in (3.1) equals the sum in (2.2) except for corrections which are exponentially small in relation to the partition function. RUELLE [4] has shown that when p is not on the boundary of M

$$\hat{Z}_N = \hat{\lambda}_0^N \pm \hat{\lambda}_1^N \pm \hat{\lambda}_2^N \pm \cdots \tag{3.2}$$

where $\hat{\lambda}_0 = \lambda_0$. We propose to employ equations (3.1) and (3.2) when $p = 1/4$.

We can compute $\hat{\lambda}_m(D,p)$ in some special cases. Consider first $D = 0$, $0 \leqslant p \leqslant 1/4$.

There are $2^N - 1$ cycles of length N, so that

$$\hat{Z}_N(D,p) = 2^N - 1 \quad . \tag{3.3}$$

Now consider the limit of large D. Only cycles with small $|df^N/dz|$ will contribute to the sum (3.1). The fixed point has the smallest derivative, thus

$$\hat{Z}_N(D,p) \approx \mu^{-DN} \tag{3.4}$$

so that $\hat{\lambda}_0 \approx (1 + \sqrt{1-4p})^D$ for large D. Aside from these special cases the computation must be done numerically.

We will now report the results of our numerical work. The convergence of some of our results is far worse than that shown in Table 1. We will discuss the reliability of our results and the possible cause of poor convergence. Two principal results of which we are quite confidant are

$$D_F(p) = D^* - x_0\sqrt{1-4p} + O(1-4p) \tag{3.5}$$

where $D^* = 1.083$ and $x_0 = 0.15$, and

$$\hat{\lambda}_0 = (1 + \sqrt{1-4p})^{-D} + O(1-4p) \tag{3.6}$$

when $D \geqslant D^*$. A third result, valid for $D < D^*$, is

$$\hat{\lambda}_0 = 1 + A(D^* - D)^y + O(\sqrt{1-4p}) \tag{3.7}$$

where $A = -0.434$ and $y = 1$. Equations (3.6) and (3.7) show that $\hat{\lambda}_0$ is a singular function of D at D^*. By analogy with statistical mechanics we will analyze this singularity as if it were a singluarity in some thermodynamic quantity as a function of inverse temperature. Thus, assume

$$\hat{\lambda}_0(D,p) - 1 = (D^* - D)^a \psi\left(\frac{D^*-D}{(1-4p)^b}\right) \quad . \tag{3.8}$$

Equation (3.6) requires $\psi(x) \sim x^{-a}$ for $x \to -\infty$ and $ab = 1/2$. Equation (3.7) requires $\psi(x) \sim A$ for $x \to +\infty$ and $a = y$. Assume that x_0 is a zero of $\psi(x)$. By equations (2.4) and (3.8) we have

$$D_F(p) = D^* - x_0(1-4p)^{1/2y} \tag{3.9}$$

which is consistent with (3.5) if $y = 1$.

4. Numerical Difficulties

Our numerical procedure is similar to that used to study small p. The essential difference is that we must allow subtraction as well as addition of eigenvalues in fitting $\hat{Z}_N$. Padé analysis on the zeta function [1,4] leads us to conclude that the spectrum is still discrete, real, and positive, and that the partition function takes the form

$$\hat{Z}_N = \hat{\lambda}_0^N + 1 - \hat{\lambda}_2^N + \hat{\lambda}_3^N - \hat{\lambda}_4^N + \cdots \quad . \tag{4.1}$$

Equation (4.1) suggests that the measurement of D^* should improve exponentially with L as we saw in Section 2. In Table 2 we see that instead we have $1/N$ type corrections. Our preliminary evidence based on fits to (4.1) shows that when $D < D_F$ the value of $\hat{\lambda}_0$ converges exponentially while all other $\hat{\lambda}_m$ approach 1 with $1/N$ corrections. When $D > D_F$ all eigenvalues approach 1 with $1/N$ corrections.

Table 2. Convergence of D_F when $p = 1/4$. The fit is to (4.1) and N is the largest size lattice used in the fit. The third column includes corrections up to $(1/N)^3$

N	D_F	D_F corrected
14	1.073217	1.095506
15	1.073734	1.089383
16	1.074229	1.086428
17	1.074688	1.084867
18	1.075107	1.083814
19	1.075487	1.083152

These numerical difficulties lead us to speculate that the spectrum may actually be continuous between 0 and 1 with an isolated $\hat{\lambda}_0 > 1$ when $D < D^*$. The Padé analysis could easily miss this fact. It is interesting to note that $1/N$ decays have been observed in other dynamical systems which are hyperbolic except at isolated points [12].

Acknowledgement

We wish to thank L.P. Kadanoff for many stimulating discussions. This work was supported in part by NSF Grant #80-20609 at the University of Chicago. We acknowledge the support of a W.R. Harper fellowship (M.W.) and an R.R. McCormick fellowship (S.J.S.).

References

1. D. Ruelle, J. Ergodic Theory and Dynamical Systems 2, 99 (1982).
2. R. Bowen, Publ. Math. IHES 50, 11 (1979).
3. M. Widom, D. Bensimon, L.P. Kadanoff, S.J. Shenker, J. Stat. Phys. to appear (1983).
4. D. Ruelle, Invert. Math. 34, 231 (1976).
5. R. Bowen, *Equilibrium States and the Ergodic Theory of Anosov Diffeomorphism*, Springer (Berlin) Lecture Notes in Math. 470 (1975).
6. B. Mandelbrot, Annals of the N.Y. Academy of Sciences 357, 249 (1980).
7. For an excellent review article on theorems of complex mappings see H. Brolin, Arkiv For Matematik 6, 103 (1965).
8. D. Ruelle, *Thermodynamic Formalism*, Addison-Wesley:Reading (1978).
9. Y. Oono and Y. Takahashi, Preprint (1983).
10. R. Pandit and M. Wortis, Phys. Rev. B25, 3226 (1982).
11. E. Fradkin, O. Hernandez, B.A. Huberman, and R. Pandit, Nuclear Physics B215, 137 (1983).
12. J. Machta, Preprint (1983).

Dynamical System Related to an Almost Periodic Schrödinger Equation

M. Kohmoto

Department of Physics and the Materials Research Laboratory
University of Illinois at Urbana-Champaign
Urbana, IL 61801, USA

A dynamical map is obtained from a class of the almost periodic Schrödinger equation. The potentials are constant except for steps at special points.

1. Introduction

There has been much interest in the almost periodic Schrödinger equation and its finite-difference analogue [1]. Since Floquet's theorem holds only for periodic systems, there is an interesting localization problem even in one dimension. There is a tendency for the spectrum to be a Cantor set, i.e., a closed set with no isolated points and whose complement is dense.

A lot of attention has been paid to the study of dynamical systems and much progress has been made in recent years. Some dynamical systems(strange attractor, Smale,horseshoe etc.) also have Cantor sets. Therefore it is desirable to have some connection between these two different areas of physics so that the theories of the dynamical systems can be applied to problems of condensed matter physics.

Recently, KOHMOTO et al. [2] and OSTLUND et al. [3] showed that the transfer matrices of a special almost periodic Schrödinger equation obey the following recursion relation:

$$M_{\ell+1} = M_{\ell-1}M_{\ell}, \tag{1}$$

where M_ℓ is a 2 x 2 real matrix with unit determinant.

The purpose of this article is to review the derivation of this dynamical system and also to give a more general class of the equations than reported previously.

2. Transfer Matrix for the Almost Periodic Model and its Recursion Relation

The almost periodic (discrete) Schrödinger equation in one dimension is written as

$$\psi_{n+1} + \psi_{n-1} + V\,(n\omega)\psi_n = E\psi_n, \tag{2}$$

where V is periodic,i.e.,$V(t+1) = V(t)$, and ω is an irrational number.

It is traditional to introduce a transfer matrix in one-dimensional problems:

$$\Psi_{n+1} = M(n\omega)\Psi_n, \tag{3}$$

where $\Psi_n = \begin{bmatrix} \psi_n \\ \psi_{n-1} \end{bmatrix}$ and M is the transfer matrix given by $M(t) = \begin{bmatrix} E-V(t) & -1 \\ 1 & 0 \end{bmatrix}$.

The matrix satisfies $M(t+1) = M(t)$ and $\det M(t) = 1$. Higher order transfer matrices are also defined,

$$\Psi_{n+k} = M^{(k)}(n\omega)\Psi_n, \tag{4}$$

where

$$M^{(k)}(t) = M(t+(k-1)\omega)\cdots M(t+\omega)M(t). \tag{5}$$

For an almost periodic system, the evaluation of $M^{(k)}$ for a large value of k is very delicate. The system tries to repeat itself on many length scales. However, it fails to do so and the degree of the failure depends on the incommensurability and the length scale.

We use a recursion relation to calculate $M^{(k)}$. This follows the spirit of the renormalization group theory. Divide k into two integers k_1 and k_2, then the transfer matrix $M^{(k)}(t)$ is written as

$$M^{(k)}(t) = M^{(k_1)}(t+k_2\omega)\, M^{(k_2)}(t). \tag{6}$$

In order to implement the recursive structure in this equation, write $k = F_{\ell+1}$, $k_1 = F_{\ell-1}$ and $k_2 = F_\ell$; and moreover $M^{(F_\ell)} \equiv M_\ell$.

$$M_{\ell+1}(t) = M_{\ell-1}(t+F_\ell\omega)\, M_\ell(t), \tag{7}$$

where F_ℓ is a Fibonacci number defined by $F_{n+1} = F_{n-1} + F_n$ and $F_o = F_1 = 1$.

One of the main obstacles to calculate the transfer matrix using (7) is the presence of $F_\ell\omega$ on the right-hand side. This gives the dependence of M_ℓ on its argument. The first step to overcome this difficulty is to choose an irrational number ω such that $F_\ell\omega$ is close to an integer. Note that $M_n(t+1) = M_n(t)$, so the argument of M_n is defined on a circle. The best choice of ω is the inverse of the golden mean $\omega^* = \frac{\sqrt{5}-1}{2} = 0.618\ldots$, since it satisfies

$$F_\ell\omega^* = F_{\ell-1} - (-\omega^*)^{\ell+1}. \tag{8}$$

The difference between $F_\ell\omega^*$ and an integer $F_{\ell-1}$ becomes geometrically small for large ℓ's. The recursion (7) is then written as $M_{\ell+1}(0) = M_{\ell-1}(-(-\omega^*)^{\ell+1})M_\ell(o)$, where t has been set to be 0 for simplicity.

The second step is to construct potentials V(t) which give

$$M_{\ell-1}(-(-\omega^*)^{\ell+1}) = M_{\ell-1}(0), \tag{9}$$

so that we have the simple recursion relation (1) which does not contain arguments. The condition (9) can be simply satisfied by a constant potential where the incommensurability is lost and the problem is solved trivially. In order to have a non-trivial model, we must allow a potential to have steps. Then, the potential takes a discrete set of values. Recall that (5) is rewritten as

$$M_{\ell-1}(t) = M(t+(F_{\ell-1}-1)\omega^*)\cdots M(t+\omega^*)M(t). \tag{10}$$

From this we learn that discontinuities must lie outside the following intervals:

$$I_\ell : [n\omega^*,\ n\omega^*-(-\omega^*)^{\ell+1}];\ n=0,1,2,\cdots,F_{\ell-1}-1, \tag{11}$$

in order that (9) be satisfied. The points $n\omega^*$ are always on the edges of the intervals which lie on either side of those points depending on whether ℓ is odd or even. In order to control the intervals it is useful to write (11) in terms of $m = F_{\ell-1}-n$. A little algebra using (8) gives

$$I_\ell :\ [-m\omega^*-(-\omega^*)^\ell, -m\omega^*-(-\omega^*)^\ell(\omega^*)^2];\ m=1,2,\cdots F_{\ell-1}. \tag{12}$$

Each of these intervals has a length $\omega^{*\ell+1}$ and the edges approach the points $-m\omega^*$ as ℓ becomes large. The points $-m\omega^*$, $m=1,2,\cdots,F_{\ell-1}$, are clearly outside the "bad" intervals I_ℓ, so we can have discontinuities at those points. KOHMOTO, KADANOFF and TANG [2] chose the simplest possible potential which has two discontinuities at $-\omega^*=(\omega^*)^2$ (mod 1) and $-2\omega^* = -(\omega^*)^3$ (mod 1), i.e.,

$$V(t) = \begin{cases} V_o & \text{for } -\omega^*<t\leq-\omega^{*3} \\ V_1 & \text{for } -\omega^{*3}<t\leq\omega^{*2}. \end{cases} \tag{13}$$

For this potential, it can easily be shown that the recursion (1) holds for all $\ell\geq1$.

Note that the two discontinuities could have been placed on different points. By choosing an appropriate original site for the transfer matrix, one discontinuity can always be placed at $-\omega^*$. Suppose the other discontinuity is at $-m_o\omega^*$, then the recursion (1) holds for $\ell\geq\ell_o$, where ℓ_o is the smallest integer which satisfies $F_{\ell_o-1}\geq m_o$. In other words, at some earlier steps, the simple recursion (1) does not hold because transfer matrices of different arguments are needed. At the step ℓ_o, the original potential with the discontinuities at $-\omega^*$ and $-m_o\omega^*$ can be regarded as having been renormalized to the potential (13) with appropriate values of V_o and V_1.

The results described above can be easily generalized to potentials which have more than two discontinuities at points $-k\omega^*$ $(k=1,2\cdots)$. Those points are dense on the circle. The intervals between the discontinuities are always commensurate with ω^*, i.e., $n\omega^*$ mod 1 $(n=1,2,\cdots)$.

3. Concluding Remarks

The trace of the transfer matrix M_ℓ gives the following mapping problem [2]:

$$(x_{\ell+1},\ y_{\ell+1}, z_{\ell+1}) = (2x_\ell y_\ell - z_\ell, x_\ell, y_\ell), \tag{14}$$

where $x_\ell = 1/2 \operatorname{Tr} M_\ell$, $y_\ell = x_{\ell-1}$ and $z_\ell = x_{\ell-2}$. There is a conserved quantity for this map given by

$$I = x_\ell^2 + y_\ell^2 + z_\ell^2 - 2\, x_\ell y_\ell z_\ell - 1. \tag{15}$$

This determines a two-dimensional manifold on which the dynamical system (14) is defined. The manifold is non-compact and is simply connected for $I > 0$ [4].

The cycles of the map were studied by KADANOFF [5]. KOHMOTO and OONO [4] found homoclinic and heteroclinic points for the map. This explains the Cantor set behavior of the energy spectrum of the almost periodic Schrödinger equation. From fixed point analysis one can derive the scaling behavior of the spectrum which has previously been found numerically [6].

Acknowledgement

Most of the works reported here are results of stimulating conversations with L. P. Kadanoff. It is a pleasure to thank him for most enjoyable collaboration. I have had useful discussions with F. Delyon. This work is supported by the NSF under grant DMR-80-20250.

References

1. For reviews, see, e.g., B. Simon: Adv. Appl. Math. 3, 463 (1982); R. Johnson: preprint.
2. M. Kohmoto, L. P. Kadanoff and C. Tang: Phys. Rev. Lett. 50, 1870 (1983).
3. S. Ostlund, R. Pandit, D. Rand, H.J. Schellnhuber and E. Siggia: Phys. Rev. Lett. 50, 1873 (1983).
4. M. Kohmoto and Y. Oono: University of Illinois preprint.
5. L.P. Kadanoff: University of Chicago preprint.
6. M. Kohmoto: University of Illinois preprint.

Mean Field Hausdorff Dimensions of Diffusion-Limited and Related Aggregates

K. Kawasaki and M. Tokuyama

Department of Physics, Faculty of Science, Kyushu University 33
Fukuoka 812, Japan

Abstract

A mean field argument is presented for the Hausdorff dimension of the diffusion-limited aggregates in analogy with the Flory theory of polymer chains. The concept of ideal aggregate is introduced. The results are favorably compared with those of computer simulations. Other related aggregates are also briefly discussed.

1. Introduction

The diffusion-limited aggregation (DLA) model invented by WITTEN and SANDER [1] is perhaps the simplest nontrivial model of growth by accretion that exhibits interesting critical behavior. The current interest on the model is focused on obtaining the Hausdorff dimension D [2] of the aggregate as a function of the topological dimension d. So far the major source of information on D comes from computer simulations, especially by MEAKIN [3].In addition there is one paper of real space renormalization group [4] and one attempt for mean field theory by MUTHUKUMAR [5]. The latter author obtained the remarkable result $D=(d^2+1)/(d+1)$. Unfortunately, however, Muthukumar's derivation remains totally mysterious to us. For further details of the field, see the article by Nauenberg in these Proceedings.

2. Mean Field Theory

Here we would like to present a simple heuristic argument for obtaining D. We will be led by the idea that the optimum aggregate configuration is the result of two competing effects, namely, the tendency to maximize the number of configurations and the screening of diffusion field by the grown aggregate. We will also draw analogy from the Flory mean field theory of polymer chain [6] where the competition of maximizing the number of chain configurations and the excluded volume effects plays a decicive role. The major difference of course is the fact that DLA is specifically a nonequilibrium phenomenon and hence a quantity like free energy has no meaning. Nevertheless we believe that our physical idea of optimization is correct and we introduce a variational function F to realize this physical idea. We consider an aggregate grown on a seed particle which consists of N particles and has the radius of gyration R, so that $N=R^D$ in some dimensionless unit where we ignore all the unimportant coefficients throughout this paper. The average number density inside the aggregate is $\rho=N/R^d$. Then the screening length becomes $\ell=\rho^{-1/2}$ for small particle number density, i.e., for $D<d$. This screening length divides the aggregate into blobs of the size ℓ^d and we can define the blob

number density inside the aggregate as $\rho_b = \ell^{-d} = \rho^{d/2}$. The function F now consists of two parts

$$F = F_S + F_I \, . \tag{1}$$

F_S which takes care of the screening is thus taken to be

$$F_S = R^d \rho_b^2 = R^{d-d^2} N^d \, . \tag{2}$$

This is an analogue of the excluded volume energy $R^d\rho^2$ of the polymer chain. For the purpose of finding F_I in (1) which is the analogue of the ideal chain free energy R^2/N, we introduce the concept of ideal aggregate. This is the aggregate formed by accretion where the screening is totally absent. This means that particles diffusing in from the outer space can freely penetrate the aggregate and can be deposited at any site adjoining the sites already deposited with the equal probability in the spirit of mean field theory, where unlimited multiple occupancy of each site is permitted. This is a kind of super-compact aggregate in which unlimited multiple occupancy of each site has the effect of endowing an extra dimension to the aggregate, which thus has the dimensionality d+1. A concrete example of the ideal aggregate is described in the Appendix. Therefore we postulate

$$F_I = R^{d+1}/N \, . \tag{3}$$

The same result (3) can be obtained also for a somewhat different aggregation model in which particles are deposited on the aggregate surface and the growth in size occurs by a Gaussian random walk process [7]. Balancing (2) and (3) or optimizing the sum of (2) and (3) with respect to N we finally find

$$N = R^D \tag{4}$$

where D is given by the following Muthukumar form:

$$D = (d^2+1)/(d+1) \, . \tag{5}$$

For large d, we find $D \simeq d-1$ which agrees with the result of NAUENBERG and co-workers [8] and is also obtained by setting F_S, (2), equal to unity.

The result (5) is compared with those obtained by computer simulations in Table 1 and the agreement is impressive.

Table 1. D(th) is from (5). D(co) is from computer simulations by MEAKIN [3].

d	D(th)	D(co)
2	5/3	1.71
3	5/2	2.5
4	3.4	3.3
5	4.3··	4.2
6	5.285	5.3

3. Discussion

The screening term (2) takes the form $R^d\rho^\zeta$ with $\zeta = d$. It is thus of some interest to see whether other aggregation models [9] can be described as a competition of two effects by optimizing the function F of the following form:

$$F = R^d \rho^\zeta + R^{d+1}/N \tag{6}$$

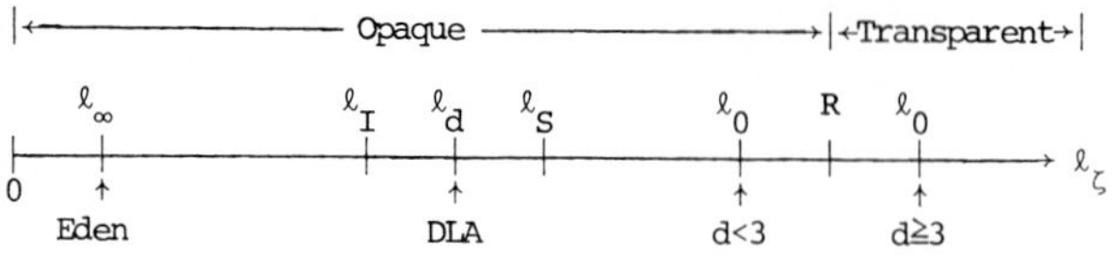

Fig. 1 Screening length of various aggregate models. Here ℓ_I and ℓ_S correspond to the invasion aggregate (D=1.89, p=0.94 and ζ=8.1) and the Sawada aggregate (D=1.39, p=0.7 and ζ=0.64), respectively

where the first term represents the "screening" effect with the "screening" length ℓ_ζ given by

$$\ell_\zeta=\rho^{-1/2}\,. \tag{7}$$

Here ζ is a measure of compactness of the aggregate and may be called the compactness parameter. Optimization then gives

$$D=(1+\zeta d)/(1+\zeta) \tag{8}$$

and thus

$$\ell_\zeta=R^{(d-1)/2(1+\zeta)}\,. \tag{9}$$

Following [1], let us define the opacity p by $\ell_\zeta/R=R^{-p}$. Then $p=(2\zeta+3-d)/2(1+\zeta)=(D+2-d)/2\leq 1$. If $p>0$, the aggregates are opaque and if $p<0$, they are transparent. The most compact aggregate is obtained for $\zeta=\infty$, for which we find $D=d$, $p=1$ and $\ell_\infty=O(1)$, corresponding to the Eden aggregate [10]. The least compact aggregate with $\zeta=0$ gives $D=1$ and $p=(3-d)/2$. There are other aggregate models besides DLA with intermediate compactness [9]. Figure 1 gives a schematic representation of each model with different values of ℓ_ζ [13,9]. For instance the invasion aggregate [11,9] and the Sawada aggregate with sufficiently large tip priority factors [12], both for d=2, give, respectively, $\zeta\simeq 8.1$ and $\zeta\simeq 0.64$. At this moment, however, we cannot provide concrete physical arguments for these particular values of ζ, and usefulness of the compactness parameter in general cases remains to be seen.

One of us (K.K.) would like to thank M. Nauenberg for useful discussions. This work was partly supported by the Scientific Research Fund of the Ministry of Education.

APPENDIX

We consider a lattice with a seed particle on the lattice site at the origin. Diffusing particles from the outer space are deposited on the z nearest-neighbor sites of the seed at time t=1. At time t=2 another set of diffusing particles are deposited on all the nearest-neighbor sites of each of the already deposited sites allowing multiple occupancy. The process then continues indefinitely to form an ideal aggregate. One can readily find that the number of particles $n_j(t)$ deposited at time t on the lattice site j which can be connected to the origin by at least c_j nearest-neighbor bonds is given by

$$n_j(t)=\begin{cases}1+z_j+z(t-c_j-1) & \text{for } t\geq c_j+1\\ 1 & \text{for } t=c_j\\ 0 & \text{for } t<c_j\end{cases} \tag{A.1}$$

where z_j is the number of occupied nearest-neighbor sites of the site j at the time $t=c_j$. For this model it is clear that $N(t)=\Sigma_j n_j(t)$ and $R=[\Sigma_j c_j{}^2 n_j(t)]^{1/2}/N(t)^{1/2}$ scale respectively as t^{d+1} and t and hence $N=R^{d+1}$.

One can generalize the model in an obvious way by assuming random partial occupations with the equal probability of the nearest-neighbor sites of the site already occupied with the result $N=R^{d+1}$ remaining unchanged.

References

1. T.A. Witten and L.M. Sander: Phys. Rev. Lerr. 47, 1400(1981) and preprint(1982).
2. B.B. Mandelbrot: The Fractal Geometry of Nature (W.H. Freeman, San Francisco 1982).
3. P.Meakin: Phys. Rev. A27, 604, 1495, 2316 (1983).
4. H. Gould, F. Family and H.E. Stanley: Phys. Rev. Lett. 50, 686(1983).
5. M. Muthukumar: Phys. Rev. Lett. 50, 893(1983).
6. P. Flory: Principles of Polymer Chemistry (Cornell University Press, Ithaca, N.Y. 1971);
   P.G. de Gennes: Scaling Concepts in Polymer Physics (Cornell University Press, Ithaca, N.Y. 1979).
7. M. Tokuyama and K. Kawasaki: preprint (1983).
8. M. Nauenberg: Phys. Rev. B28, 449(1983)
   M. Nauenberg, R. Richter and L.M. Sander: Phys. Rev. B28, 1649(1983).
9. H.E. Stanley, F. Family and H. Gould: J. Poly, Sci.(1983).
10. M. Eden, Proc. Fourth Berkeley Symposium on Mathematical Statistics and Probability, J. Neyman (ed.), Vol. IV, p.223 (University of California Press, Berkeley, 1961).
11. R. Chandler, J. Koplik, K. Lerman and J.F. Willemsen: J. Fluid. Mech. 119, 249(1982).
12. Y. Sawada, S. Ohta, M. Yamazaki and H. Honjo: Phys. Rev. 26A, 3557(1982).
13. P.A. Rikvold, Phys. Rev. A26, 647(1982).

Part III

Onset of Chaos

Stability of the Scenarios Towards Chaos

P. Coullet

Mécanique Statistique, Laboratoire de Physique de la Matière Condensée
Parc Valrose, F-06034 Nice Cedex, France

For the last ten years, the study of chaotic behaviors generated by deterministic dynamical systems has received an increasing interest among Physicists and Mathematicians. A fascinating aspect of this field of research is its interdisciplinarity. The deep reason for this universality lies in the very nature of the subject: The Mathematics of Time [1]. The study of simple dynamical systems, such as the iterations of mappings or the ordinary differential equations thus appears as a study of these Mathematics of time. Two kinds of problems can be easily identified. The first one deals with the transition towards complex behaviors. The second one is concerned by the chaotic behaviors themselves. In this paper, we will only discuss some general aspects of the first class of problems. In fact, these problems are in some sense simpler to study. They generally lead to universal scenarios [2] by which a periodic or quasiperiodic signal progressively looses its regularity, as an external constraint varies. These problems of transition bear a striking analogy with critical phenomena. The common analytical tool is the renormalization group [3].

In this paper, we show how to use this technics to obtain some information on the structural stability of various scenarios. Among all the scenarios we can imagine, only a small number has been numerically or experimentally observed. The notion of stability, introduced via the renormalization group analysis, could give some light on this experimental fact. The organization of the paper is the following. We first recall the method of renormalization using the period-doubling route [4] to chaos as a pedagological example. The notion of stability will then be introduced on a very simple example : the effect of a small noise on this cascade [5] [6]. The second part of this paper will be devoted to study the stability of two scenarios : a cascade of period doubling for tori [7] [8][9] and a cascade of symmetry breaking [10] [11]. To conclude, we will give some comments on the other scenarios and their stabilities.

I The EFFECT of NOISE on the CASCADE of PERIOD-DOUBLING BIFURCATIONS : AN EXAMPLE of an INSTABILITY for a SCENARIO

We first recall the main steps of the renormalization group analysis of the cascade of period-doubling bifurcations. This way to analyze this phenomenon has been successfully used by M.J. FEIGENBAUM [4] . Independently, C. TRESSER and myself [12] rediscovered it later. Rigourous results have been given by several authors [13] . The simplest dynamical system which exhibit a period-doubling cascade is given by the iteration of non-invertible maps on an interval [14] . We can, without inconvenients, restrict ourselves to mappings $f(x)$ of the interval $[-1,+1]$ of the form :

$$f(x) = 1 + ax^2 + bx^3 + \dots \quad \text{with} \quad a < 0. \qquad (1)$$

The starting point of the renormalization group anlysis is the constatation that when f has a periodic orbit of period 2, its second iterate $f^{(2)}(x) = f(f(x))$, near the periodic points, looks like f (see Fig. 1)

In particular, since zero is also a critical point of $f^{(2)}$, one can try to compare $f^{(2)}$ around zero to f . This is simply achieved by the transformation :

$$Nf = -\frac{1}{\alpha} f\, f(\alpha x) \qquad (2)$$

where $\alpha = -f(1)$ is a scaling factor choosen to make possible the comparison between f and Nf, since then $f(0) = Nf(0) = 1$. This transformation is the elementary operation of the renormalization group. It divides by two the period of even periodic orbits in the same way that the KADANOFF-WILSON [3] transformation divides by a factor $s > 1$ the correlation length in critical phenomena. Having in hand such a transformation, it is quite natural to look for a fixed point $g(x)$:

$$g(x) = -\frac{1}{\lambda} g(g(\lambda x)) \qquad \text{where } \lambda = -g(1). \qquad (3)$$

This equation introduced by CVITANOVIC and FEIGENBAUM has been numerically solved [4] .In the generic case, it leads to a quadratic function

$$g(x) = 1 - 1.527\, x^2 + \dots \qquad (4)$$

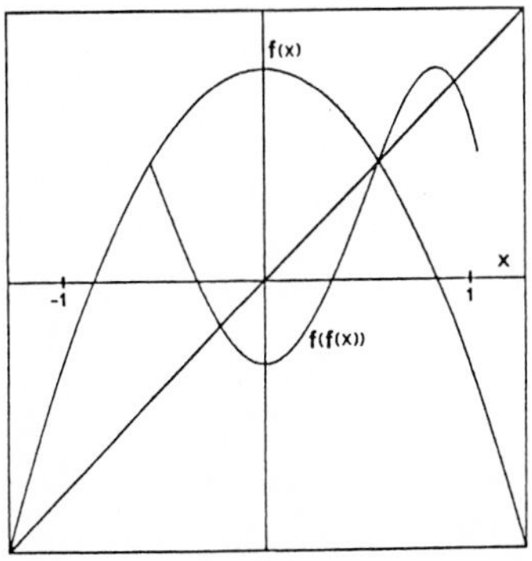

Fig. 1 : Illustration of the Renormalization group operation

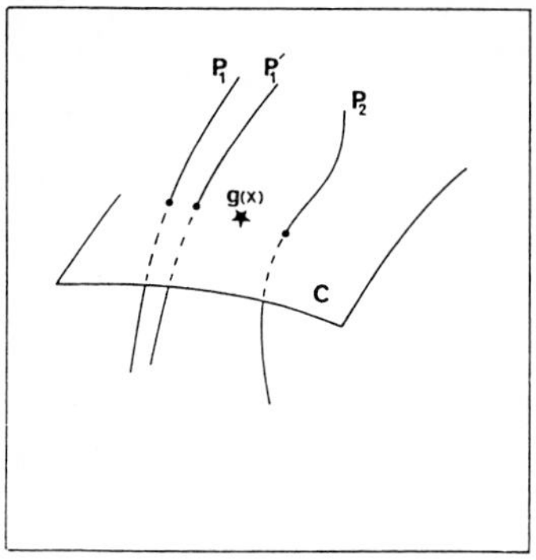

Fig. 2 Topology of a codimension 1 unstable fixed point

The stability of this fixed point is then investigated through the spectrum of the linear part of the operator

$$T_g e(x) = N(g + e)\ (x) - Ng(x). \tag{5}$$

Let us note D_gT this operator. As usual, we look for a basis of function $e_i(x)$ which diagonalize DT_g. A general perturbation $e(x)$ of $g(x)$ is then expanded on this basis

$$e(x) = \sum_i t_i\ e_i(x). \tag{6}$$

In critical phenomena the t_i's are known as scaling fields[15]. The eigenvalue equation writes as

$$DT_g\ e_i = \delta_i\ e_i \tag{7}$$

when the eigenvalues are inside the unit circle, the corresponding scaling fields are said irrelevant. They are marginal or relevant if the eigenvalues have moduli equal or greater than one. As in critical phenomena, we are only interested by marginal and relevant perturbations. In our case, we have only one relevant eigenvalue $\delta_1 = 4.669$ ... associated with an even perturbation $e_1(x) = \tilde{e}_1(x^2)$. The corresponding scaling field is the deviation of the external constraint r from its critical value r_c . The various critical exponents are simply computed as functions of λ and δ_1 . For example, the period of superstable cycle of period 2^n is given by :

$$T_n \sim (r_c - r_n)^{-\nu} \tag{8}$$

where $\nu = \text{Log } 2 / \text{Log } \delta_1$.

The fixed point $g(x)$ is said codimension 1 unstable since, in the space of maps f, it has a stable manifold of codimension 1 and an unstable one of dimension 1. This geometric situation has been illustrated on the Fig. 2 .

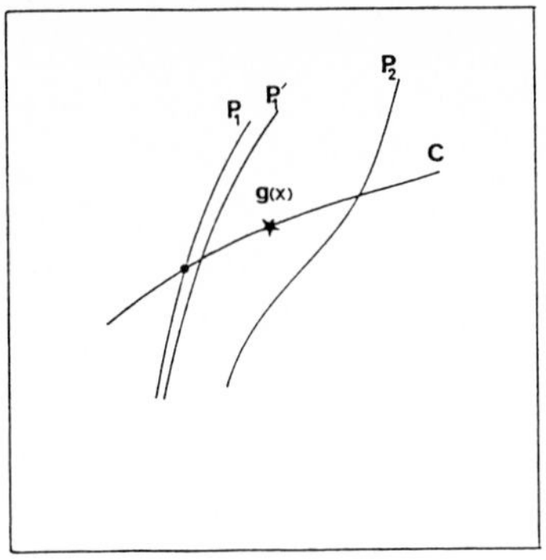

Fig. 3 Topology of a codimension 2 unstable fixed point

P_1, P_1', P_2 represent three paths corresponding to three typical variations of parameters in f. P_1 is a path for which the transition occurs. P_1' is a small perturbation of P_1. Since the critical manifold C (stable manifold of g) has a codimension 1, P_1' will intersect it again. We will observe for both P_1 and P_1' the same kind of transition described by the same critical behavior. The corresponding scenario is stable. P_2 represent an other generic path, but at distance from P_1 for which the transition is still present. In opposition, the Fig. 3 represents an hypothetical fixed point of codimension 2 (two relevant scaling fields).

In this case a generic perturbation P_1' of P_1 do not cross the critical manifold. The transition disappears. We will say that the corresponding route to chaos is an unstable one. Nevertheless, since P_1' is close to P_1, we will observe following P_1' critical effects which are reminiscent of the proximity of the critical manifold of g. However, when we do not know this critical manifold, a generic path P_2 is not likely to pass near it. In this case, we do not expect to observe the corresponding scenario in a typical one-parameter experiment. To illustrate this second situation, let us consider a simple example of unstable scenario.

The unavoidable presence of external noise in real experiments is simply modelled on the iteration of f by the addition of a noise to the mapping at each iteration step[16] . The effect of this noise has been investigated through a renormalization group analysis [5] [6]. A perturbation of the form $\xi e(x)$ is sought as an eigenvector of $D_g T$. ξ represents a random variable characterized by an even distribution on [-1,1]. It has been found [5] [6] that such a perturbation is relevant with an eigenvalue $\delta_2 = 6.619 \ldots$ The corresponding scaling field is the intensity of the noise. Thus the renormalization amplify the noise. We have now in hand a fixed point of codimension two, and consequently, following our definition, an unstable scenario. This instability manifests itself mainly at the vicinity of the bifurcation points in the cascade (see Fig. 4).

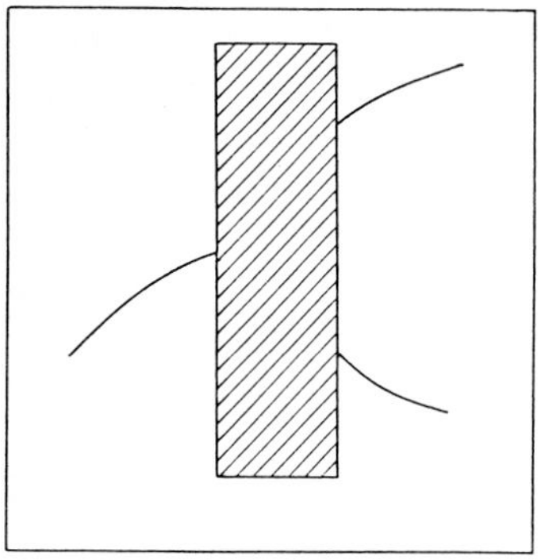

Fig. 4 Carricature of a fuzzy pitchfork bifurcation

The bifurcations become in some sense fuzzy. As a consequence, the last steps of the cascade will always disappear in this fuziness. Nevertheless, since it is quite natural in experiments to try to decrease as much as possible the effect of the external noise, we will always observe the period-doubling scenario. This natural constraint give back some stability to this scenario and crossover effects are thus expected to occur [5] [6] . More generally, an unstable scenario conveniently constrained in order to make as small as possible the relevant scaling fields, except the one associated to the cascade itself, will again become stable.

II TWO UNSTABLE SCENARIOS : A CASCADE of PERIOD-DOUBLING BIFURCATIONS for TORI and a CASCADE of SYMMETRY-BREAKING BIFURCATIONS

1° Doubling of Torus

Recently several authors [7] [8] [9] have described a route to chaos via a cascade of period doubling of tori. A doubling of torus appears on a Poincaré map as a doubling of the invariant circle which represents the section of a quasiperiodic motion on a torus. The common observation of these authors is the unavoidable interruption of the cascade by the appearance of chaotic behaviors localized around the tori. This observation is a clear symptom of the instability of this scenario. To make more precise this conclusion, let us consider [7] the general third-order system

$$\partial_t x = F_r(x) \tag{9}$$

where r represents a real parameter and $x = (x^1, x^2, x^3)$ $F_r = (F_r^1, F_r^2, F_r^3)$. To every periodic orbit (9) one can associate an invariant torus by the device of adjoining to (9) the equation

$$\partial_t\theta = \omega \qquad (10)$$

where θ represent an angular variable. The dynamical system (9)-(10) is now defined in the phase space $\mathbb{R}^3 \times T$, where T is the circle of length 2π . Suppose now that (9) when r varies in some interval $[\, r_{min} , r_c]$ presents a cascade of period doubling for a periodic orbit. Then (9)-(10) will exhibit the corresponding sequence for tori. The troubles come when one try to couple (9) and (10). What does happen when one considers the system

$$\begin{aligned} \partial_t x &= F_r(x) + \varepsilon g(x, \theta) \\ \partial_t &= \omega + \varepsilon h(x, \theta) . \end{aligned} \qquad (11)$$

Various numerical experiments on such systems show [7] the phenomenon of the interruption of the cascade. The instability of this scenario has an intrinsic origin and is related to the fact that a bifurcation of torus is by itself unstable. Rather restrictive conditions have been given [17] to insure the existence of a clean transition. When these conditions are not exactly satisfied, we get a fuzzy bifurcation, as it is possible to show it explicitely in some case[18] . The domain of fuzziness extends itself on each side of the "clean" transition point when the invariant tori are not enough normally hyperbolic. The existence of this fuzziness is the intrinsic analog of the extrinsic fuzziness generated by the presence of an external noise in the case of the usual period-doubling cascade. The analogy is now clear enough. We generically can expect an interruption of the cascade and various critical behaviors for small ε (crossover). For example, at the critical point of the uncoupled system, the Liapunov exponent which measures the quantity of noise generated by the dynamical system will behave as ε^{γ} , where ε is the strength of the coupling, which is now a relevant scaling field and $\gamma = \mathrm{Log}2 \,/\, \mathrm{Log}\, \delta_2$, where δ_2 is the corresponding relevant eigenvalue. The instability associated to this scenario is more inconvenient that the previous one since the stabilizing constraint is no longer natural. It consists to uncouple two oscillators when their natural tendency is to couple. One can conclude that this scenario is not likely to occur in one-parameter experiments on autonomous systems expected near special points where, for example, several instabilities are competing [19] . The invariant torus will generally undergo a more stable route to chaos.

2° Cascade of symmetry breaking

The route to chaos we are going to describe has been studied by several authors [10] [11] . It has been observed in ordinary differential equations [10] and partial derivative equations [11] as well having a symetry property "à la Lorenz"

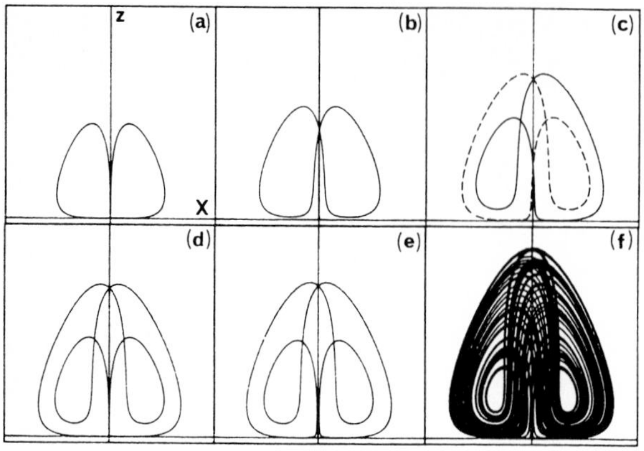

Fig. 5. A numerical investigation of the system -12) with α= 1.8, β= - 0.07 , ξ = .5 displays : (a) a pair of stable homoclinic orbits for μ =0.07607...,(b) a stable symmetric periodic orbit for μ= 0.05, (c) a pair of stable orbits which are mutual images under S for μ = 0.034, (d) a pair of stable homoclinic orbits for μ=0.03218..., (e) a stable symmetric periodic orbit for μ = 0.0321, (f) seemingly a strange attractor for μ = 0.02 . All these patterns are projected on a plane XOz orthogonal to the stable manifold of the flow linearized at 0 and containing the z axis (X = 5x-y).

[20] . The corresponding scenario is unstable and has a codimension 2 , The symetry being the stabilizing constraint which allows the scenario to be stable. Figure 5 represents this scenario as it has been numerically observed on the differential system

$$\begin{aligned} \partial_t x &= \alpha x - \alpha y \\ \partial_t y &= - 4\alpha y + xy + \mu x^3 \\ \partial_t z &= - \xi\alpha z + xy + \beta z^2 \end{aligned} \qquad (12)$$

as we decrease μ.α,β and ξ being constant parameters. This system has the symetry, S : $(x,y,z) \to (-x , -y , z)$.

Using the work of GUCKENHEIMER and WILLIAMS [21]on the geometrical Lorenz flow,we have modelized the return map of system displaying this cascade of homoclinic bifurcations as

$$\begin{aligned} f_r &= r - |x|^\xi && \text{for} \quad x < 0 \\ f_r &= -r + |x|^\xi && \text{for} \quad x > 0 . \end{aligned} \qquad (13)$$

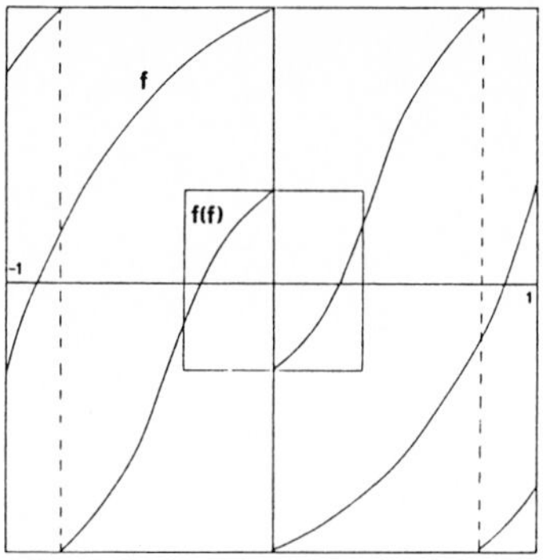

Fig. 6 Illustration of the renormalization group operation for the symetry-breaking cascade

The parameter ξ turns out to be simply the ratio of the eigenvalues involved in the homoclinic trajectory [21] . $\xi = -\lambda_1 / \lambda_3$ ($\lambda_1 < 0$, $\lambda_3 > 0$). We are interested here in the case $\xi > 1$. A simple scaling allows to normalize f_r such that $f_r(0_-) = 1$ and $f_r(0_+) = -1$.

We can formulate the renormalization operation as suggested on Fig. 6

$$N' f = \frac{1}{\alpha} f(f(\alpha x))$$

where $\alpha = f(1)$ is a positive scaling factor. It is then easy to show that $k(x) = -\mathrm{sgn}x\, g(x)$ is a fixed point of this transformation, g being the fixed point of the period-doubling case. We found that the spectrum of DT'_k , the linearization of N' around k(x) contains two relevant eigenvalues. One is the usual eigenvalue $\delta_1(\xi)$ associated with the scaling field $(r-r_c)$. The other $\delta_2(\xi) > \delta_1(\xi)$ is associated with the strength s of a non-symmetric perturbation as a scaling field. In the case of $\xi = 1$, $\delta_1 = 2$, $\delta_2 = \infty$. The conclusions are now the same as previously. A slight breaking of the symmetry lead to some fuzziness in the homoclinic in the bifurcation as it is possible to anticipate using rigourous results [22] on some non-symmetrical mappings. We get, as expected, for small s various critical behaviours. For example, the Liapunov exponent is given by

$$l = s^{\gamma} \; L\left((r-r_c)/s^{\Delta}\right) \quad \text{where} \quad \gamma = \mathrm{Log}2/\mathrm{Log}\,\delta_2 \qquad \Delta = \mathrm{Log}\delta_1 / \mathrm{Log}\,\delta_2 \, . \qquad (15)$$

The renormalization group analysis of the cascade comes from a joint work with P. COLLET and C. TRESSER.

III CONCLUSION

Using the notion of stability introduced here, all the known scenarios with the exception of the period-doubling cascade and the intermittency [23] appear as unstable routes to chaos. Indeed, we have restricted our discussion to self-similar route for which a renormalization group analysis can be done. This is not an important restriction since this is the majority of smooth route to chaos and even intermittency fall in this class [24] [25] . We also have to include in this class a codimension two scenario consisting in a cascade of frequency locking leading to chaos via a quasiperiodic behavior [26][27][28], some transitions involving three oscillators [29] .

As a conclusion we could conjecture that in real and numerical experiments most of the stable scenarios have been identified. It remains the naturally constrained systems which will probably exhibit new routes to chaos. For the unstable scenarios themselves, one can always try to identify it by applying appropriate external constraints on a system, in the same way that one study polycritical points in phase transition.

REFERENCES

1 S. Smale : The Mathematics of Time (Springer Verlag, 1980)

2 J.P. Eckmann : Rev. Mod. Phys. 53, 643 (1981)

3 K.G. Wilson : Rev. Mod. Phys. 55, 583 (1983)

4 M.J. Feigenbaum : J. Stat. Phys. 19, 25 (1978)
M.J. Feigenbaum : J. Stat. Phys. 21, 669 (1979)

5 J.P. Crutchfield, N. Nauenberg and J. Rudnick: Phys. Rev. Lett. 46, 933 (1981)

6 B. Shraiman, C.E. Wayne and P.C. Martin: Phys. Rev. Lett. 46, 935 (1981)

7 A. Arnéodo, P. Coullet and A.E. Spiegel: Phys. Lett. 94A, 1 (1983)

8 V. Franceschini: Physica 6D, 285 (1983)

9 K. Kaneko : "Doubling of Torus" preprint (1983)

10 A. Arnéodo, P. Coullet and C. Tresser : Phys. Lett. 81A, 197 (1981)

11 Y. Kuramoto and S. Koga : Phys. Lett. 92A, 1 (1982)

12 P. Coullet and C. Tresser : J. de Physique, colloque 39, C5-25 (1978)
C. Tresser and P. Coullet : C.R.Acad. Sc. Paris 287, 577 (1978)

13 J.P. Eckmann "Routes to chaos with special emphasis on period doubling" Les Houches Summer school 1981 Chaotic Behavior in Deterministic Dynamical Systems, Ed. G. IOOSS, HELLEMAN, STORA (North Holland 1983)

14 P. Collet and J.P. Eckmann : Iterated maps on an interval as dynamical systems (Birkhäuser, Boston 1980)

15 S.K. Ma : Modern theory of critical phenomena (Benjamin, Reading, Mass. 1976)

16 B.A. Huberman and J. Rudnick: Phys. Rev. Lett. 45 , 154 (1980)

17 A. Chenciner and G. Iooss: Arch. Rational Mech. Anal. 69, 109 (1979)

18 "Bifurcation de difféomorphismes de $\mathbb{R}^2$ au voisinage d'un point fixe elliptique" in Les Houches Summer school 1981 Chaotic Behavior in Deterministic Dynamical Systems , Ed. IOOSS, HELLEMAN, STORA (North Holland 1983)

19 P. Coullet:"Chaotic behaviors in the unfolding of singular vector fields" to be published in the Proceedings of the "Workshop on Common Trends in Particle and Condensed Matter Physics" at Les Houches (1983)

20 E.N. Lorenz: J. Atmos. Sci. 20, 130 (1963)

21 J. Guckenheimer and R.F. Williams : Publ. Math. IHES 50, 307 (1979)

22 C. Tresser : C.R. Acad. Sci. Paris 296 , 729 (1983)

23 P. Manneville and Y. Pomeau : Phys. Lett. 75 A, 1 (1979) ; Commun Math. Phys. 74, 74 (1980 ; Physica 1D , 219 (1980)

24 J.E. Hirsch, N. Nauenberg and D.J.Scalapino : Phys. Lett. 87A , 391 (1982)

25 B. Hu, and J. Rudnick : Phys. Rev. Lett. 48, 1645 (1982)

26 H. Daido : Prog. Theor. Phys. 68, 1935 (1982)

27 M.J. Feigenbaum, L.P. Kadanoff and S.J. Shenker: Physica 5D, 370 (1982)

28 D. Rand, S. Ostlund, J.P. Setna and E.D. Siggia : Phys. Rev. Lett. 49, 132 (1982)

29 J.P. Setna and E.D. Siggia "Universal transition in a dynamical system forced at two incommensurate frequencies" Preprint 1983

Functional Renormalization-Group Equations Approach to the Transition to Chaos†

B. Hu

Department of Physics, University of Houston, Houston, TX 77004, USA* and Theoretical Division and Center for Nonlinear Studies, Los Alamos National Laboratory, Los Alamos, NM 87545, USA

ABSTRACT

The functional renormalization-group equations approach to the study of the universal scaling properties of period doubling and intermittency is reviewed. The differential-equation method of obtaining the exact solutions for intermittency is explained in detail.

I. Introduction

In recent years there has been much interest in the study of the transition to chaos in nonlinear systems [1]. Various scenarios have been proposed [2]. In particular, the discovery of universality [3] in the period-doubling route to chaos has attracted a great deal of attention. The formulation of the universality theory in terms of functional renormalization-group equations [3] has become an important and standard method of gleaning useful information at the onset of chaos.

In this lecture I will give a pedagogical introduction to the functional renormalization-group equations approach to the transition to chaos, using period doubling and intermittency [4-6] as examples. In Sec. II a simple method of deriving the functional renormalization-group equations for period doubling is illustrated. In Sec. III the differential-equation method of obtaining the exact solutions [7] to these functional equations in the case of intermittency is shown in detail. Finally some concluding remarks are given in Sec. IV.

II. Functional Renormalization-Group Equations for Period Doubling

In the study of period doubling, Feigenbaum discovered two universal numbers α and δ, which measure respectively the amount of rescaling and the rate of bifurcation. These numbers depend only on the order of the local maximum and not on the details of the function studied.

---

†Invited talk presented at the Sixth Kyoto Summer Institute on Chaos and Statistical Mechanics (1983).

*Permanent address.

To provide a theoretical explanation of these numbers, Cvitanović and Feigenbaum discovered a functional equation that serves as the basis of the universality theory. A simple derivation of this equation can be given as follows. Since the basic composition rule for the periods in period doubling is

$$2^n + 2^n = 2^{n+1} \ , \tag{1}$$

the corresponding composition rule for the functions is

$$f^{2^n} \circ f^{2^n} = f^{2^{n+1}} \ . \tag{2}$$

If one assumes that as $n \to \infty$ the sequence of functions f^{2^n} approaches asymptotically a universal function $f^*(x)$ at the period-doubling accumulation point under rescaling by α

$$f^{2^n}(x) = (-\alpha)^{-n} f^*((-\alpha)^n x) \ , \tag{3}$$

then the functional composition rule implies

$$\begin{aligned} f^{2^n} \circ f^{2^n} &= (-\alpha)^{-n} f^*((-\alpha)^n (-\alpha)^{-n} f^*((-\alpha)^n x)) \\ &= (-\alpha)^{-n} f^*(f^*((-\alpha)^n x)) \ . \end{aligned} \tag{4}$$

Since

$$f^{2^{n+1}} = (-\alpha)^{-(n+1)} f^*((-\alpha)^{n+1} x) \ , \tag{5}$$

we have

$$(-\alpha)^{-(n+1)} f^*((-\alpha)^{n+1} x) = (-\alpha)^{-n} f^*(f^*((-\alpha)^n x)) \ . \tag{6}$$

The functional equation for the universal fixed-point function then follows:

$$f^*(f^*(x)) = -\frac{1}{\alpha} f^*(-\alpha x) \ . \tag{7}$$

To fix the scale, we set

$$f^*(0) = 1 \ , \tag{8}$$

which implies

$$\alpha = -\frac{1}{f^*(1)} . \tag{9}$$

Equation (7) has the same import as a recursion relation in the renormalization-group theory of critical phenomena. However, we now have to deal with a functional equation instead, and the fixed point of this functional equation is a function rather than a point as in critical phenomena.

A simple method to solve Eq. (7) is to use polynomial approximation. I will illustrate the method by employing the lowest-order polynomial approximation:

$$f^*(x) = 1 - ax^2 . \tag{10}$$

Substituting it into Eq. (7) and keeping only quadratic terms, we have

$$\begin{aligned} f^*(f^*(x)) &= 1 - a(1-ax^2)^2 \\ &\cong 1 - a(1-2ax^2) \\ &= (1-a) + 2a^2x^2 , \end{aligned} \tag{11}$$

and

$$\begin{aligned} -\frac{1}{\alpha} f^*(-\alpha x) &= -\frac{1}{\alpha}[1-a(-\alpha x)^2] \\ &= -\frac{1}{\alpha} + a\alpha x^2 . \end{aligned} \tag{12}$$

Equating equal powers, we obtain two equations for the two unknowns a and α:

$$1 - a = -\frac{1}{\alpha} , \tag{13}$$

$$2a^2 = a\alpha . \tag{14}$$

Solving these two equations gives

$$a = \frac{1+\sqrt{3}}{2} \cong 1.366 , \tag{15}$$

$$\alpha = 1+\sqrt{3} \cong 2.732 . \tag{16}$$

Compared to the numerical value, $\alpha = 2.503$, the result is not bad. Higher-order approximations will give more and more accurate values.

Now that we have derived the functional equation determining α, we can proceed to derive an equation for δ. Consider a small perturbation to the fixed-point function $f^*(x)$:

$$f(x) = f^*(x) + \varepsilon h(x) . \tag{17}$$

Then

$$\begin{aligned} f(f(x)) &= f(f^*(x)+\varepsilon h(x)) \\ &= f^*(f^*(x)+\varepsilon h(x)) + \varepsilon h(f^*(x)+\varepsilon h(x)) \\ &\cong f^*(f^*(x))+\varepsilon[f^{*\prime}(f^*(x))h(x) + h(f^*(x))] . \end{aligned} \tag{18}$$

The eigenvalue equation for $h(x)$ is therefore

$$f^{*\prime}(f^*(x))h(x) + h(f^*(x)) = -\left(\frac{\lambda_\varepsilon}{\alpha}\right) h(\alpha x) . \tag{19}$$

The leading eigenvalue of λ_ε is δ. For period doubling, it has been proved that there is a unique relevant eigenvalue $\delta > 1$.

Finally, since chaotic phenomena by their very nature may appear noisy, it is important to take the effect of external noise [8,9] into account. Consider a stochastic perturbation to the fixed-point function:

$$f(x) = f^*(x) + \xi g(x) , \tag{20}$$

where ξ is a Gaussian stochastic variable. Then

$$\begin{aligned} f(f(x)) &= f(f^*(x)+\xi g(x)) \\ &= f^*(f^*(x)+\xi g(x)) + \xi' g(f^*(x)+\xi g(x)) \\ &\cong f^*(f^*(x)) + \xi f^{*\prime}(f^*(x))g(x) + \xi' g(f^*(x)) \\ &= -\frac{1}{\alpha} f^*(-\alpha x) + \xi''[f^{*\prime 2}(f^*(x))g^2(x) + g^2(f^*(x))]^{\frac{1}{2}} . \end{aligned} \tag{21}$$

In going to the last line, we have made use of the fact that ξ and ξ' are independent random variables, and ξ'' is also a random variable. The stochastic eigenvalue equation follows immediately

$$f^{*\prime 2}(f^*(x))g^2(x) + g^2(f^*(x)) = \left(\frac{\lambda_\xi}{\alpha}\right)^2 g^2(\alpha x) . \tag{22}$$

The functional renormalization-group equations (7), (19) and (22) constitute the cornerstones of the universality theory.

III. Exact Solutions for Intermittency

A. Intermittency

Period doubling is based on the pitchfork bifurcation, in which a 2^n-cycle gives rise to a 2^{n+1}-cycle when the eigenvalue (i.e., the derivative in one dimension) of f^{2^n} passes through -1. Beyond the period-doubling limit point, various odd cycles appear via the tangent bifurcation, where bifurcation takes place when the eigenvalue passes through +1 instead. The tangent bifurcation was first invoked by Manneville and Pomeau to explain the phenomenon of intermittency. To see how this comes about, let us consider, for example, the third iterate of the logistic map $x_{n+1} = r\, x_n(1-x_n)$. The 3-cycle is born when $f^3(x)$ is tangent to the 45° line (Fig. 1), and let us denote this parameter value by r_3. To see what happens in the vicinity of r_3, let us focus our attention on only one of the elements of the 3-cycle (Fig . 2a). For $r \gtrsim r_3$, the iterates of the map, after an initial transient, settle down to a 3-cycle, which then undergoes the usual period-doubling bifurcation. However, for $r \lesssim r_3$, even though there is no stable 3-cycle, suc-

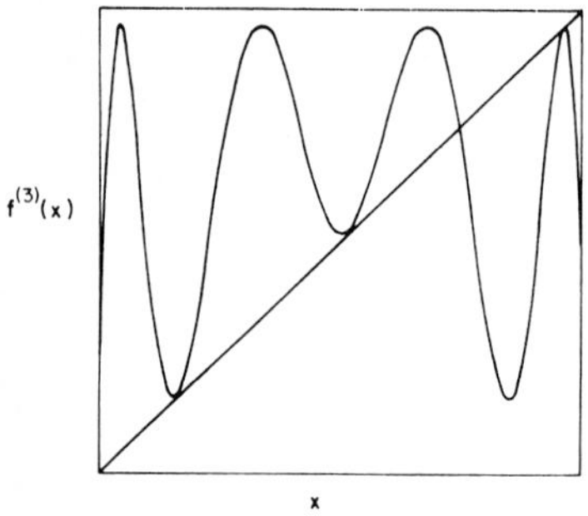

Fig. 1 The third-iterated map $f^{(3)}(x)$ at r_3

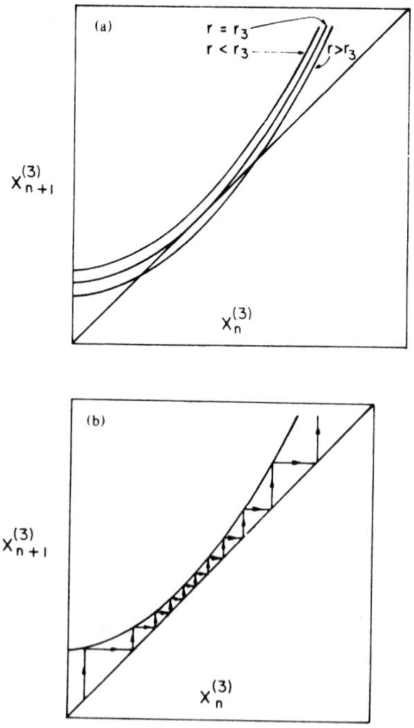

Fig. 2 (a) Tangent bifurcation near r_3 region
(b) Slow passage through the channel

cessive third iterates of the map pass more or less regularly through the channel region (Fig. 2b). This corresponds to regular or laminar flow. After passing through the channel region, the iterates move wildly until they return to this neighborhood or a similar one near the other two elements of the 3-cycle. This phenomenon of regular flow interrupted by intermittent bursts is called intermittency.

Recently it was found [4-6] that the length of laminarity ℓ, in the presence of noise of amplitude σ, takes on a universal scaling form

$$\ell(\varepsilon,\sigma) = \varepsilon^{-(1-1/z)} f(\sigma/\varepsilon^{(z+1)/2z}) , \tag{23}$$

where $\varepsilon = r_3 - r$. Most remarkably, HIRSCH, NAUENBERG and SCALAPINO [6] later found that these scaling properties and the exponents can be simply derived by employing the same functional equations (7), (19) and (22), with a mere change of sign for α and the boundary condition appropriate to the tangency condition:

$$f^*(0) = 0 \ , \ f^{*\prime}(0) = 1 \ . \tag{24}$$

Thus the functional renormalization-group equations provide a unified treatment of both period doubling and intermittency. Recently a complete set of exact solutions to these equations has been found by HU and RUDNICK [7]. The methods of obtaining these exact solutions will be the subject matter of the next subsection.

B. Exact Solutions

The first hint that exact solutions to the functional equations are possible came from series analysis [7]. The first few terms of the series expansions, enough to reveal the pattern of the entire series, are given below:

$$f^*(x) = x + ax^z + \frac{z}{2!} a^2x^{2z-1} + \frac{z(2z-1)}{3!} a^3x^{3z-2} + \cdots \tag{25}$$

$$h(x) = 1 + \frac{z}{2!} ax^{z-1} + \frac{z(3z-2)}{3!} a^2x^{2(z-1)} + \frac{z(3z-2)(4z-3)}{4!} a^3x^{3(z-1)} + \cdots \tag{26}$$

$$g^2(x) = 1 + \frac{2z}{2!} ax^{z-1} + \frac{2z(4z-2)}{3!} a^2x^{2(z-1)} + \frac{2z(4z-2)(5z-3)}{4!} a^3x^{3(z-1)} + \cdots . \tag{27}$$

The interested reader is invited to write down the general terms and resum them.

There are other more powerful methods [7] to obtain the exact solutions. However, a very illuminating method is to use a differential-equation formulation [7]. In Eq. (25), if we consider an infinitesimal change, $a \to dt$, and neglect all the higher-order terms, then we obtain a simple differential equation

$$\frac{dx}{dt} = x^z \ . \tag{28}$$

We will first show how to find the rescaling factor α of Eq. (7). Define

$$w(\tau) = \alpha x(t(\tau)) \ , \tag{29}$$

where

$$t(\tau) = 2\tau \ . \tag{30}$$

The equation for w is

$$\frac{dw}{d\tau} = (2\alpha^{z-1})w \ . \tag{31}$$

The invariant nature of the fixed-point equation dictates that

$$2\alpha^{z-1} = 1 \tag{32}$$

so that the equations for x(t) and w(τ) are the same under rescaling. We have therefore

$$\alpha = 2^{1/(z-1)} \ . \tag{33}$$

We now proceed to find the fixed-point function $f^*(x)$. Integrating Eq. (28)

$$\int_x^{x'} x^{-z}\,dx = \int_0^a dt \ , \tag{34}$$

we obtain immediately the fixed-point function

$$x' = f^*(x) = [x^{-(z-1)} - (z-1)a]^{-1/(z-1)} \ . \tag{35}$$

That the global solution can be obtained from the local differential equation is of course well known in the theory of Lie groups. Since intermittency is based on the tangent bifurcation, the transition is continuous. The underlying invariance group is simply the one-parameter translation group. The solutions to the functional equation, which expresses the recursion relation between x(t+τ) and x(t), can therefore be simply obtained by integrating the differential equation.

Now that we have obtained the exact solution to Eq. (7), it is relatively straightforward to find the exact solutions to the eigenvalue equations (19) and (22). Let us first try to solve the deterministic eigenvalue equation. Consider a perturbation to Eq. (28):

$$\frac{dx}{dt} = x^z + \varepsilon x^n \ . \tag{36}$$

We will again first find the eigenvalues λ_ε. In terms of the variable $w(\tau)$, Eq. (36) reads

$$\frac{dw}{d\tau} = w^z + \varepsilon(\alpha^{z-n})w^n . \tag{37}$$

It has exactly the same form as Eq. (36) except for a factor of α^{z-n}. The eigenvalues of Eq. (19) are therefore

$$\lambda_\varepsilon = 2^{(z-n)/(z-1)} . \tag{38}$$

We proceed to find the eigenfunctions $h(x)$. Under an infinitesimal change $x \to x + \varepsilon y$, the variation y satisfies the differential equation

$$\frac{dy}{dt} = zx^{z-1}y + x^n . \tag{39}$$

Letting $y = x^z v$, we obtain the differential equation for v:

$$\frac{dv}{dt} = x^{n-z} . \tag{40}$$

Integrating this equation

$$\int_0^{v'} dv = \int_0^a dt\; x^{n-z}$$

$$= \int_x^{x'} dx\; x^{-z}\; x^{n-z} , \tag{41}$$

we get

$$v' = \frac{1}{2z-n-1} \{[x^{-(z-1)}-(z-1)a]^{(2z-n-1)/(z-1)} - x^{-(2z-n-1)}\} . \tag{42}$$

Writing $y' = x'^z v'$, we obtain the eigenfunctions $h(x) = y'$ of the deterministic eigenvalue equation:

$$h(x)$$
$$= \frac{1}{2z-n-1} [x^{-(z-1)}-(z-1)a]^{-z/(z-1)}\{x^{-(2z-n-1)}-[x^{-(z-1)}-(z-1)a]^{(2z-n-1)/(z-1)}\} . \tag{43}$$

Finally let us try to solve the stochastic eigenvalue equation. Consider a stochastic perturbation to Eq. (28)

$$\frac{dx}{dt} = x^z + \varepsilon x^n \xi , \tag{44}$$

where $\xi(t)$ is a Gaussian random variable satisfying

$$\langle\xi(t)\xi(t')\rangle = \delta(t-t') . \tag{45}$$

Due to the singular nature of this random variable, we have to exercise some care in trying to find the stochastic eigenvalues λ_ξ. It will be clearer to rewrite the stochastic differential Eq. (44) as a finite difference equation:

$$\frac{x(t+\Delta)-x(t)}{\Delta} = x^z + \varepsilon x^n \xi_\Delta , \tag{46}$$

where

$$\langle \xi_\Delta^2 \rangle = \frac{1}{\Delta} . \tag{47}$$

In terms of the variable w, Eq. (46) becomes

$$2\left[\frac{w(\tau+\Delta')-w(\tau)}{\Delta}\right] = w^z + \varepsilon w^n \alpha^{z-n} \xi_\Delta . \tag{48}$$

It can be seen easily that Δ and ξ_Δ are related to Δ' and $\xi_{\Delta'}$ by

$$\Delta = 2\Delta' , \tag{49}$$

$$\xi_\Delta = 2^{-\frac{1}{2}} \xi_{\Delta'} . \tag{50}$$

Eq. (46) therefore becomes

$$\frac{w(\tau+\Delta')-w(\tau)}{\Delta'} = w^z + \varepsilon(\alpha^{z-n} 2^{-\frac{1}{2}}) w^n \xi_{\Delta'} . \tag{51}$$

This stochastic difference equation for $w(\tau)$ is the same as that for $x(t)$ except for the factor $\alpha^{z-n}2^{-\frac{1}{2}}$, which gives the stochastic eigenvalues

$$\lambda_\xi = 2^{(z-2n+1)/(z-1)} . \tag{52}$$

We now proceed to find the eigenfunctions of the stochastic eigenvalue equation. Under an infinitesimal change, $x \to x + \varepsilon y$,

$$\frac{dy}{dt} = zx^{z-1} y + x^n \xi . \tag{53}$$

Writing $y = x^z v$, we obtain the stochastic differential equation for v:

$$\frac{dv}{dt} = x^{n-z} \xi . \tag{54}$$

The solution to this stochastic differential equation is

$$v' = [\int_0^a x^{2(n-z)} dt]^{\frac{1}{2}} \xi' , \tag{55}$$

where ξ' is another random variable. Working out the integral, we get

$$v' = \frac{1}{\sqrt{3z-2n-1}} \{x^{-(3z-2n-1)} - [x^{-(z-1)} - (z-1)a]^{(3z-2n-1)/(z-1)}\}^{\frac{1}{2}} \xi' . \tag{56}$$

Writing $y' = x'^{z} v'$, we obtain the eigenfunctions, $g^2(x) = y'^2$, of the stochastic eigenvalue equation

$$g^2(x)$$

$$= \frac{1}{3z-2n-1} [x^{-(z-1)} - (z-1)a]^{-2z/(z-1)} \{x^{-(3z-2n-1)} - [x^{-(z-1)} - (z-1)a]^{(3z-2n-1)/(z-1)}\} . \tag{57}$$

This completes the derivation of the exact solutions to the functional equations for intermittency in the one-dimensional case. Exact solutions for intermittency in two dimensions have also been obtained [7]. This differential-equation method is applicable to any higher dimensions.

IV. Concluding Remarks

The functional renormalization-group equations provide a powerful approach to the study of the universal scaling properties of dynamical systems at the onset of chaos. Unfortunately not much is known about these equations or their solutions. It was initially quite surprising that a complete set of exact solutions could be found for such complicated functional equations in the case of intermittency. The connection between the differential-equation formulation and the functional equations is intriguing. It will be a very important task in the field of the transition to chaos that study of the properties of these functional equations and the analytic methods of solving them be vigorously pursued.

Acknowledgments

The work reported here was done in collaboration with Professor Joseph Rudnick and supported in part by the University of Houston. I would like to thank Professor Y. Kuramoto for his invitation to lecture at the Kyoto Summer Institute and his gracious hospitality. The manuscript was written while I was visiting the Theoretical Division and the Center for Nonlinear Studies at the Los Alamos National Laboratory in the summer of 1983. The warm hospitality extended to me by Drs. P. Carruthers and D. K. Campbell is also gratefully acknowledged.

References

1 For a recent review and references, see, for example, B. Hu, Phys. Rep. 91, 233 (1982)

2 J.-P. Eckmann, Rev. Mod. Phys. 53, 643 (1981)

3 M. J. Feigenbaum, J. Stat. Phys. 19, 25 (1978); 21, 669 (1979)

4 P. Manneville and Y. Pomeau, Phys. Lett. 75A, 1 (1979); Physica 1D, 219 (1980); Y. Pomeau and P. Manneville, Commun. Math. Phys. 74, 189 (1980)

5 J.-P. Eckmann, L. Thomas and P. Wittwer, J. Phys. A14, 3153 (1981)

6 J. E. Hirsch, B. A. Huberman and D. J. Scalapino, Phys. Rev. A25, 519 (1982) J. E. Hirsch, M. Nauenberg and D. J. Scalapino, Phys. Lett. 87A, 391 (1982)

7 B. Hu and J. Rudnick, Phys. Rev. Lett. 48, 1645 (1982); Phys. Rev. A26, 3035 (1982); "Differential-Equation Approach to Functional Equations: Exact Solutions for Intermittency," University of Houston preprint (1983)

8 J. P. Crutchfield, M. Nauenberg and J. Rudnick, Phys. Rev. Lett. 46, 933 (1981)

9 B. Shraiman, C. E. Wayne and P. C. Martin, Phys. Rev. Lett. 46, 935 (1981)

Collapse of Tori in Dissipative Mappings

K. Kaneko

Department of Physics, Faculty of Science, University of Tokyo, Hongo, Bunkyo-ku
Tokyo 113, Japan

1. Introduction

Transition from torus to chaos with frequency lockings has been extensively studied in recent years. Most of these works, however, are restricted only to the study of phase motion of a torus, which is based on a one-dimensional circle map. (see,e.g. [1]) In this paper, we use higher dimensional mappings to study the instability of amplitude motion of a torus. Contents are as follows: (A) Oscillation of torus in connection with the damped oscillation of an unstable manifold (§2 i) . (B) Onset of chaos through "fractalization" of torus (§2 ii) . (C) Doubling of torus stops by a finite number of times. The mechanism of the interruption of doubling is explained by the instability against a perturbation of the direct product state of coupled logistic and torus maps. (D) Three-torus persists for a finite perturbation, which loses its stability through lockings and chaos appears. Thus, the picture by RUELLE and TAKENS [11] does not always hold (§4).

2. Oscillation of Torus

i) Connection with the damped oscillation of an unstable manifold ——
Various two-dimensional mappings show a transition "cycle → (Hopf bifurcation) →

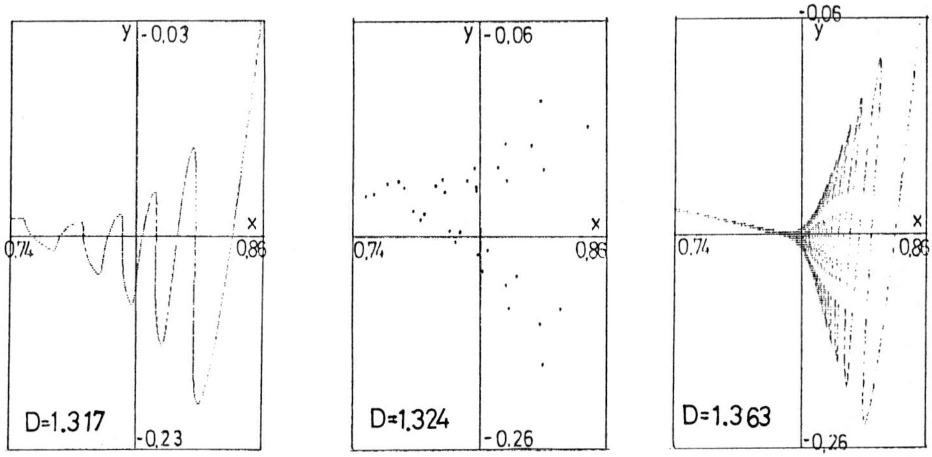

Fig. 1 A part of the attractor of the map with A = 0.12

torus → (distortion) → lockings → chaos" [2,3]. We consider here a typical model which shows this behavior, that is a two-point delayed logistic map [3,7,12],

$$x_{n+1} = Ax_n+(1-A)(1-Dy_n^2) \quad , \quad y_{n+1} = x_n \quad . \tag{1}$$

At D = 1/(1-A), this map shows a Hopf bifurcation and a torus appears. Transition to chaos occurs via lockings as the nonlinearity D is increased. There appears an oscillatory behavior of torus before the chaos appears. (See Fig. 1a (torus) and 1b (locking) and 1c (chaos) for the attractor.) The oscillation (or distortion) of a torus has also been observed in a coupled logistic map [2]. We also note that the attractor in Rayleigh-Bénard convection shows a quite similar behavior [4].

The mechanism of the oscillation is understood as follows. We consider a locking which appears at a different but close value of a bifurcation parameter. When the locking occurs, periodic saddles also appear. If the unstable manifold of the saddles crosses to a stable manifold of a node, it crosses infinite times. The situation is similar to the homoclinic oscillation in conservative mappings, but the oscillation in our case is damped since the mapping is dissipative. We calculated numerically an unstable manifold of a saddle with period 4 for the map (1). It shows an oscillation just like Figs. 1a) or 1c). Thus, oscillation of the torus can be regarded as a manifestation of the oscillation of an unstable manifold. The damping ratio of the oscillation is given by the eigenvalue of a matrix $J(x_1,y_1)\cdots J(x_p,y_p)$ where J is a Jacobi matrix and (x_j,y_j) is a periodic point. The oscillation, therefore, can be seen typically, if the stability of a periodic point along the amplitude direction is small. In the map (1), this is satisfied for small A.

ii) "Fractalization" of Torus

In usual two-dimensional mappings, oscillation of torus is masked by lockings. Here, we propose a modulation model which has no lockings, i.e.,

$$\begin{cases} x_{n+1} = f(x_n)+\varepsilon h(\theta_n) \\ \theta_{n+1} = \theta_n+c \quad , \quad c = \text{irrational (mod 1)} \quad . \end{cases} \tag{2}$$

The modulation map has been investigated in doubling of torus and stability of three-torus. Here, we take $f(x) = -ax+bx^2$, $h(\theta) = \sin(2\pi\theta)$ and $c = (\sqrt{5}-1)/2$. When $\varepsilon = 0$, the torus is a straight line ($x = 0$). As ε is increased, the torus oscillates more and more strongly, till it collapses at $\varepsilon = \varepsilon_c$ and chaos emerges for $\varepsilon > \varepsilon_c$. In Fig. 2, the attractors are shown. It seems that the torus loses its smoothness at $\varepsilon = \varepsilon_c$.

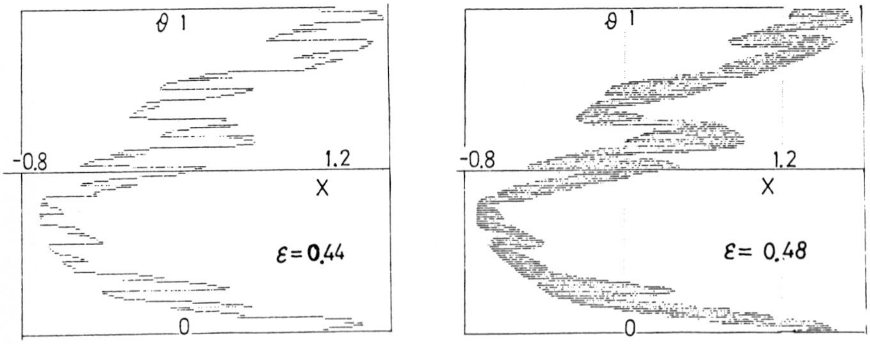

Fig. 2 Attractor of the map (2) with a = b = 1 and $c = (\sqrt{5}-1)/2$. ($\varepsilon_c \simeq 0.472$)

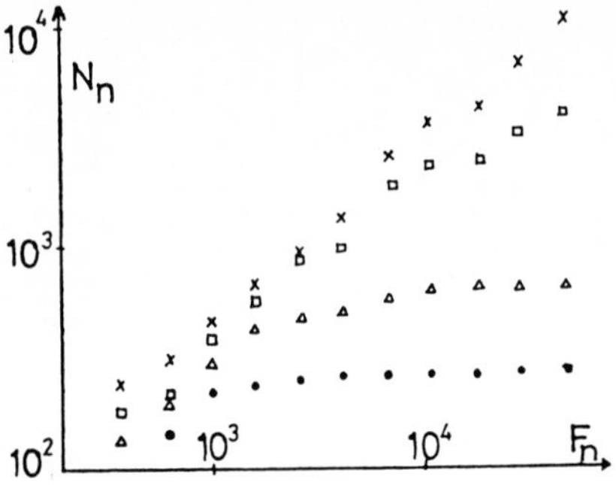

Fig. 3 The number of extrema N_n as a function of F_n. The parameters a and b are both 1 and ε = 0.45 (•), 0.46 (△), 0.465 (□) and 0.47 (x)

In order to study the oscillatory behavior in more detail, we introduce the equation of an invariant curve. The invariant curve $x = g(\theta)$, if it exists, must obey the functional equation

$$g(\theta+c;\ 1) = f(g(\theta))+\varepsilon h(\theta) \quad , \tag{3}$$

where $(x;\ 1)$ denotes x(mod 1). If f(x) is linear (b = 0), Eq. (3) is easily solved to give

$$g(\theta) = \varepsilon(\sin 2\pi(\theta-c)+a\sin 2\pi\theta)/\{(1+a\cos 2\pi c)^2+a^2\sin^2 2\pi c\} \quad . \tag{4}$$

For $b \neq 0$, it is not easy to solve (3). We search for a solution numerically by iterating the equation

$$g_n(\theta+c;\ 1) = f(g_n(\theta))+\varepsilon h(\theta). \tag{5}$$

For numerical convenience , we replace c by $c_n = F_{n-1}/F_n$, where F_n is the Fibonacci sequence [5]. We made 5000 iterations of Eq. (5) for c_n. The convergence is very fast for $\varepsilon \ll \varepsilon_c$. It becomes slower as ε approaches ε_c, and the iteration (5) does not converge for $\varepsilon \lesssim \varepsilon_c$ within the above steps. We calculated the number of extrema N_n, by counting the integer i ($1 \leq i \leq F_n$) such that $\{g(c_n(i+1);\ 1)-g(c_n i;\ 1)\} \times \{g(c_n i;\ 1)-g(c_n(i-1);\ 1)\} < 0$. If the curve is smooth, the number N_n approaches a constant value as n is increased, while it grows infinitely in proportion to n^α, if the curve is fractal [6]. The number N_n is shown in Fig. 3 as a function of F_n. Small scale structures are increasingly excited as ε approaches $\varepsilon_c \simeq 0.472$. We also measured the length of the invariant curve at $\varepsilon \simeq \varepsilon_c$, by changing the scale [6], which also supports that the torus becomes fractal at $\varepsilon \simeq \varepsilon_c$. Detailed results on the critical phenomena at this transition will be reported elsewhere [7].

3. Doubling of Torus

In the two-dimensional mapping of §2 i), the instability of amplitude direction is interrupted by lockings. For higher dimensional mappings, however, the doubling of torus itself is also possible, which was reported by the author and by ARNÉODO et al.[8,9]. The models in [8] are three-point or four-point delayed logistic maps and some modified ones. These models show only a finite number of doublings before chaos appears. Thus, there arises a question whether the direct product state of the torus map and Feigenbaum's fixed point function [10] is stable against a perturbation or not. To answer this question, we performed a simulation of the map

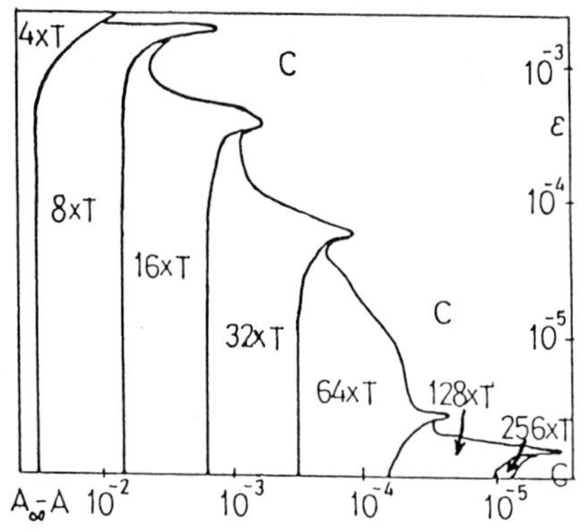

Fig. 4 Phase diagram of the map (6) with the perturbations (I) and $c = (\sqrt{5}-1)/2$. The transverse axis denotes $(A_\infty - A)$, where A_∞ is the value of the onset of chaos for the map $x_{n+1} = 1-Ax_n^2$. The notations c and $n \times T$ represent chaos and n ⊗ torus respectively. This diagram is obtained by the calculations of the first and second Lyapunov exponents

$$x_{n+1} = 1-Ax_n^2+\varepsilon g(x_n,y_n) \quad , \quad y_{n+1} = y_n+c+\varepsilon h(x_n,y_n) \quad (\text{mod } 1) \tag{6}$$

where g and h are perturbations and c is fixed at an irrational number. We mainly studied the following two cases; (I) $g(x,y) = \sin(2\pi y)$ and $h(x,y) = x$ (II) $g(x,y) = \sin(2\pi y)$ and $h = 0$ (modulation model). In Fig. 4, the phase diagram for the case (I) with $c = (\sqrt{5}-1)/2$ is given, which shows the instability of the direct product state. The number of doubling decreases as the perturbation increases. The qualitative behavior does not change for other values of c and for the case (II).

When the doubling stops and chaos appears, oscillation of the torus is remarkably seen. Thus, a possible mechanism of the onset of chaos is considered to be the fractalization of torus in §2 ii). For the modulation model (II), we can construct a one-dimensional map $x_{n+1} = \tilde{g}(x_n)$ by iterating the map (6) F_n times, if $c = (\sqrt{5}-1)/2$ is replaced by F_{n-1}/F_n. The map $\tilde{g}(x)$ shows an intermittent-like burst at the onset of chaos. Since $\tilde{g}(x)$ satisfies the Schwarzian condition, the cause of the interruption of the doubling will be the multi-modal property of $\tilde{g}(x)$ [3].

4. Fates of Three-Torus

In 1971, RUELLE and TAKENS [11] proposed a new picture for the onset of turbulence. They have shown that a three-torus (we use this word for a quasiperiodic state with three incommensurate frequencies) is unstable against some perturbations. Their

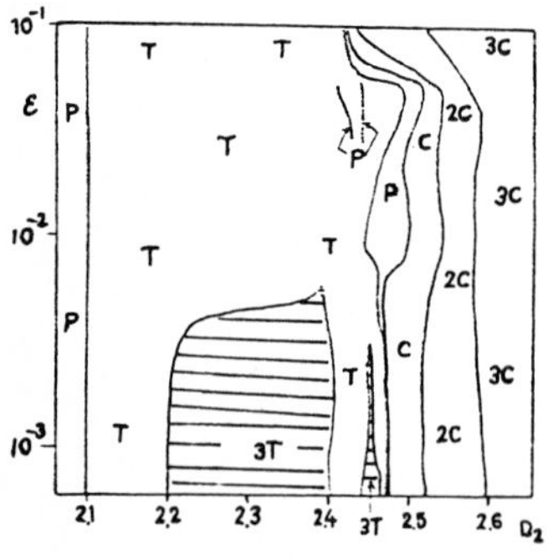

Fig. 5 Phase diagram of the map (7) with $D_1 = D_2+0.1$, $h_1 = z_n-z_{n-1}$ and $h_2 = x_{n-1}-x_n$. The notations P, T, 3T, C, and nC denote periodic state (cycle), torus, three-torus, chaos and hyperchaos with n positive Lyapunov exponents, respectively. This diagram is obtained through the calculations of four Lyapunov exponents

argument, however, does not include the consideration on the measure of such perturbations. Thus, it is an open problem whether the transition to chaos by their mechanism is physically observable or not. We made a simulation of the coupled map [13]

$$\begin{cases} x_{n+1} = Ax_n+D_1x_{n-1}(1-x_{n-1})+\varepsilon h_1(x_n,x_{n-1},z_n,z_{n-1}) \\ z_{n+1} = Az_n+D_2z_{n-1}(1-z_{n-1})+\varepsilon h_2(x_n,x_{n-1},z_n,z_{n-1}) \end{cases} \tag{7}$$

to study this problem. If the perturbation does not exist ($\varepsilon = 0$), each two-dimensional mapping shows a transition "fixed point → torus → locking → chaos". For $A = 0.4$, Hopf bifurcation occurs at $D = 3-2A = 2.2$ and chaos appears for $D \gtrsim 2.59$. Thus, the direct product state such as torus ⊗ cycle, torus ⊗ torus, chaos ⊗ torus exists corresponding to the values D_1 and D_2, for $\varepsilon = 0$.

We add perturbations $h_1 = z_n-z_{n-1}$ and $h_2 = x_{n-1}-x_n$ to see the stability of the direct product states. In Fig. 5, the phase diagram is shown, where the attractors are classified according to the sign of Lyapunov exponents. As is seen in this figure, three-torus exists for finite perturbations. We also studied other types of perturbations such as $h_1 = (z_n-z_{n-1})^2$ or $h_1 = |z_n-z_{n-1}|^{1/2}$ (similar for h_2). The qualitative features of the phase diagram do not change for these perturbations. We note that the collapse of three-torus for large perturbations is not due to the emergence of chaos, but due to the complete locking to torus. The chaos in our model is born only out of a locking to cycle which again is born of a locking to torus. Thus, the picture by RUELLE and TAKENS [11] is not observed in our model.

If our interest is restricted only to the phase motion of the three-torus, a further simplification of the model may be possible. Thus, the coupled circle map $\theta_{n+1} = \theta_n+d_1+h_1(\theta_n,\phi_n)$ (mod 1), $\phi_{n+1} = \phi_n+d_2+h_2(\theta_n,\phi_n)$ (mod 1) will be relevant to study the stability of the three-torus [12,13]. The result of this map also confirms the above statements about the map (7).

The locking to torus from a three-torus forms a "double devil's staircase", in the sense that the number of the elements to construct a staircase is two. The properties of the "double devil's staircase" can also be analyzed by the coupled circle map [13].

5. Discussion

In this paper we have investigated various instabilities of tori in dissipative mappings. Especially, fractalization of torus will give a new critical phenomenon at the onset of chaos. Renormalization group approach in the wave number space will be important to study this critical phenomenon. Doubling of tori is also a mechanism for the onset of chaos though the cascade cannot continue infinitely. Stability of a three-torus is an interesting problem. Since it is stable for finite perturbations in our simulations, it will be possible and important to construct a dissipative version of the KAM theory. We have studied the stability of the various direct product states using the method of coupled maps. The hyperchaos as a direct product state of chaos seems to be most stable in our simulations. Thus, it may be interesting to consider the fully developed turbulence as a direct product state of a lot of chaos, instead of Landau's picture on turbulence.

References

1. K. Kaneko, to appear in Turbulence and Chaotic Phenomena in Fluids (Ed. T. Tatsumi, North-Holland Pub.) and references cited therein
2. K. Kaneko, Prog. Theor. Phys. 69, 1427 (1983)
3. K. Kaneko, to be published
4. P. Bergé, to appear in Physica Scripta
5. J.M. Greene, J. Math. Phys. 20, 1183 (1979)
6. B.B. Mandelbrot, The Fractal Geometry of Nature (Freeman, San Francisco, 1982)
7. K. Kaneko, to be submitted to Phys. Lett. A
8. K. Kaneko, Prog. Theor. Phys. 69, 1806 (1983)
9. A. Arnéodo, P.H. Coullet, and E.A. Spiegel, Phys. Lett. 94A, 1 (1983)
10. M.J. Feigenbaum, J. Stat. Phys. 19, 25 (1978); 21, 669 (1979)
11. D. Ruelle and F. Takens, Comm. Math. Phys. 20, 167 (1971)
12. C. Grebogi, E. Ott, J.A. Yorke, Phys. Rev. Lett. 51, 339 (1983)
13. K. Kaneko, "Fates of Three-Torus I", submitted to Prog. Theor. Phys.

Periodic Forcing Near Intermittency Threshold - Resonance and Collapse of Tori

H. Daido†

Research Institute for Fundamental Physics, Kyoto University
Kyoto 606, Japan

I. Introduction

There are a variety of problems concerning chaos. One of the interesting and important ones would be to examine how chaos responds to external perturbations. In particular the response of chaos near its onset is interesting since such a baby of chaos is generally speaking expected to respond in more or less universal ways. In fact, as to stochastic perturbations, recently many studies have been devoted to such problems near onset of chaos mainly via period doubling and intermittency [1]. On the other hand, the response to deterministic perturbations, i.e., periodic forcing seems still to remain an open problem.

Among various ways of onset of chaos, intermittency [2] is particularly interesting. As the threshold is approached, intermittent chaos becomes more and more similar to purely periodic oscillations. Then, how does the response of an intermittent state compare to that of a periodic state which has been well studied so far (see,e.g.,[3], [4]) ? Moreover, needless to say, intermittency is one of the typical routes to chaos, so that in this connection it may be a natural question to ask what kind of routes to chaos are produced by periodic forcing near its onset. Trying to answer these questions, some results are reported in the following of numerical and analytical investigations into the behavior of chaotic systems near intermittency threshold under periodic perturbation.

II. The Lorenz Model [5]

First of all some numerical results [6] are described here on the forced Lorenz model: $\dot{x}=\sigma(y-x)$, $\dot{y}=-y-xz+(r+a\cos\omega t)x$, $\dot{z}=-bz+xy$, where $\sigma=10$ and $b=8/3$. In the absence of forcing ($a=0$), it is known [7] that the model exhibits intermittent chaos for $r>r_c=166.06\cdots$ which is type 1 according to the classification by Pomeau and Manneville [2]. The results of two cases are presented.

Case 1

In this case the Rayleigh number r is chosen just beyond r_c, i.e., $r=166.07$, and the driving frequency ω is varied around twice the natural frequency ($2\omega_n \sim 11$) with $a=1$ fixed. This corresponds to investigating the 1/2 resonance. The results are summarized as follows: a fairly narrow region of period 2 entrainment, $\omega_\ell \leq \omega \leq \omega_r$, exists where $\omega_\ell=11.2059\cdots$ and $\omega_r=11.2173\cdots$. Around this region intermittent oscillations are found roughly over $\sim 10.5<\omega<\sim 11.4$ beyond which oscillations are fully chaotic.

† Fellow of the Japan Society for the Promotion of Science

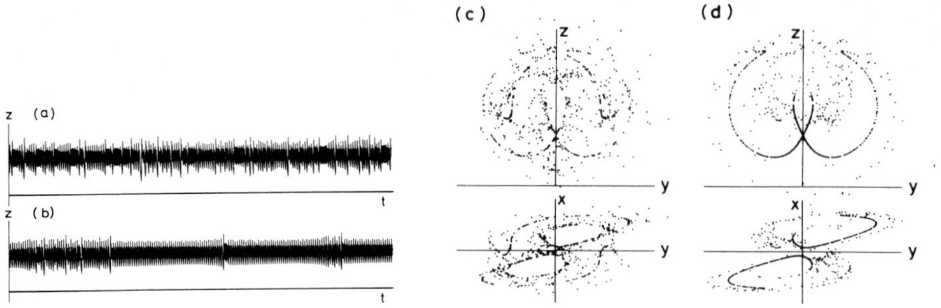

Fig.1 Typical wave forms of z and stroboscopic phase portraits for $|x|\leq 60$, $|y|\leq 100$, $80\leq z\leq 250$. (a), (c) $\omega=10$ (b), (d) 11 for r=166.07, a=1

Figure 1 shows some typical examples of the wave forms and the stroboscopic phase portraits. From these figures it is clear that a kind of order develops near the entrainment region. Namely, a pair of one-dimensional manifolds emerge there to result in the appearance of intermittency which the wave form clearly indicates. Now the character of the instabilities of the entrainment at ω_r and ω_ℓ should be noted. For ω_r one of the Floquet multipliers was found to become unity, so that the intermittency is type 1 just beyond ω_r. On the other hand, for ω_ℓ a pair of complex multipliers has absolute value unity, which implies that the intermittency is type 2 just below ω_ℓ. In this way it turns out that periodic forcing can produce several different kinds of intermittency. An important new feature of the resonance discussed so far is that instead of a torus one-dimensional manifolds accompanied by chaotic bursts appear in the stroboscopic phase portrait outside the entrainment region. Qualitatively similar results were obtained for some smaller values of a. For details see [6].

Case 2

In this case the Rayleigh number is set below r_c, i.e., r=166, and the amplitude a is varied from zero with $\omega=10$ fixed (see Fig.2). For $0<a<a_c=0.0577\cdots$ quasi-periodic oscillations (torus) are found, while for $a>a_c$ a chaotic state appears. Figure 2(d) shows that the stroboscopic phase portrait consists of a ghost of torus and

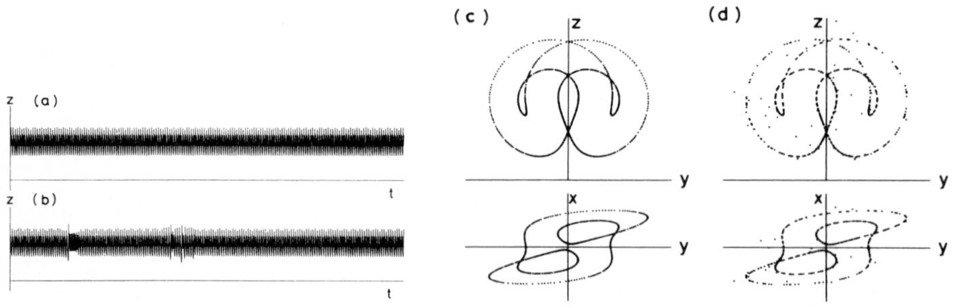

Fig.2 Typical wave forms of z and stroboscopic phase portraits for $|x|\leq 60$, $|y|\leq 100$, $80\leq z\leq 250$. (a) a=0.054 (b) 0.059 (c) 0.0576 (d) 0.058 for r=166, $\omega=10$

chaotic bursts for $a>a_c$. Therefore what happens at the threshold a_c is an intermittent transition from torus to chaos, which is one of the fundamental routes to chaos, and in fact its candidates already exist in some experimental results of Bénard convection [8], [9], [10]. Hereafter this type of phenomena will be called *torus intermittency*.

III. A Two-Dimensional Dissipative Map [11], [12]

In order to investigate in detail torus intermittency and related problems I introduce the following map:

$$\theta_{n+1} = \theta_n + \Omega + \alpha X_n , \qquad \text{(mod. 1)}$$

$$X_{n+1} = X_n + cX_n^2 + \varepsilon + a \cos 2\pi\theta_n \qquad \text{(mod. 2)}$$

which describes typically the dynamics of angle θ and radius X on a suitable Poincaré section. For a=0 the second equation reduces to a map which has been used to study type 1 intermittency and for which the threshold is $\varepsilon=0$. Roughly speaking the constant Ω denotes the ratio of the driving frequency to the natural one from the viewpoint of the previous section. Hereafter c is fixed to be 1.75.

Pair annihilation of tori

The first task is to consider the mechanism of the appearance of torus intermittency. For this purpose an example of this route to chaos in the map is presented in Fig. 3(a), (b) where a is varied with other constants fixed. Subsequent figures (c),

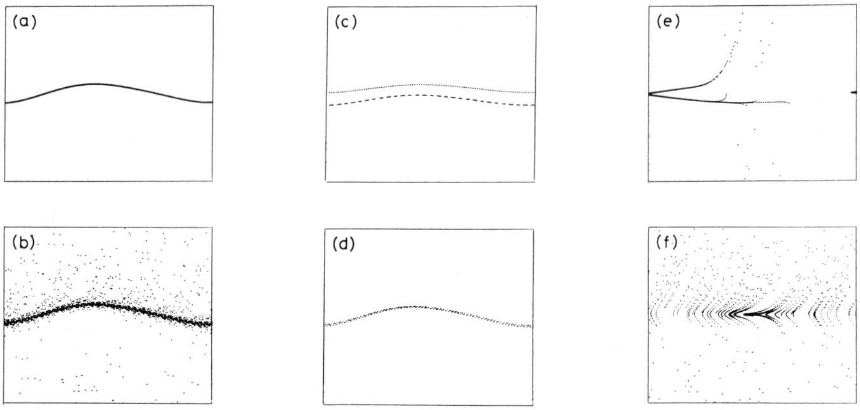

Fig.3 Onset of intermittency from torus ((a), (b), (e)), and pair annihilation of tori ((c), (d)), and collapse of unstable fixed points, (f) . Each window covers $0\leq\theta<1$ in abscissa and $-1\leq x\leq 1$ in ordinate. (a) a=0.203916 (b) 0.208 (c) 0.1 (d) 0.203 for $\varepsilon=-0.01$, $\Omega=\sqrt{2}-1$, $\alpha=0.01$. ($a_c=0.203916\cdots$). In (c), (d) only 100 points are plotted on each torus. (e) $\varepsilon=-0.00984242$ ($\varepsilon_c=-0.00984242417\cdots$) (f) 0.01 for a=0.01, $\Omega=0$, $\alpha=0.1$

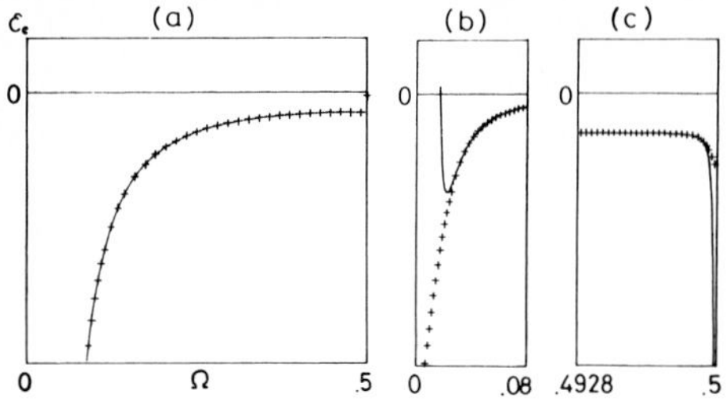

Fig.4 Comparison between theory (real line) and numerical experiment (+) for $\varepsilon_c(\Omega)$ (α=0, a=0.01). The bottom levels: (a) -3×10^{-4} (b) -8×10^{-3} (c) -1.5×10^{-4}. (b) and (c) show the neighborhoods of Ω=0 and 1/2, respectively

(d) clarify the mechanism of this transition. Namely for $a<a_c$ a pair of tori exist: one is stable (downside one) and the other unstable. Moreover, as a is varied towards a_c, both approach each other to collide finally for $a=a_c$ and disappear for $a>a_c$. Therefore it turns out that the onset of torus intermittency takes place via pair annihilation of stable and unstable tori.

Shift of the intermittency threshold

One of the interesting and important problems as to the effect of perturbation is to see how the location of the threshold is affected by perturbation. (For the case of stochastic perturbations see [1] and references therein.) For this purpose ε has to be chosen as a bifurcation parameter since its critical value ε_c is then the shift of the intermittency threshold due to periodic forcing. Using the functional equation $\phi(\theta+\Omega+\alpha\phi(\theta))=\phi(\theta)+c\phi(\theta)^2+\varepsilon+a\cos2\pi\theta$ satisfied by torus $X=\phi(\theta)$, it is possible to obtain for small a

$$\varepsilon_c = \varepsilon_1(\Omega)a^2 + \varepsilon_2(\Omega)a^4 + \cdots ,$$

where $\varepsilon_1(\Omega)=(-1/4)(c+2\pi\alpha\sin2\pi\Omega)/(1-\cos2\pi\Omega)$ and $\varepsilon_2(\Omega)=(c^3/64)(3+4\cos2\pi\Omega)/\{(1-\cos2\pi\Omega)^2(1-\cos4\pi\Omega)\}$ (for α=0 to avoid a lengthy expression). A comparison is made in Fig.4 between the theoretical ε_c and that obtained by numerical experiments. An excellent agreement is found except very near Ω=0 and 1/2. Important points to be noticed are that the shift always occurs towards the negative side and that this effect is enhanced near resonance, in particular, in the neighborhood of Ω=0.

Intermittency via saddle connection

To see what happens where the perturbation theory breaks down, I now discuss the particular case of Ω=0 ($\alpha\neq0$) as an example. The intermittent transition from torus to chaos takes place in this case as well, but it is based on a different new mechanism. Figures 5(a),(b) show the phase plane structures just below and beyond ε_c. Notice first that a pair of fixed points exist for $-a<\varepsilon<a$ one of which is unstable focus and the other saddle. For ε immediately below ε_c a stable torus exists downside near the saddle and its manifolds. Then, at the threshold, this torus and the downside branches of the manifolds merge to form a saddle connection. Finally just

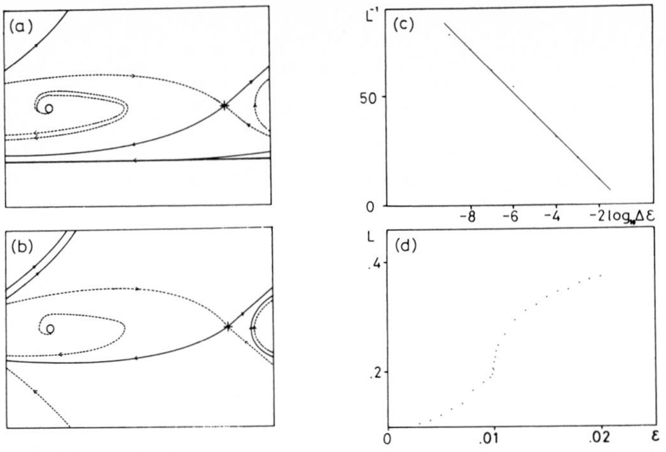

<u>Fig.5</u> The $\Omega=0$ case. (a), (b): schematic illustrations of the phase plane structure near x=0 for $\varepsilon<\varepsilon_c$, (a), and $\varepsilon>\varepsilon$, (b) . o: unstable focus. +: saddle. The broken and real lines are the stable and unstable manifolds of the saddle, respectively. The thick line in (a) is the stable torus. (c), (d): the Lyapunov number vs. ε ($\Delta\varepsilon=\varepsilon-\varepsilon_c$ in (c)) for a=0.01, $\alpha=0.1$

beyond ε_c a narrow channel from downside to upside opens. Figure 3(e) shows an example of the phase portrait in this region. The Lyapunov number L is found to grow like $L\propto -1/\ln(\varepsilon-\varepsilon_c)$ which is in clear contrast with the square root law observed in the pair annihilation case (see Fig.5(c)). This route to chaos provides another new type of intermittent transition from torus to chaos.

Cliff

A peculiar phenomenon, which I call *cliff*, is found far beyond the threshold ε_c discussed above. Namely, the Lyapunov number $L(\varepsilon)$ seems to have an infinite slope for $\varepsilon=a=0.01$ (see Fig.5(d)) where the unstable focus and the saddle collapse (see Fig.3(f)). In fact I observed numerically $L-L(\varepsilon=a)\simeq\pm|\varepsilon-a|^{\gamma}$, where $\gamma\sim0.5$ for $\varepsilon<a$ while $\gamma\sim0.7$ for $\varepsilon>a$. Usually, as a function of bifurcation parameter, the Lyapunov number has a singularity at a transition point to chaos. In the case of cliff, however, the singularity appears in the chaotic region. This implies that a sudden enhancement or enfeeblement of chaos takes place as a parameter is varied. Correspondingly, a sharp change is found to occur in power spectra [12].

IV. Conclusion and Remarks

In conclusion I have shown that intermittent systems can exhibit a variety of phenomena under periodic perturbation. It is important to note that *external periodic forcing is not necessarily important to observe these phenomena*. Even in the absence of forcing such phenomena may be observed provided that at least two internal oscillating degrees of freedom exist and interact each other in a proper way. In particular a couple of examples of the intermittent transition from torus to chaos have been presented. These are based on radius dynamics rather than angle dynamics, but I believe that they have physical significance no less important than other various routes [13].

References

1. For review see J.P. Crutchfield, J.D. Farmer, and B.A. Huberman, Phys. Rep. 92, 47 (1982).
2. Y. Pomeau and P. Manneville, Commun. Math. Phys. 74, 189 (1980).
3. T. Kai and K. Tomita, Prog. Theor. Phys. 61, 54 (1979).
4. K. Tomita and H. Daido, Phys. Lett. 79A, 133 (1980).
5. E.N. Lorenz, J. Atm. Sci. 20, 130 (1963).
6. H. Daido, Prog. Theor. Phys. 70, 879 (1983).
7. P. Manneville and Y. Pomeau, Physica 1D, 219 (1980).
8. J.P. Gollub and S.V. Benson, J. Fluid Mech. 100, 449 (1980).
9. J. Maurer and A. Libchaber, J. Phys. Lett. 41, L515 (1980).
10. M. Sano and Y. Sawada, in these proceedings.
11. H. Daido, preprint (RIFP-525): submitted to Prog. Theor. Phys. (August 1983).
12. H. Daido, to be published.
13. Relevant papers in these proceedings.

Perturbation Theory Analysis of Bifurcations in a Three-Dimensional Differential System

T. Shimizu

Division of Physics, Faculty of Engineering, Kokushikan University, Setagaya
Setagaya-Ku, 154 Tokyo, Japan

1. Introduction

For the appearance of chaos some typical scenarios, which include the inverted bifurcation, the period-doubling bifurcation, intermittent chaos and a transition from a torus to a strange attractor, have been investigated experimentally and theoretically[1]. Among these scenarios, in particular, a transition of a torus to a strange attractor is of great interest in the study of the connection between chaos in dissipative systems and in conservative systems. In one- and two-dimensional difference systems several important properties have been found. The whole aspect of the scenario, however, seems not yet to be clear.

In this study we propose a three-dimensional differential system[2]:

$$\ddot{x}(t) - \varepsilon a\dot{x}(t) + bx(t)(x(t)^2 - 1 + m(t)) = 0, \qquad (1)$$
$$\dot{m}(t) - \varepsilon(cm(t) - fx(t)^2 - r\dot{x}(t)^2/(1 - m(t))^2) = 0,$$

where a, b, c, f and r are positive parameters and ε is a smallness parameter. This three-dimensional system has a sheet-type attractor (like the Lorenz attractor) for $\varepsilon < 0$. In particular, this system coincides with the Lorenz model for $r = 0$[3]. On the other hand, this system has a torus-type attractor and exhibits a transition of a torus to a strange attractor for $\varepsilon > 0$. In this study we pay attention to the latter case. For this purpose we put these parameters as $a = 1.0625$, $b = 0.0625$, $c = 0.25$, $f = 1.75$ and $\varepsilon = 0.001$. The parameter r is chosen as a bifurcation parameter. Since this system has a smallness parameter ε, we can solve (1) by using a perturbation theory. The main purposes of this study are to give an asymptotic solution of (1) and to discuss the mechanism of onset of chaos and the transition of the torus to the strange attractor by using the asymptotic solution.

2. Asymptotic solution

Let us first try to find an asymptotic solution of (1) by using Kuzmak's method. Hereafter we consider only the case of $m(t) < 1$, which is possible by choosing an initial value $m(0)$ as $m(0) < 1$. We introduce two time scales τ and ω :

$$\tau = \varepsilon t, \qquad d\omega/dt = \phi(\tau), \qquad (2)$$

where $\phi(\tau)$ is determined in the course of analysis. Then the time derivatives with respect to t are transformed according to

$$d/dt = \varepsilon\partial/\partial\tau + \phi(\tau)\partial/\partial\omega,$$
$$d^2/dt^2 = \varepsilon^2\partial^2/\partial\tau^2 + \varepsilon\phi'(\tau)\partial/\partial\omega + 2\varepsilon\phi(\tau)\partial^2/\partial\tau\partial\omega + \phi(\tau)^2\partial^2/\partial\omega^2, \qquad (3)$$

where the prime denotes differentiation with respect to τ. We assume that there exist asymptotic representations for x and m of the form

$$x(t) = x_0(\tau,\omega) + \varepsilon x_1(\tau,\omega) + \dots, \qquad m(t) = m_0(\tau,\omega) + \varepsilon m_1(\tau,\omega) + \dots. \tag{4}$$

Substituting (3) and (4) into (1) and equating the coefficients of ε^0 and ε^1 to zero, we get

$$\phi(\tau)\partial m_0/\partial\omega = 0, \tag{5}$$
$$\phi(\tau)^2\partial^2 x_0/\partial\omega^2 + bx_0(x_0^2 - 1 + m_0) = 0, \tag{6}$$
$$\partial m_0/\partial\tau + \phi(\tau)\partial m_1/\partial\omega = cm_0 - fx_0^2 - r(\phi(\tau)/(1 - m_0))^2(\partial x_0/\partial\omega)^2, \tag{7}$$
$$\phi(\tau)^2\partial^2 x_1/\partial\omega^2 + b(3x_0^2 - 1 + m_0)x_1 = -\phi'(\tau)\partial x_0/\partial\omega - 2\phi(\tau)\partial^2 x_0/\partial\tau\partial\omega + a\phi(\tau)\partial x_0/\partial\omega - bx_0 m_1 = 0. \tag{8}$$

We assume that $\phi(\tau) \neq 0$. Eq.(5) implies that m_0 is a function of τ only. Multiplying (6) by $\partial x_0/\partial\omega$, we get

$$\phi^2(\partial x_0/\partial\omega)^2/2 + bx_0^4/4 - b(1 - m_0)x_0^2/2 = E_0(\tau), \tag{9}$$

where E_0 may be a function of τ. This means that in the time scale ω the motion can be represented as the oscillation in the fourth-order potential, because E_0 and m_0 are constant on this time scale. If $E_0 > 0$, the motion extends over both the right valley $x_0 > 0$ and the left valley $x_0 < 0$. For $E_0 < 0$ the motion is localized in one of two valleys. On the time scale τ, however, m_0 and E_0 can change. Then the motion becomes very complicated, as will be shown later. It is convenient to introduce the Jacobi elliptic functions sn(u,k), cn(u,k) and dn(u,k) and the Jacobi zeta function zn(u,k) with modulus k. Let us denote the complete elliptic integral with modulus k of the first kind and of the second kind by K(k) and E(k), respectively. According to the sign of E_0 the solution of (9) can be expressed by

$$x_0 = A(\tau)\,\mathrm{cn}(K(\nu)\omega,\nu) \quad \text{for } E_0 > 0 \quad \text{case(I)}$$
$$x_0 = B(\tau)\,\mathrm{dn}(K(\nu)\omega,\nu) \quad \text{for } E_0 < 0 \quad \text{case(II)} \tag{10}$$

where the modulus ν and the amplitudes A and B may be functions of τ. Substituting (10) into (9) and using the property of Jacobi elliptic functions, we find that the energy E_0, the amplitude A (or B) and ϕ can be represented in terms of ν and m_0,

$$E_0 = b(1 - \nu^2)(\nu(1 - m_0)/(2\nu^2 - 1))^2, \quad A = \pm(2\nu^2(1 - m_0)/(2\nu^2 - 1))^{1/2},$$
$$\phi = (b(1 - m_0)/(2\nu^2 - 1))^{1/2}/K(\nu), \quad \text{case(I)}$$
$$E_0 = b(\nu^2 - 1)((1 - m_0)/(2 - \nu^2))^2, \quad B = \pm(2(1 - m_0)/(2 - \nu^2))^{1/2},$$
$$\phi = (b(1 - m_0)/(2 - \nu^2))^{1/2}/K(\nu), \quad \text{case(II)}. \tag{11}$$

Therefore we can determine the lowest order approximation x_0 completely, if we know the time evolutions of ν and m_0. The time evolution equations for ν and m_0 can be derived from the secular conditions.

After some calculations we get the following asymptotic solution in cases (I) and (II).

Case(I) $E_0 > 0$ and $\phi_1 > \alpha\varepsilon$

$$x(t) \doteqdot x_0(\tau,\omega) = A(\tau)\,\mathrm{cn}(K(\nu)\omega,\nu),$$
$$p(t) \doteqdot \phi(\tau)\partial x_0(\tau,\omega)/\partial\omega = -\phi(\tau)A(\tau)K(\nu)\,\mathrm{sn}(K\omega,\nu)\,\mathrm{dn}(K\omega,\nu), \tag{12}$$
$$m(t) \doteqdot m_0(\tau) + \varepsilon m_1(\tau,\omega)$$
$$dm_0/d\tau = cm_0 - (A^2 fG_1(\nu) + 2\nu^2 brf_1(\nu)/(2\nu^2 - 1)^2), \tag{13}$$
$$d\nu/d\tau = N_1(\nu,m_0)/D_1(\nu,m_0), \tag{14}$$
$$N_1(\nu,m_0) = ((3/2)(dm_0/d\tau)/(1 - m_0) + a)f_1 - (f\nu^2 + r\nu^2(2\nu^2 - 1)\phi^2K^2/(1 - m_0)^2) * (G_2 - G_1^2) - r\nu^4\phi^2K^2(G_1G_2 - G_3)/(1 - m_0)^2,$$

$$D_1(\nu,m_0) = -2(\nu^2+1)f_1/((2\nu^2-1)\nu) - (1+2\nu^2)/(3\nu) + 2(1+\nu^2)G_1/(3\nu),$$
$$m_1(\tau,\omega) = -(A^2/\phi K\nu^2)(f + r\phi^2K^2(2\nu^2-1)/(3(1-m_0)^2))zn(K\omega,\nu) - (r/(3(1-m_0)^2)\phi^2K^2\nu^2\, dn(K\omega,\nu)sn(K\omega,\nu)cn(K\omega,\nu)),$$

$$\omega(t) = (1/\varepsilon)\int_{\varepsilon t_0}^{\varepsilon t}\phi_1(\tau)d\tau + \omega(t_0), \qquad \phi_1(\tau) = (b(1-m_0)/(2\nu^2-1))^{1/2}/K(\nu),$$

case(II) $E_0 < 0$ and $\phi_2 > \alpha\varepsilon$

$$x(t) \doteqdot x_0(\tau,\omega) = B(\tau)dn(K(\nu)\omega,\nu)$$
$$p(t) \doteqdot \phi(\tau)\partial x_0(\tau,\omega)/\partial\omega = -\phi(\tau)B(\tau)K(\nu)cn(K\omega,\nu)sn(K\omega,\nu), \tag{15}$$
$$m(t) \doteqdot m_0(\tau) + \varepsilon m_1(\tau,\omega),$$
$$dm_0/d\tau = cm_0 - fB^2(1-\nu^2+\nu^2G_1(\nu)) - 2brf_2(\nu)/(2-\nu^2)^2, \tag{16}$$
$$d\nu/d\tau = N_2(\nu,m_0)/D_2(\nu,m_0), \tag{17}$$
$$N_2(\nu,m_0) = ((3/2)(dm_0/d\tau)/(1-m_0) + a)f_2 - (f\nu^4 + r\nu^6\phi^2K^2/(1-m_0)^2) * (G_2 - G_1^2) - r\nu^6\phi^2K^2(G_1G_2 - G_3)/(1-m_0)^2,$$
$$D_2(\nu,m_0) = 3\nu f_2/(2-\nu^2) + \nu^3(1-G_1),$$
$$m_1(\tau,\omega) = -(B^2/\phi K)((f + r\phi^2K^2(2-\nu^2)/(3(1-m_0)^2))zn(K\omega,\nu) - (r/(3(1-m_0)^2))\phi^2K^2\nu^2 dn(K\omega,\nu)sn(K\omega,\nu)cn(K\omega,\nu),$$

$$\omega(t) = (1/\varepsilon)\int_{\varepsilon t_0}^{\varepsilon t}\phi_2(\tau)d\tau + \omega(t_0), \quad \phi_2(\tau) = (b(1-m_0)/(2-\nu^2))^{1/2}/K(\nu),$$

where $f_1(\nu)$, $f_2(\nu)$, $G_1(\nu)$, $G_2(\nu)$ and $G_3(\nu)$ are defined by

$$f_1(\nu) = 1 - \nu^2 + (2\nu^2-1)G_1(\nu) - \nu^2G_2(\nu),$$
$$f_2(\nu) = \nu^4(G_1(\nu) - G_2(\nu)),$$
$$G_1(\nu) = (E(\nu)/K(\nu) - 1 + \nu^2)/\nu^2,$$
$$G_2(\nu) = (1/3)((1-\nu^2)/\nu^2 + (4 - 2/\nu^2)G_1(\nu)),$$
$$G_3(\nu) = (1/5)(3(1-\nu^2)G_1(\nu)/\nu^2 + (8 - 4/\nu^2)G_2(\nu)).$$

Therefore the time dependence of $m_0(\tau)$ and $\nu(\tau)$ can be determined by solving simultaneous equations (13) and (14) for case(I) and eqs.(16) and (17) for case(II). By substituting the solutions into (11), we can study the time development of $E_0(\tau)$, $A(\tau)$ (or $B(\tau)$) and $\phi(\tau)$. In (12) and (15) we retained the first-order term m_1, because the lowest-order approximation of the time derivative of $m(t)$ is of order of ε. It should be noted here that the asymptotic forms (12) and (15) depend on the history of the motion because of $\omega(t)$.

So far we discussed the asymptotic solution of (1) for positive and negative E_0. Near $E_0 = 0$, however, the asymptotic solution ceases to be valid, because $E_0 = 0$ implies that $\nu = 1$ in (11). If ν approaches 1, ϕ will vanish. Therefore ϕ can be of order of ε at some value ν near 1. Then the separation between two time scales τ and ω can not be discriminated. Therefore for $\phi(\tau) < \alpha\varepsilon$, where α is an appropriate value between 1 and $(1/\varepsilon)$, we employ two time scales τ and η instead of τ and ω

$$\tau = \varepsilon t, \qquad d\eta/dt = \psi(\tau) \tag{18}$$

and assume that the lowest-order approximation $x_0(\tau,\eta)$ takes the form

$$x_0(\tau,\eta) = C(\tau)cn(\eta+\theta(\tau),1) = C(\tau)dn(\eta+\theta(\tau),1) = C(\tau)/\cosh(\eta+\theta(\tau)). \tag{19}$$

Substituting (19) into (9) and using the property of Jacobi elliptic functions, we get

$$C(\tau) = \pm\sqrt{2(1-m_0)}, \quad \psi(\tau) = \sqrt{b(1-m_0)}.$$

The lowest-order term m_0 of m is given by

$$m_0(\tau) = M\exp(c\tau), \tag{20}$$

where M is a constant coefficient. In the case where ν is nearly equal to zero but not exactly equal to zero, we can not neglect the deviation from the approximate solution (19). Therefore we carried out perturbation calculations up to the first order in ε and we get the following form :

case(III) $\quad \phi_1 < \alpha\varepsilon$ and $\phi_2 < \alpha\varepsilon$

$$
\begin{aligned}
x(t) = {} & \pm(\sqrt{2(1 - m_0)}/\cosh(\eta+\theta) + \varepsilon(\sinh(\eta+\theta)/\cosh^2(\eta+\theta))((\sqrt{2/b}/(6(1 - m_0)) \\
& \times(2f - a + 2br/5 + (a - 2f -3c/2)m_0)\sinh^2(\eta+\theta) - (2r\sqrt{2b})/(15(1 - m_0)) \\
& \times\log(\cosh(\eta+\theta))) + \varepsilon F\sinh(\eta+\theta)/\cosh^2(\eta+\theta) + \varepsilon G(- 3/(2\cosh(\eta+\theta) \\
& + \cosh(\eta+\theta)/2)), \\
p(t) = {} & \pm(- \sqrt{2b}(1 - m_0)\sinh(\eta+\theta)/\cosh^2(\eta+\theta) + \varepsilon(- (cm_0/\sqrt{2}(1 - m_0))/\cosh(\eta+\theta) \\
& + (\sqrt{2}/6\sqrt{(1 - m_0)})(2f - a + 2br/5 + (a - 2f - 3c/2)m_0)(2 + \cosh^2(\eta+\theta)) \\
& \times\sinh^2(\eta+\theta)/\cosh^3(\eta+\theta) + (2\sqrt{2}br/15\sqrt{(1 - m_0)})((2 - \cosh^2(\eta+\theta)) \\
& \times\log(\cosh(\eta+\theta)) + \sinh^2(\eta+\theta))/\cosh^3(\eta+\theta) + F\sqrt{b(1 - m_0)}(2 - \cosh^2(\eta+\theta)) \\
& /\cosh^3(\eta+\theta) + G(\psi(3\sinh(\eta+\theta)/(2\cosh^2(\eta+\theta)) + \sinh(\eta+\theta)/2) + 3\sqrt{b(1 -m_0)} \\
& \sinh(\eta+\theta)/(2\cosh^2(\eta+\theta))), \\
m(t) = {} & m_0 - (\varepsilon/\sqrt{b(1 - m_0)})(2f(1 - m_0)\tanh(\eta+\theta) + (2br/3)\tanh^3(\eta+\theta)),
\end{aligned}
\tag{21}
$$

where m_0 is given by (20) and

$$
\begin{aligned}
\eta(t) = {} & - (\sqrt{b}/\varepsilon c)(2\sqrt{(1 - M)} + \log(|(1 -\sqrt{1 - M})/(1 + \sqrt{1 - M})|) \\
& + (\sqrt{b}/\varepsilon c)(2\sqrt{(1 - m_0)} + \log(|(1 - \sqrt{1 - m_0})/(1 + \sqrt{1 - m_0})|), \\
\theta(\tau) = {} & \theta(0) - (3\sqrt{b}/2\sqrt{2})G\tau.
\end{aligned}
$$

In (21) F, G and M are constants. It should be noted here that $x(t)$ and $p(t)$ have the terms which grow exponentially as $\sinh(\eta+\theta)$ or $\cosh(\eta+\theta)$.

On the face of it the perturbation method discussed above may seem to have no advantage, because the set of eqs.(13) and (14) or eqs.(16) and (17) looks more complicated than the model equation (1). This method, however, has an advantage in the following point. By means of this method we can reduce the starting 3-dimensional problem to a 2-dimensional one in the (E_0, m_0) plane in cases(I) and (II) and for case(III) we can get the asymptotic form (21), which makes the analysis easier. Consequently we can clearly understand the origin of chaos in this model and study the structure of the strange attractor, as will be shown later.

3. Periodicity, quasi periodicity, phase locking and chaos

We are now in a position to discuss the set of equations (13) and (14) or eqs.(16) and (17). Let us consider the motion in the (E_0, m_0) plane instead of that in the (ν, m_0) plane, because the motion is quite different according to the sign of E_0. We solved eqs.(13) and (14) or eqs.(16) and (17) numerically. According to the magnitude of r the following cases are discriminated.

(A) $r > r_1 \doteqdot 2.26$. In this parameter region there exists a stable fixed point (E_c, m_c) in the (E_0, m_0) plane, where $E_c > 0$. Therefore E_0, ν, m_0, A and ϕ tend to constant values for large t. If we substitute these constant values into (12), the solution (12) gives a stable periodic orbit in the three-dimensional $(x,\ P=\dot{x},\ m)$ space.

(B) $r_1 > r > r_2 \doteqdot 1.81$. At $r = r_1$ the fixed point (E_c, m_c) loses its stability and there appears a stable limit cycle around it, whose shape is shown in Fig.1. Since the limit cycle locates in the region of $E_0 > 0$, the asymptotic solution in the steady state is given by (12). The functions E_0, m_0, ν, A and ϕ change periodically in the steady state. Let us denote the period of the oscillation by T. Since the period of cn function is $4K(\nu)$, it turns out that for large t the

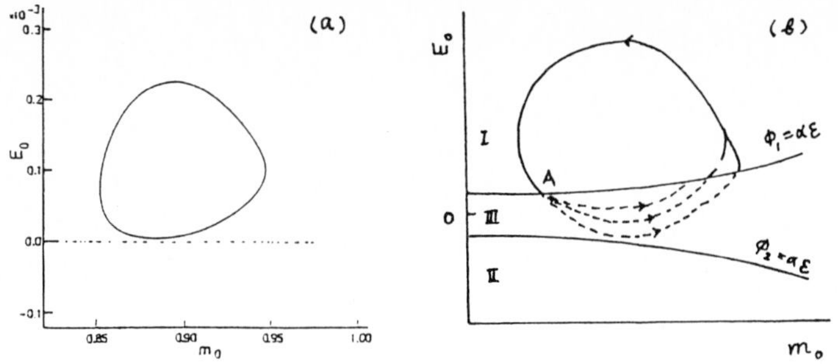

Fig.1 Limit cycle on the (E_0, m_0) plane. (a) r = 1.81 (b) schematic in case (C)

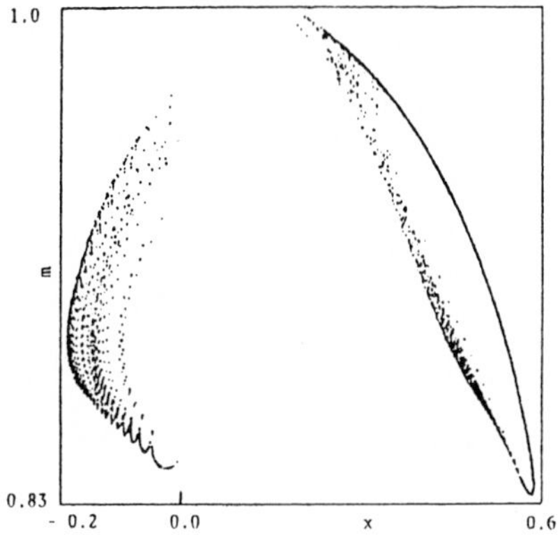

Fig.2 Poincaré map on the p=0 plane at r = 1.5

motion is periodic in the case of $(\omega(t+T) - \omega(t))/4$ = rational, and quasi-periodic in the case of $(\omega(t+T) - \omega(t))/4$ = irrational. In the latter case the solution (12) corresponds to a torus-type attractor in the (x,p,m) space.

(C) $r_2 > r > r_3 \doteqdot 1.71$. The size of the limit cycle grows gradually, if r is decreased. At $r = r_2$, the trajectory in the stationary state enters into the region (III) which is surrounded by two lines $\phi_1 = \alpha\varepsilon$ and $\phi_2 = \alpha\varepsilon$. In this region the trajectory which enters into the region (III) from the region (I) gets out into the region (I). Therefore the asymptotic solution of the system can be described in terms of both the solution (I) and the solution (III). The two solutions should be connected smoothly on the line $\phi_1 = \alpha\varepsilon$. This condition for connection determines the arbitrary coefficients F, G and M in the solution (III). The trajectory in the region (III) can not be illustrated on the (E_0, m_0) plane, because the solution (III) can not be described by E_0 and m_0 as can be seen in (21). In Fig.1b the dotted line in the region (III) means only the existence of the solution (III) and the shape of the curve has no special meaning. At the point A in Fig.1b each trajectory has the same values of E_0 and m_0, because of the limit cycle. However, the phase ω is, in general, different in each return, because it depends on the history of the motion. Since the solution (III) has terms which grow exponentially, the difference of ω yields quite different trajectory in the region (III). The system, in general, becomes chaotic. In some parameter region, however, there appears a stable limit cycle which extends over two regions (I) and (III). Then the phase ω has also the same value at the point A in each return. Then the trajectory is periodic, which means the so-called phase-locking phenomena. In this region the torus is destroyed.

(D) $r_3 > r$. The trajectory in the stationary state begins to extend over three regions (I), (II) and (III). Therefore three solutions (I), (II) and (III) must be connected successively: (I) → (III) → (II) → (III) → (I)→.... in the same manner discussed in (C). When E_0 is negative, the motion is confined to the right or left valley. The motion is deterministic and the connection formula determines one of two valleys. However, it is very difficult to predict to which valley the trajectory is confined in each process, because the solutions (I) and (II) depend on the history of the motion. A small difference of initial values causes a great difference in the future. That is, the system exhibits the orbital instability in cases (C) and (D). To confirm it, we calculated the largest Lyapunov number of the system by computer. The largest one in cases (C) and (D) was positive (for example, 0.0008 at $r = 1.5$). The smallness of this value comes from the fact that the time development of this system is very slow because of the smallness parameter ε. In this sense this motion may be called chaotic. On the other hand, the largest Lyapunov number in case (B) was zero, as it should be. The Poincaré map on $p=0$ plane at $r = 1.5$ is plotted in Fig.2. The theoretical result, which is calculated with the help of the asymptotic solution, is in a good agreement with those of the computer simulation.

References

1. R.H.G.Helleman, in Fundamental Problems in Statistical Physics V, E.G.D.Cohen ed., (North-Holland, Amsterdam, 1980) p.165.
2. T.Shimizu and A.Ichimura, Phys. Letters 91A, 52 (1982).
3. T.Shimizu, Physica 97A, 383 (1979).

Part IV

One-Dimensional Mappings

Noise-Induced Order - Complexity Theoretical Digression

I. Tsuda

Bioholonics Project, Research Development Cooperation of Japan, Nissho build. 5F
4-14-24 Koishikawa, Tokyo 112, Japan

K. Matsumoto

Department of Physics, Kyoto University, Kyoto 606, Japan

1. Introduction

We are living in a fluctuating environment and in a finite universe. From this fact, also in chaos research, we are obliged to consider chaos from a standpoint of finitism. As is well known, in one-dimensional mappings, a binary coding is possible [4],[5]. Usually, the left-hand side of the extremum of the mapping is coded "0" and otherwise "1". One can get a string consisting of 0 or 1 corresponding to one orbit. That one has an infinite string is equivalent to that the position on an attractor is determined with an infinite precision. Moreover, it takes an infinitely long time to produce that type of infinite string. However, we are in a finite universe, that is, we have just finite time to observe any system.

Therefore, in very natural manner, a heat bath is introduced in all figures but certain finite figures [6],[7]. This situation is very similar to the 2nd kind of uncertainty introduced by B. B. Mandelbrot in the 1960s [8]. Anyhow, we are obliged to have a finite precision because of having only finite time to observe, even in the case without external perturbations. In the case with external perturbations, the noise level naturally gives the upper bound of the length of the symbols, because one cannot get much more information about the initial condition of the orbit even if one knows the longer string of symbols. Then, we naturally inquire what the influence of finite precision is. To investigate this, we consider chaotic systems under some external random perturbation.

2. Observation of Noise-Induced Order[1]

There are several contributions to the research concerning the character of chaos influenced by external noise in the case of the logistic model [9] . In the logistic model, external noise induces the transition from the periodic behavior to the chaotic behavior. There appears a broadening of the invariant density and of the power spectrum and an increase in the Lyapunov number.

We study here other types of one-dimensional maps which exhibit a very different response from the one obtained in the logistic model. The new transition from chaotic behavior to ordered behavior is observed [3] . The transition to the order is indicated by sharpening of power spectrum, abrupt decrease of entropy, appearance of negative Lyapunov number, and localization of orbit.

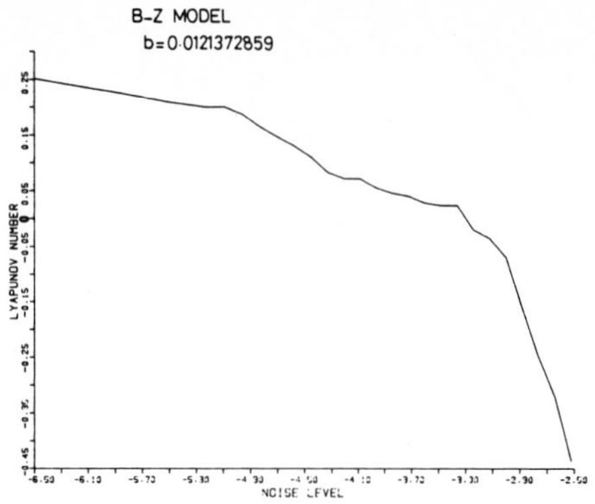

Fig. 1. Lyapunov number vs. noise

We investigate the models with additive noise equally distributed in the interval$[-\sigma, \sigma,]$where σ is the maximum value of the noise level.

One of the models is the B-Z map:

$$f(x)=((x-0.125)^{\frac{1}{3}}+0.50607357)\exp(-x)+b \qquad \text{for } x<0.3,$$

$$f(x)=0.121205692(10x\exp(-\frac{10}{3}x))^{19}+b \qquad \text{for } x\geqq 0.3,$$

where b is the controlled parameter which corresponds to the flow rate in real experiments.

Our first observation is that the originally positive Lyapunov number changes to a negative one as the noise level is increased. This is shown in Fig. 1.

To check whether or not our phenomenon is observable in experiments, we study the power spectrum. The results are shown in Fig. 2. In Fig. 2 a) the power spectrum in the case without noise is shown. Fig. 2 b) is the case where the noise level is 0.01. For relatively large noise, the sharp peak which implies some kind of order appears. An important characteristic of spectrum shown in

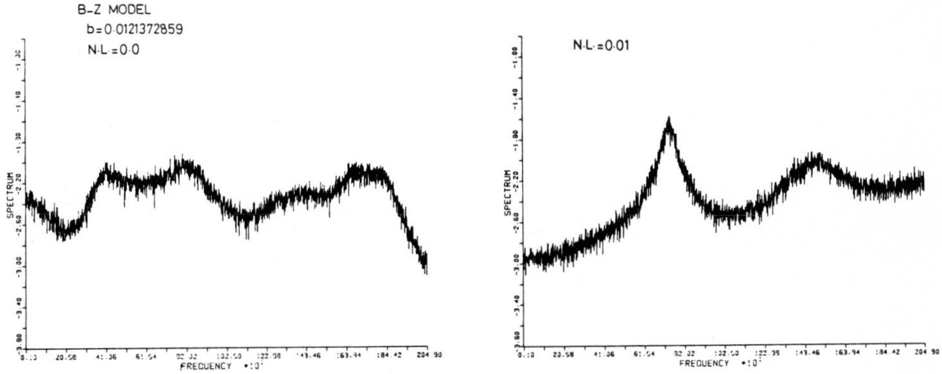

Fig. 2. (a) Power spectrum (FFT) in the case without noise
(b) Power spectrum (FFT) in the case with noise level 1.0×10^{-2}

Fig. 2 b) is that this figure of spectrum cannot be distinguished from the figure for the map in which a 7-periodic orbit is superstable with the same noise level, except for a slight shift of the primary peak of the spectrum. This is the reason why we call this phenomenon noise-induced order.

We also observe the nature of the orbit and the invariant density. About these see reference [1].

The numerical calculations presented above show definitely that the chaos is unstable against external random perturbations in some class of maps.

To trace this type of instability to its origin, we construct two families of models each having a parameter which controls the steepness of the slope on the left-hand side of the map.

One of them is

$$f(x)=0.8\ (0.3)^{-\alpha} \times e^{\alpha} x^{\alpha} \exp(-\frac{\alpha}{0.3}x)+b.$$

The other map is

$$f(x)=\text{const.} \times\{\arctan[\beta(x-0.2)]+\arctan(0.2\beta)\}/[1+(2x)^{19}]+b,$$

where const. is chosen so as to give

$$\max(f(x))=0.8.$$

A parameter α or β in each model controls the steepness of the slope on the left-hand side of the map.

For these maps, we calculate the same features as those for the B-Z map. When α and β are small, namely, the left slope of the map is less steep, the transition noise-induced order is not observed. On the other hand, when α and β are large, namely, the left slope is large enough, the transition is observed.

To investigate this phenomenon further, we use symbolic dynamics approach and calculate the entropy. We associate a symbol 0 with the left-hand side of the extremum and 1 with the right-hand side of the extremum. Regarding the 0-1 sequence of an orbit as the product of an information source, we can easily calculate its entropy.

We assume that the symbolic dynamics is a Markov process of degree n, i.e., n+1-th symbol depends only on the last n symbols. Then we calculate the entropy H of a Markov process

$$H=-\sum_{i,j} P_i P_{ij} \ln(P_{ij}),$$

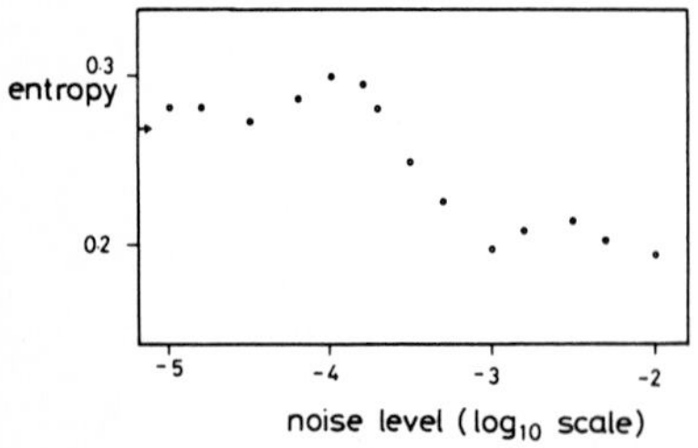

Fig. 3. Entropy vs. noise. The arrow Indicates the value of entropy in the case which is noiseless but for round-off error (10^{-16})

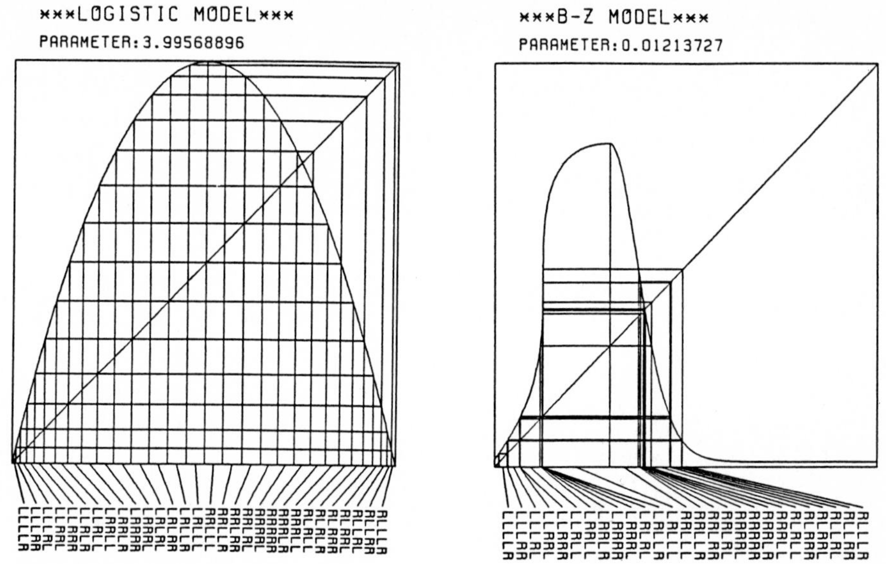

Fig. 4. 4-th refinement of a Markov partition in (a) logistic map and (b) B-Z map

where P_{ij} is the transition matrix of the Markov process and P_i the stationary probability.

As shown in Fig. 3, the entropy abruptly decreases as the noise level is increased. This abrupt decrease of the entropy in the B-Z model indicates clearly that out of many possibilities a few orbits survive the noise.

3. A Global Character of Maps[2]

We observe the difference in Markov partition of the B-Z map and of the logistic map(See Fig. 4). In the logistic map, the interval [0,1] is divided into almost equal refinement. On the other hand, in the B-Z map, that division produces the extremely small partitions and comparatively large partitions. We discussed the entropy decrease using the above facts. Our calculation of entropy is based on the reformalized Markov partition. Applying noise, the partitions with smaller size than the noise level disappear.

4. Relation to the Algorithmic Complexity

Here, we discuss noise-induced order from the viewpoint of algorithmic complexity [10]. We note that there are two effects of external noise. On the one hand, the noise decreases the number of orbits $-\Delta N_p$ caused by the decrease of the number of partitions, and on the other hand, the noise increases the number of orbits by leak ΔN_a. Roughly speaking, the former is responsible for the decrease of the entropy and the latter is responsible for the increase of the entropy. Therefore, there exists competition between ΔN_a and $-\Delta N_p$.

If ΔN defined by $\Delta Na-\Delta Np$ is positive, then the entropy increases as a result. On the other hand, if ΔN is negative, the entropy decreases. The latter case corresponds to the noise-induced order. The logistic model is in the former case.

Moreover, we connect the quantity ΔN with the algorithmic complexity.

The algorithmic complexity $A(S^n)$ of a binary string S^n of length n is defined by the minimum size of a computer program which generates the above string.

For a periodic string, say, {01011 01011 01011..... 01011}, we write a computer program for this example with PASCAL:

```
FOR I:=1 TO M DO                                              (p1)
    WRITE(`01011`) .
```

(This is not necessarily the shortest program.)
For a periodic string, the algorithmic complexity is proportional to the number of figures of M in binary representation, that is,

$$A(S^n) \propto \log_2 M \sim \log_2 n (\text{as } n \gg 1).$$

On the other hand, a random string cannot be expressed as the output of a program much shorter than the string itself. Therefore, a minimum size of a program is

```
WRITE(`011001.....1010001`) .                                 (p2)
```

In this case, the algorithmic complexity is proportional to the length of the string S^n, that is,

$$A(S^n) \propto n.$$

We write this relation $A(S^n)=cn$.

Let N be the maximum number of strings of length n which can be generated by the program with fixed length m in binary representation.

There are 2^m programs of length m. Some of these are not real programs and some do not type the string of length n. But when n is large and the order of m is proportional to the order of n, almost all programs type random strings of length n, since the most part of these programs is the string itself (the expression in inverted commas in (p2)). Thus we have $N \sim 2^m$. Then, we obtain the relation $N \sim 2^{cn}$. Here N is the total number of strings of complexity cn.

Now we can consider the quantity ΔN in connection with the algorithmic complexity. The noise changes the number of possible strings of fixed length n. Let N' be the number of binary strings coded from the noisy orbits by the method in §2 and N from the noiseless orbits. The c'n and cn are the corresponding algorithmic complexities. This change is represented by

$$\Delta N \equiv N'-N \sim 2^{c'n}-2^{cn} (=2^{A'(S^n)}-2^{A(S^n)}) .$$

Then, the noise-induced order is expressed as the change of the algorithmic complexity in orbits. In this interpretation, for the B-Z map, the algorithmic complexity in the case with noise becomes smaller than the one in the case without noise, i.e., $c' < c$. On the other hand, for the logistic map, we obtain the inequality $c' > c$.

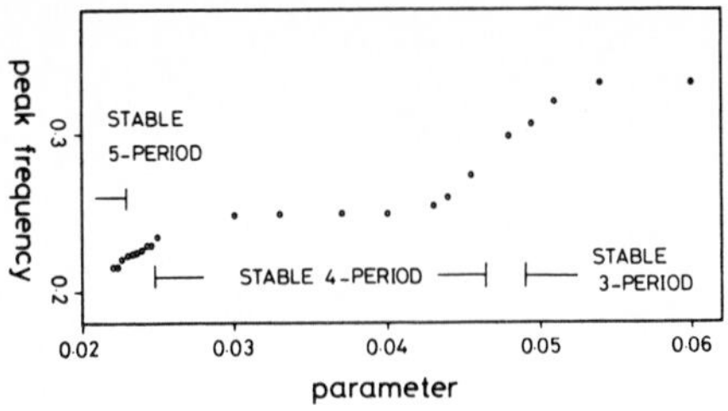

Fig. 5. The frequency of the peak in the power spectrum vs. the bifurcation parameter b in the B-Z map. Noise level is 1.0×10^{-2}, but the noise level dependence of the frequency is not noticeable

5. Summary and Discussions

The transition from chaos to order was observed as the noise level was increased. We call it noise-induced order. This kind of transition was attributed to the steepness of the map. We investigated the global character of maps in terms of symbolic dynamics and attributed the origin of the noise-induced order to the nonuniformity in the width of partitions. Moreover, we discussed the noise -induced order from a viewpoint of algorithmic complexity. The inequality $c' < c$ obtained in the case of noise-induced order implies a surprising fact that the noise shortens the computer program. This aspect seems to be very important in considering an information processing in the brain of human beings.

Finally, we add some comments from a practical point of view. The transition which was observed here in one-dimensional maps may be observed in flow systems. Lozi proposed a concept 'Motif' to construct a systematic three-dimensional model which is supposed to explain the bifurcation structures found in several experiments of the Belousov-Zhabotinsky reaction[11]. Applying his idea to the present problem, that the noise-induced order can appear also in the flow systems, is expected to be proved mathematically.

The next problem is formulated as follows:Suppose that a periodic solution with noise is obtained in some experiment. Then how do we decide whether the system is chaotic or periodic in the noiseless case, i.e., the ideal case? As shown in Fig. 5, if the chaos exists in the ideal case, the peak of the power spectrum is continuously shifted as the bifurcation parameter is continuously varied. On the other hand, if the periodic solution exists in the ideal case, either other peaks, for example, those of subharmonics can appear or, alternatively, the primal peak of the spectrum stays in the same position as the bifurcation parameter is continuously varied. Thus, we can distinguish the above two cases by studying the bifurcation at the same noise level.

References

1. K. Matsumoto and I. Tsuda, J. Stat. Phys. 31, 87(1983);ibid. 33, December (1983)
2. K. Matsumoto, to appear in J. Stat. Phys.
3. The transition to ordered behavior by noise has been independently found in the literature:G. Mayer-Kress and H. Haken, in "Evolution of chaos and order", ed. by H. Haken, Springer Series in Synergetics(1982);in Physica D (1983)
4. V. M. Alekseev and M. V. Yakobson, Physics Reports, 75, 287(1981)
5. J. P. Crutchfield and N. H, Packard, Int. J. Theor. Phys. 21, 433(1982)

6. R. Shaw, in "Chaos and order in Nature" ed. by H. Haken, Springer Series in Synergetics. Vol. 11, 218(1981)
7. I. Tsuda. Thesis(in Japanease) 1982
8. B. B. Mandelbrot, J. Polit. ECON. LXXI, 421(1963)
9. G. Mayer-Kress and H. Haken, J. Stat. Phys. 26, 149(1981);J. P. Crutchfield , J. D. Farmer, and B. A. Huberman, (Preprint);J. P. Crutchfield and B. A. Huberman, Phys. Lett. 77A, 407(1980)
10. R. J. Solomonoff, Information Control, 7, 224(1964), A. N. Kolmogorov, Problems of Information Transmission, 1, 1(1965), G. J. Chaitin, J. Assoc. Computing Machinary, 13, 547(1966);Sci. Amer. May (1975)
11. R. Lozi, Thèse (in French) 1983 (University of Nice, Dept. of Mathematics)

Symbolic Dynamics Approach to Intermittent Chaos – Towards the Comprehension of Large Scale Self-Similarity and Asymptotic Non-Stationarity

Y. Aizawa and T. Kohyama

Department of Physics, Kyoto University
Kyoto 606, Japan

1. Introduction

One of the most important problems brought forward by the study of chaos is "What is probability?". The probabilistic world is not completely disorder, but rather a strongly ordered one which is ruled over by the self-similarity law. The order in chaos must be also studied in the same context as self-similarity. On one side of the small scale limit, the strong order in chaos appears as the Cantor structure such as the Smale's horse shoe, and on the other side the large scale order comes to appear as the intermittent phenomena. Our main concerns are to find out the variety of stochastic processes generated by the dynamical law, and to bridge the gaps between probability and dynamics. Especially, the aim of this report is to elucidate the theoretical basis of the large scale self-similarity law immersed in the intermittent chaos.

The statistical properties of intermittency are related to such large scale universality that appears in the attractive domain of stable distributions and/or in the renormalization group theory. For instance, as the result of such universality the long time behaviors of intermittency do not sensitively depend on the way of the coarse-graining. This remarkable feature enables us to adopt the symbolic dynamics approach, but here an important point is that the long-range order of the intermittency is not understood from the simple Markovian process. Indeed in this paper, two types of intermittency are explained with the use of one-dimensional maps; one is identified by a countable infinite Markov chain and the other by a non-Markovian renewal process.

The large scale order is correlated to the 1/f spectral distribution and to the Pareto-Zipf distribution of the laminar phase residence time. Interrelations between both hyperbolic laws, which were discussed many years ago by Mandelbrot [1] from a purely probabilistic viewpoint, are elucidated from a chaotic dynamics viewpoint. In the following sections, we will refine and extend the Manneville's excellent scenario [2]. The universality of the 1/f sepctrum is guaranteed by the stability of the Pareto-Zipf law under the external noise.

One more purpose is to classify the 1/f spectral phenomena in the intermittent chaos. Assuming the following asymptotic form ($x \simeq 0$) of the one-dimensional map $F: x \to x$ $(0 \leq x \leq 1)$,

$$F(x) \simeq x + Ax^B + \varepsilon \quad (\varepsilon: \text{small parameter}) \quad , \tag{1-1}$$

the laminar state residence time distribution P(m) becomes

$$P(m) \sim m^{-\beta} \qquad (\beta = B/(B-1)) \ , \tag{1-2}$$

and the low-frequency power sepctrum $S(\omega)$ ($\omega = 2\pi f$)

$$S(\omega) \sim \omega^{-\nu} \qquad (\nu = 3 - \beta) \ . \tag{1-3}$$

Here the appropriate symbolic dynamics plays an important role. The relations among the indices B, β, and ν are asymptotically exact. For the case of $B \geq 2$ the σ-finite invariant density is asymptotically $\delta(x)$ at $\varepsilon = 0$, but for the finite $\varepsilon \to 0_+$ practically there occurs the 1/f sepctrum. The cases of $B \leq 2$ and $B > 2$ are respectively called the infrared crisis and the infrared catastrophe. The case of the weak catastrophe for $B > 2$ may be called the asymptotic non-stationary chaos, since the infrared catastrophe usually appears in the non-stationary self-similar processes [3]. Such non-stationary chaos must be characterized by the Allan variance [4] and by the A entropy instead of the Kolmogorov-Sinai invariant.

2. Intermittency of the Λ model

The intermittency of the Lorenz model is typically described by the 1-dim. map of Fig.1-(a). By taking the symbolic state (0 and 1) as given in the figure, the topological feature of a chaotic orbit is determined by a sequence; the admissible words (1), (10),$\cdots$, (10$\cdots$0) (Fig.1-(b)) [5]. The number of these admissible words depends on the bifurcation parameters. Here our discussion is limited to $n = 2$ ($r \simeq 138.0$). The onset of chaos ($r > r_c = 137.976$) is characterized by the fractal dimensions (Fig.1-(c)), where D_A and D_H are the dimensions of strange attractor and of the chaotic time series. The transition probabilities among three admissible states (1), (10) and (100) are identified by a simple Markovian approximation as,

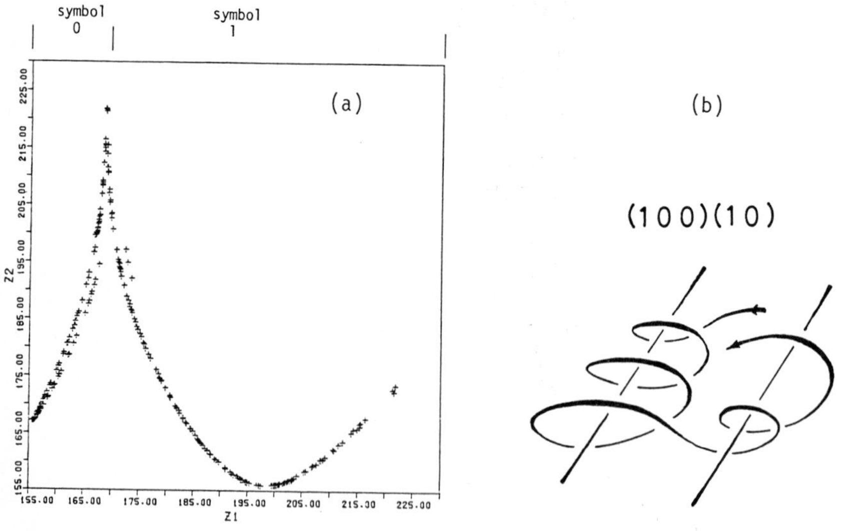

Fig.1 (a) Lorenz plot ($\dot{x} = 16(y-x)$, $\dot{y} = rx-y-xz$, $\dot{z} = -4z+xy$, $r=138$)
(b) Topological aspect for a symbolic sequence

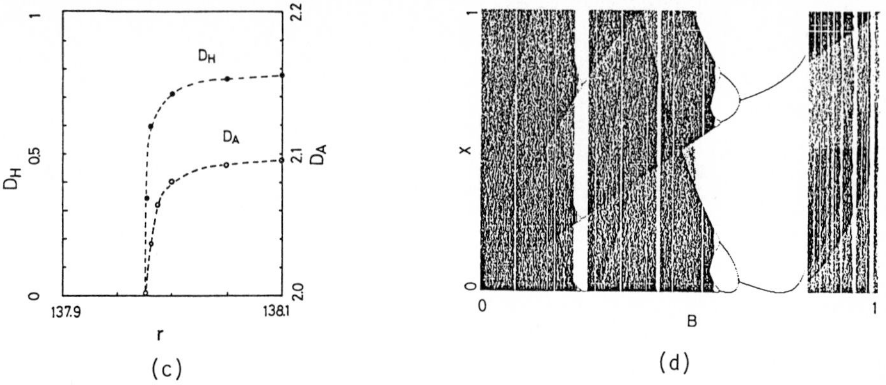

Fig.1 (c) Fractal dimensions, (d) Global bifurcations of the Λ model

$$\begin{array}{cccc} & (1) & (10) & (100) \\ (1) & 0.45, & 0.41, & 0.14 \\ (10) & 0.68, & 0.26, & 0.06 \\ (100) & 0.06\varepsilon^{\mu}, & 0.56\varepsilon^{\mu}, & 1-0.62\varepsilon^{\mu} \end{array} \quad \rightarrow \quad \begin{array}{ccc} 0.45, & 0.41, & 0.14 \\ 0.68, & 0.26, & 0.06 \\ 0, & 0, & 1 \end{array} \qquad (2\text{-}1)$$

$$(\varepsilon \equiv r - r_c > 0, \text{ with } \mu \simeq 0.32) \qquad (\varepsilon \lesssim 0 \text{ ; window}) .$$

The window phenomena ($r \leq r_c$) are understood as the decomposition of the matrix into two irreducible Markov parts. The transient states (1) and (10) show the Cantor chaos (which is existing but non-observable), and the (100) state is a periodic sink.

To study the intermittency in the large, the following 1-dim. Λ model was studied precisely [6],

$$\begin{aligned} x_{n+1} &= \frac{1-B}{A}\, x_n + B \qquad (0 \leq x_n \leq A) \ , \quad \text{symbol } 0 \\ &= \frac{1}{(A-c)^2}\,(x_n - c)^2 \qquad (A \leq x_n \leq 1) \ , \quad \text{symbol } 1 \end{aligned} \qquad (2\text{-}2)$$

$$(\ B : \text{bifurcation parameter}, \quad A = 0.4, \quad c = 0.75\ ) \ .$$

The global bifurcation scheme (Fig.1-(d)) qualitatively reproduces almost all the aspects of the Lorenz system, but the index μ of the transition probabilities is nearly 0.5. The onset of the (100) intermittency is observed for $B > B_c \simeq 0.268330$.

3. Hyperbolic laws in the intermittent chaos

The essential feature of intermittency is beyond the simple Markovian identification. The remarkable non-Markovian aspects appear in the residence time distribution for each admissible state. The distributions for the burst states (1) and (10) are the exponential ones, which is certified by the numerical results

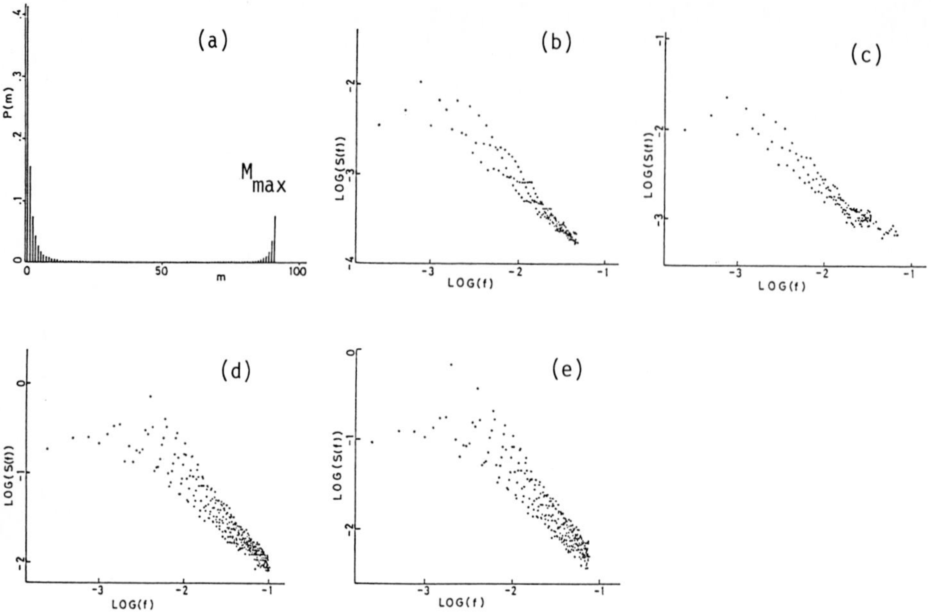

Fig.2 (a) Residence time distribution of (100) state (B=0.26835)
(b) $\sim$ (e) Power spectrum for each symbolic sequence (B=0.268331)

as well as by the theoretical caluculations. In other words, the Cantorian chaos mentioned in §2 is well identified by a certain simple Markov chain. On the other hand, however, the distribution P(m) of the laminar state (100) residence time is not exponential but is theoretically estimated as,

$$P(m) \sim \frac{\varepsilon}{a}\,[1+\tan^2\{\tan^{-1}(\sqrt{\frac{a}{\varepsilon}}\,A) - \sqrt{a\varepsilon}\;\; m\}\,] \qquad (3\text{-}1)$$

$(\ \varepsilon = B - B_c,\ \ a : \text{a certain constant}\ ).$

This is well supported by the numerical simulations (Fig.2-(a)). The second peak in P(m) at M_{max} increases to infinity as $B \to B_c$, and P(m) ($M \ll M_{max}$) obeys a hyperbolic law as $P(m) \sim m^{-2}$.

The numerical result of the spectral analysis is shown in Fig.2-(b) $\sim$ 2-(e); Fig.(b) is the result for the time series $(x_1, x_2, \cdots, x_n \cdots)$ which is generated by the Λ model, Fig.(c) is for the dyadic (0,1) sequence constructed in §2, and Fig.(d) for the triadic(0,1,2) sequence obtained as $(1) \to 0$, $(10) \to 1$, and $(100) \to 2$. The final Fig.(e) is for the new dyadic (0,1) sequence generated by the following transformation; the burst consisted of (1) and $(10) \to 1$, and the laminar state $(100) \to 0$. Namely the symbol 1 does not appear one after another. Almost the same type of 1/f spectral law is obtained for each case.

One of the interesting points is that the small oscillatory behavior is superposed on the mean 1/f slope (typically observed in Fig.2-(b)). To see the origin of this peculiar behavior, let's see the theoretical power spectrum corresponding to the case of Fig.2-(e),

$$S(\omega) \sim \frac{1}{\omega}\left\{ \frac{d_1(1+\cos(\frac{\pi\omega}{b})) + d_2\ln(\omega)\sin(\frac{\pi\omega}{b})}{G(\omega)} \right\} , \tag{3-2}$$

$$G(\omega) = 1 - 2d_1^{\,2}(1+\cos(\frac{\pi\omega}{b})) - 2d_2\ln(\omega)(d_2\ln(\omega)+1)(1-\cos(\frac{\pi\omega}{b}))$$

$$- 2d_1\sin(\frac{\pi}{b}\omega)(2d_3\ln\omega+1) ,$$

where d_i and b are certain constants. This is derived by assuming that the burst state is a point process where the residence time distribution of the laminar state obeys eq.(3-1). More general cases corresponding to Fig.2-(c) and Fig.2-(d) are analysed by an alternating renewal process of semi-Markov type, and that the results shown in §1 are unchanged; for the case of $B = 2$ in eq.(1-1) $S(\omega)$ has the log. correction as $S(\omega) \sim \omega^{-1}\cdot(\ln(\omega))^{-2}$. The small oscillation appearing in Figs.2 is explained by eq.(3-2), whose period is nearly equal to $1/M_{max}$. Since the cause of the oscillatory correction comes from the second peak of P(m) shown in Fig.2-(a), the 1/f spectrum becomes much clearer as the second peak decreases. In the Λ model the second peak at $m \simeq M_{max}$ does not disappear even in the limit of $B \to B_c$, therefore the 1/f spectrum is obtained only approximately in the restricted frequency domain where the Pareto-Zipf law ($P(m) \sim 1/m^2$) is adjustable. An exact 1/f spectral model is discussed in §5.

4. Stability of the hyperbolic laws

A random perturbation is added to the Λ model; $x_{n+1} = F(x_n) + \sqrt{\sigma}\xi_n$ where ξ_n is a white Gaussian noise ($<\xi_i\xi_j> = \delta_{ij}$). The residence time distribution of the laminar state (100) is shown in Fig.3-(a). The main peak of P(m) remains almost unchanged but the second peak near $m \simeq M_{max}$ is very unstable to the small perturbation. As the result of the discussion in §3, the small oscillations superposed on the 1/f spectrum are expected to diminish remarkably. Actually, the corresponding 1/f spectrum (Fig.3-(b)) has no oscillations. This interesting phenomenon can be theoretically explained by the fact that the noise effect becomes dominant

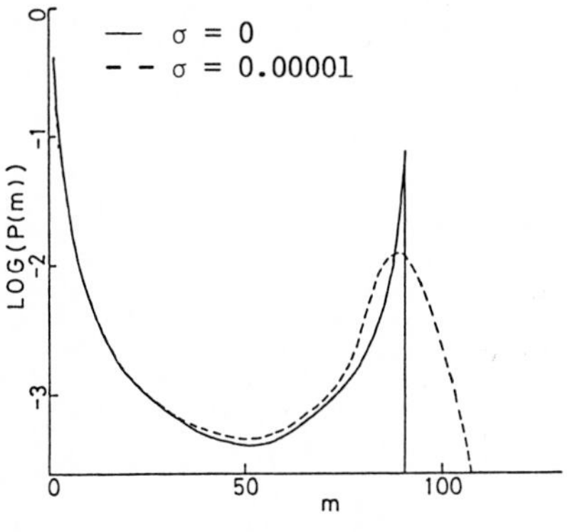

(a)

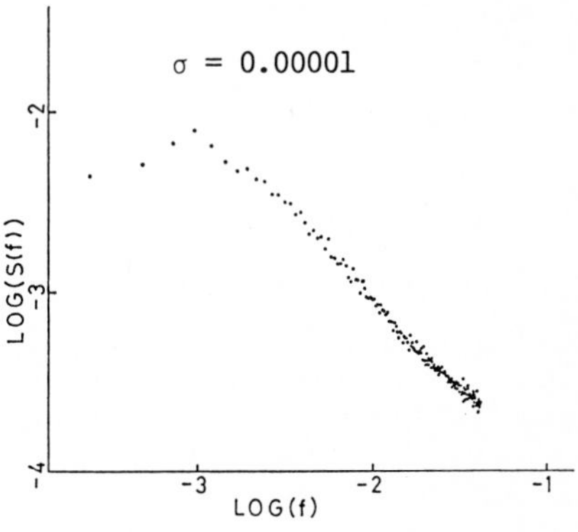

(b)

Fig.3 (a) Residence time distribution of (100) state (B=0.26835)
(b) Power spectrum for the time series $(x_1, x_2, \ldots)$ (B=0.268331)

in the region where $|F'(x)|$ is small. In other words, the laminar state with the large residence time is well randomized, so that the second peak of P(m) is easily destroyed by the perturbation. This remarkable effect may be understood as the noise recreating long-range order.

5. Asymptotic non-stationary chaos

The recurrent chaotic motion is the bounded variation, therefore the essential non-stationary process can not be observed in the usual situation. However, generally the non-stationary chaos is almost always immersed in the part of the singular measure. This is a quite similar situation to the case of the non-observable Cantorian chaos discussed in §2. The non-stationary chaos can be often observed asymptotically under a certain condition that the non-stationary singular measure competes with the weakly stable asymptotic measure in the presence of the small perturbation.

The asymptotic non-stationary choas is conventionally defined as the dynamical process which reveals the infrared weak catastrophe. From the practical viewpoint, that is characterized by the Allan variance $\sigma_A^2(\tau) \sim \tau^{2\alpha}$. Typical examples are the fractional Brownian motion $B_H(t)$ and the fractional noise $dB_H(t)/dt$ $(0 \leq H \leq 1)$ [7]. Then the index α yields as $\alpha=H$ and $\alpha=H-1$, and the corresponding power spectrum becomes as $f^{-(2H+1)}$ and $f^{-(2H-1)}$ respectively.

The modified Bernoulli system is the simplest model which generates the asymptotic non-stationary chaos.

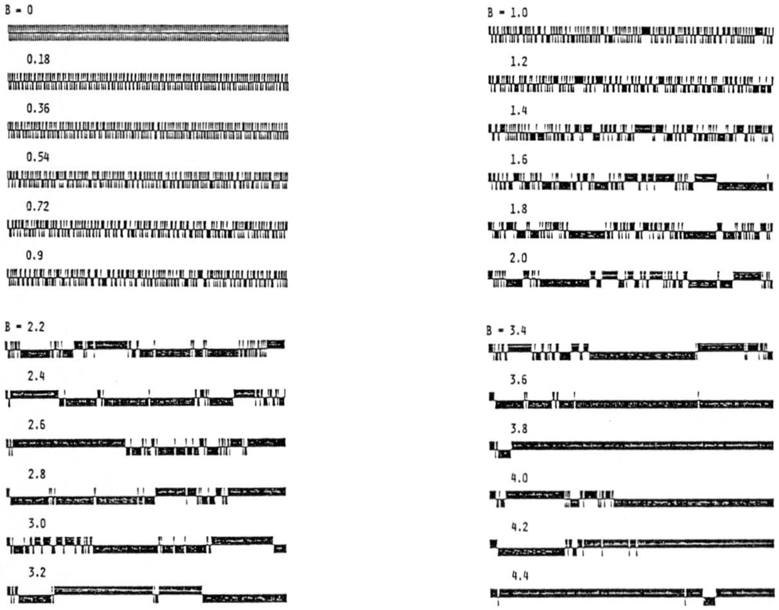

Fig.4 (a) Symbolic time sequences for various values of B

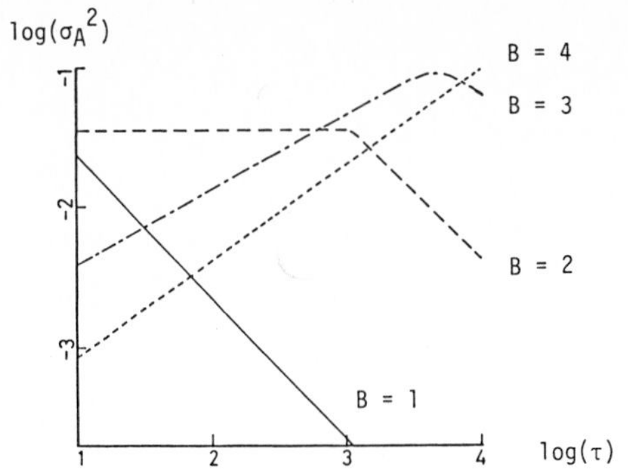

Fig.4 (b) Allan variances of the modified Bernoulli system

$$x_{n+1} = x_n + 2^{B-1}(1-2\varepsilon)x_n^B + \varepsilon, \quad (0 \le x_n \le 1/2), \quad \text{symbol } 0$$
$$= x_n - 2^{B-1}(1-2\varepsilon)(1-x_n)^B - \varepsilon, \quad (1/2 \le x_n \le 1), \quad \text{symbol } 1 \quad . \qquad (5\text{-}1)$$

(ε: small parameter)

The residence time distribution P(m) of the symbol (0) or (1) is derived as,

$$P(m) \sim \begin{cases} a^{-(1/B)m} & (B < 1) \\ b^{-m} & (B = 1) \\ m^{-B/(B-1)} & (B > 1) \end{cases} \qquad (5\text{-}2)$$

(a and b are positive constants.)

Here ε can be replaced by a small random perturbation $\xi_n(>0)$. The symbolic sequence for each parameter value of B is shown in Fig.4-(a) ($\varepsilon = 10^{-5}$). The flip-flop jump shows the strong intermittency for $B \geq 2$, and the corresponding power spectrum reveals the 1/f behavior as is summarized in §1.

The corresponding Allan variances are shown in Fig.4-(b). The flicker floor appears for the case of B=2, and its length is prolonged when the value of ε goes to zero. The positive constant slope ($2\alpha = 2\log\{\sigma_A(\tau)\}/\log\{\tau\}$) occurs before it decays as $\sigma_A(\tau) \sim 1/\sqrt{\tau}$. The index α is estimated as $\alpha \sim 1-B/2(B-1)$ for $B \geq 2$.

The asymptotic non-stationarity chaos reveals the strong long-range order, so that the Lyapunov exponent or the K-S entropy (h) can not be an available parameter. Indeed, the K-S entropy of the modified Bernoulli system is zero when ε goes to zero under the condition of $B \geq 2$, because the measure of the cylinder set is not of $o(e^{-h \cdot n})$ but of $o(n^{-\delta})$. Details will be reported elsewhere [8].

6. Summary and discussion

The symbolic dynamics approach is applied to the study of the topological entanglement of chaotic orbits. The topological features of chaos are more complex than the measure theoretical complexity, and that must be studied to elucidate the bifurcation into the chaotic phase [5][9].

In this paper we studied a mechanism of the 1/f fluctuation from a viewpoint of chaotic dynamics, and pointed out that the phase randomization based on the Pareto-Zipf law is an origin of the 1/f spectral behavior. However, many cases of the 1/f spectrum are induced by the amplitude randomization as is typically observed in the fractional Brownian motion. How possible is it to construct the dynamical model which reveals the 1/f spectrum of such amplitude origin? This remains as a future problem. Furthermore, even in the phase randomized 1/f spectral chaos there still remains a problem: how to extend the algorithmic complexity theory into the asymptotic non-stationary chaos.

references

1. B.B.Mandelbrot; IEEE Commun.Tech.March (1965),71
2. P.Manneville; J.Phys.,41 (1980),1235
3. S.Taqqu; Colloquia Mathematica Societatis Janos Bolyai 27, Random Fields, Esztergon (Hungary), (1977),1057
4. D.W. Allan; Proc. Symposium on 1/f fluctuations (July, 1977, Tokyo)
5. Y. Aizawa; Prog. Theor. Phys., 70, No.5 (1983) in press
6. T. Kohyama and Y. Aizawa; submitted to Prog. Theor. Phys.
7. B.B. Mandelbrot; SIAM Review 10, No.4 (1968) 422
8. Y. Aizawa and T. Kohyama; submitted to Prog. Theor. Phys.
9. Y. Aizawa and T. Uezu; Prog. Theor. Phys., 67 (1982), 982
   T. Uezu and Y. Aizawa; Prog. Theor. Phys., 68, No.6 (1982), 1907

Diffusion and Generation of Non-Gaussianity in Chaotic Discrete Dynamics

H. Fujisaka

Department of Physics, Kagoshima University
Kagoshima 890, Japan

1. Introduction

In this contribution we discuss the long-time dynamics of Y_t[1-4] governed by

$$Y_{t+1} = Y_t + \Delta(x_t) \ , \quad (t=0,1,2,\cdots) \quad , Y_0=0 \ , \tag{1.1}$$

where Δ is a certain steady stochastic variable generated by a chaotic variable x_t obeying a one-dimensional map

$$x_{t+1} = f(x_t) \ , \qquad (0 \leqq x_t < 1) \ . \tag{1.2}$$

The central limit theorem states that Y_t is asymptotically Gaussian as $t \to \infty$ because Y_t is expressed as the summation of the stochastic variable Δ:$Y_t=\Sigma_{s=0}^{t-1}\Delta(x_s)$. Namely, the probability distribution for Y_t is given by

$$P(y,t) \simeq Z_t^{-1}\exp[-(y-\lambda_0 t)^2/2\sigma_t] \qquad (t \to \infty) \tag{1.3}$$

[2], where λ_0 and σ_t are the average velocity and the variance of Y_t, respectively. Z_t is the normalization constant. As will be shown later, however, the above Gaussian approximation generally breaks down in an exact sense even for $t \to \infty$. The main purpose of this contribution is concerned with the violation of the Gaussian property.

Since the deviation from the Gaussian distribution is extremely weak(see section 3), we intend to enhance it. The sensitivity of the exponential function e^y on y suggests to introduce $A_t=\exp(Y_t)$ and $B(x)=\exp[\Delta(x)]$. In terms of these variables (1.1) is rewritten as

$$A_{t+1} = B(x_t)A_t \quad , \quad (A_0 = 1) \ . \tag{1.4}$$

After the above transformation we may suitably single out the non-Gaussian characteristics in the enhanced form. Eq.(1.4) is the starting dynamics for the present aim.

Note that (1.3) reduces to the log-normal distribution for A_t and that the q-order moment of A_t becomes[5,6]

$$\langle A_t^q \rangle \propto \exp[q(\lambda_0 + Dq)t] , \qquad (-\infty < q < \infty) \tag{1.5}$$

as $t \to \infty$, where $D \equiv \lim_{t\to\infty} \sigma_t/2t$ and $\langle\cdots\rangle$ means the ensemble average over $\rho(x)$, the steady distribution for x , satisfying $\rho(x)=\int_0^1 \delta(f(z)-x)\rho(z)dz$.

2. Characteristic exponents and the violation of the Gaussianity

The statistical characteristics of A_t can be discussed with the q-order moment $\langle A_t^q \rangle$. By defining the q-order characteristic exponent λ_q[6] by

$$\lambda_q \equiv q^{-1} \lim_{t\to\infty} t^{-1} \ln \langle A_t^q \rangle , \tag{2.1}$$

the moment is written as

$$\langle A_t^q \rangle = Q_t^{(q)} \exp(q\lambda_q t) , \tag{2.2}$$

where $Q_t^{(q)}$ does not contain any singular factor of the exponential form with respect to t. Noting that $\langle A_t^{2q} \rangle \geqq \langle A_t^q \rangle^2$, one finds $d\lambda_q/dq \geqq 0$, if λ_q is continuous everywhere in q. The Gaussian approximation leads to

$$\lambda_q = \lambda_0 + Dq , \quad (-\infty < q < \infty) , \tag{2.3}$$

(see (1.5)). In particular the relation (2.3) exactly holds if $\Delta(x_t)$ is exactly Gaussian. Examining the q dependence of λ_q, we can investigate the non-Gaussianity of the probability distribution for Y_t. If λ_q differs from (2.3), then the distribution turns out to have non-Gaussian components.
Generally for $q \to 0$, (2.1) reduces to

$$\lambda_q = \lambda_0 + Dq + O(q^2) , \tag{2.4}$$

where

$$\lambda_0 \equiv \lim_{t\to\infty} t^{-1} \langle \ln A_t \rangle = \langle \ln B(x) \rangle = \langle \Delta(x) \rangle \tag{2.5}$$

$$D \equiv \lim_{t\to\infty} \sigma_t/2t , \quad \sigma_t \equiv \langle (\ln A_t - \langle \ln A_t \rangle)^2 \rangle . \tag{2.6}$$

λ_0 and D have the meanings of the drift velocity and the diffusion coefficient, respectively. The generation of diffusion in chaotic systems has been recently investigated extensively by several authors[1-8]. The relation (2.4) is in agreement with the result (1.3) due to the Gaussian approximation, if $O(q^2)$ is neglected. If the relation $\lambda_q \simeq \lambda_0 + Dq$ holds till $|q|=\infty$, the Gaussian law turns out to be adequate for moments over the whole region of q. However, generally it is not true. We assume that $\lambda_{\pm\infty}$ exists, and furthermore that for $|q| \to \infty$

$$\lambda_q = \lambda_\infty - \lambda_+' q^{-1} + o(q^{-1}) , \qquad (q \to \infty) \tag{2.7}$$

$$= \lambda_{-\infty} - \lambda_-' q^{-1} + o(q^{-1}) , \qquad (q \to -\infty) , \tag{2.8}$$

where $\lambda_\pm' > 0$ from $d\lambda_q/dq \geqq 0$. An illustrative example producing (2.7) and (2.8) is discussed below. The q dependence of λ_q for $|q| \to \infty$ is highly different from

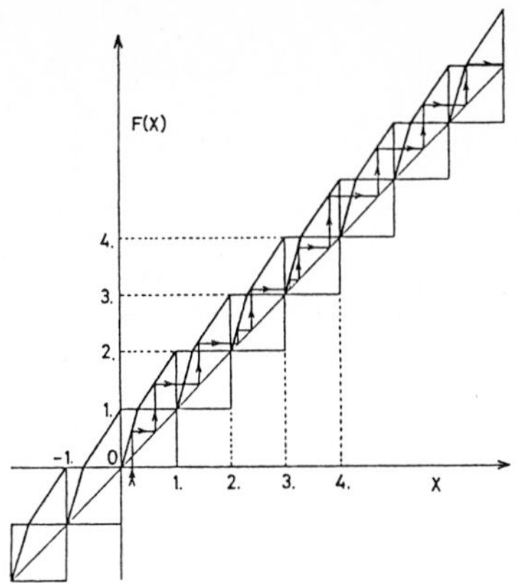

Fig.1 The example of the generalized map (2.9)(thick line).The arrow indicates the movement of the phase point X_t

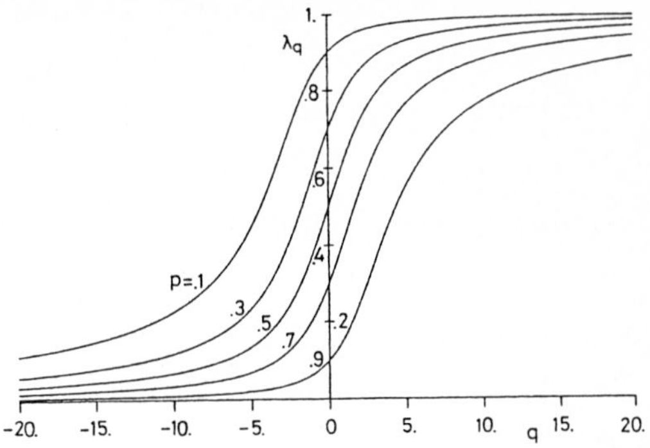

Fig.2 λ_q is drawn as the function of q for different p. We find that λ_q tends to 1(0) as $q\to\infty(-\infty)$. This indicates that the probability distribution for Y_t deviates from the Gaussian law

(2.3) and (2.4). This implies that the Gaussian approximation is inadequate to estimate large-power moments. This originates from the enhancement of the non-Gaussianity, due to the large amplitude fluctuations in $\Delta(x_t)$ which cannot be smoothed out by the summation of $\Delta(x_s)$. The existence of $\lambda_{\pm\infty}$ is guaranteed when $\Delta(x)$ is bounded. Therefore one easily observes that the Gaussian law breaks down always when $\Delta(x)$ is bounded.

One typical example described by (1.1) is the one-dimensional map

$$X_{t+1} = X_t + h(X_t) \equiv F(X_t) \tag{2.9}$$

with the one-period function $h(X):h(X+1)=h(X)$. A simple example of such a map is shown in Fig.1. After the decomposition $X_t=Y_t+x_t$ with Y_t being the box number in which the phase point X_t is located at time t,(2.9) can be rewritten into the set of (1.1) and (1.2), where $\Delta(x_t)$ is the jumping number from the Y_t-th box after unit step[1,2].

As an illustration, we consider the system (2.9)(Fig.1,[1,2]) with

$$\begin{aligned} \Delta(x) &= 0 \qquad (0 \leqq x < p) \\ &= 1 \qquad (p \leqq x < 1) \end{aligned} \tag{2.10}$$

and

$$\begin{aligned} f(x) &= x/p \qquad (0 \leqq x < p) \\ &= (x-p)/(1-p) \qquad (p \leqq x <1) . \end{aligned} \tag{2.11}$$

We can exactly solve all moments as

$$\langle A_t^q \rangle = [p+(1-p)e^q]^t , \qquad (-\infty < q < \infty) \tag{2.12}$$

and therefore

$$\lambda_q = q^{-1}\ln[p+(1-p)e^q] \ . \tag{2.13}$$

In Fig.2 λ_q is plotted as the function of q. Furthermore one easily obtains

$$\begin{aligned} &\lambda_0 = 1-p \ , \quad D = p(1-p)/2 \ , \quad \lambda_\infty = 1 \\ &\lambda_{-\infty} = 0 \ , \quad \lambda'_+ = -\ln(1-p) \ , \quad \lambda'_- = -\ln p \ . \end{aligned} \tag{2.14}$$

Evidently we observe that the q-linear characteristics break down when $|q|$ is away from zero. As will be discussed in the next section, this fact reflects the violation of the Gaussian law (1.3) in the tail regions of the probability distribution.

3. Asymptotic probability distribution

As was shown in the preceding section, there exist three typical characteristic q, ($q=0,-\infty$ and ∞). Low-order moments ($q\to 0$) are essentially determined by a global structure of the distribution(the Gaussian component) near the peak position $y=\lambda_0 t$, and are insensitive to details of tail regions. The Gaussian character of the probability distribution generally breaks down in two tails of the distribution. Precise calculation[9] tells that the crossover from the Gaussian characteristics to the non-Gaussian ones occurs near $y\simeq\lambda_{-\infty}t$ and $\lambda_\infty t$, and therefore two tails of the distribution are relevant to higher-order statistics. This fact reflects that the Gaussian distribution does not provide us with correct expressions for higher-order moments of A_t (see (1.5),(2.7) and (2.8)).

In contrast to the $\exp(-y^2/4Dt)$ decay of the Gaussian distribution($y\to\infty$),we can show[9] that the distribution producing the asymptotically correct expressions for higher-order moments decays in the form, e.g. for $y \gg \lambda_\infty t$,

$$\propto \exp[-(e^{q_+ y} - \langle A_t^{q_+}\rangle)^2/2\langle(\delta A_t^{q_+})^2\rangle] \ , \tag{3.1}$$

where q_+ stands for a certain number lying in the region $\lambda_{q_+} \simeq \lambda_\infty - \lambda'_+ q_+^{-1}$. We find that the asymptotically correct distribution decays more rapidly than the Gaussian form.

It should be noted that the break-down of the Gaussian distribution does not imply the violation of the central limit theorem. The peak position of the distribution is $\lambda_0 t$ and the peak height is $\sim 1/\sqrt{t}$. The width of the distribution which has the Gaussian form is $\sim\sqrt{t}$. On the other hand, the non-Gaussianity becomes important for $y \gtrsim \lambda_\infty t$, e.g. for the right tail. The intensity of the distribution near $y\simeq\lambda_\infty t$ is, therefore,$O(t^{-1})$, which is negligibly small as compared with the peak height. The range of the Gaussian regime ($|y-\lambda_0 t|\sim\sqrt{t}$) becomes wider and wider as $t\to\infty$. In this sense the central limit theorem holds and the probability distribution is globally approximated by the Gaussian form if the intensity of the order t^{-1} is ignored. Nevertheless tails with intensities $O(t^{-1})$ give important contributions to higher-order statistics.

4. Remarks

We have discussed the long-time statistical dynamics (1.1) or equivalently (1.4) generated by a one-dimensional map (1.2). We have shown that Y_t usually does deviate from the Gaussian process even for $t\to\infty$. This can be seen with the deviation from the q linear characteristics of λ_q.

The present analysis can be applied to a wide range of stochastic processes. Examples are the following:

(A) Consider a one-dimensional map $x_{t+1}=f(x_t)$. The Lyapunov exponent characterizing the orbital instability is given by $\lambda=\langle \ln|f'(x)|\rangle$. The quantity $\ln|f'(x_t)|$ may be called the local Lyapunov exponent and usually strongly depends on the phase point x_t. λ is just the average of such local quantities. The fluctuation effect of local Lyapunov exponents[5] can be discussed by utilizing the variation equation

$$A_{t+1} = |f'(x_t)|A_t \quad . \tag{4.1}$$

This is the same as (1.4) if we put $B(x)=|f'(x)|$. We find that the Lyapunov exponent λ is just the zero-order characteristic exponent($\lambda=\lambda_0$). For details, see FUJISAKA[5,6].

(B) The next example is seen also in the dynamics $x_{t+1}=f(x_t)$. Let us define S_t by

$$S_{t+1} = S_t + \sqrt{|f(x_t)-x_t|^2+1} \ , \tag{4.2}$$

which is the length of the x_t-t curve measured from ,e.g. t=0. This is identical to (1.1) or, after the transformation $A_t=\exp(S_t-S_0)$, $B(x)=\exp(\sqrt{|f(x_t)-x_t|^2+1})$, the above reduces to our basic equation (1.4).

(C) The most striking example may be seen in the theory of the intermittency in fully developed isotropic homogeneous turbulence[10,11]. In this case A is a certain positive definite quantity averaged over certain spatial ranges. We find that the log-normal argument[10] and the β-model approach[11] correspond to the limits $q\to 0$ and $q\to\infty$ in the present theory, respectively. The intermittency exponent μ relating the 1-st and 2-nd moments of A turns out to be expressed by

$$\mu = \lambda'_+ \ , \tag{4.3}$$

where λ'_+ is defined in (2.7), if (2.7) holds for q=1. Since the β model is proposed by a heuristic argument without any dynamical equations, we can discuss the interrelation between the log-normal theory and the β-model approach. Details will be reported elsewhere[12].

Finally we note that similar arguments on the statistical mechanics of one-dimensional maps are discussed by TAKAHASHI-OONO [13] for different purposes.

Acknowledgements

The author thanks Prof.M.Inoue for valuable discussions. This study was partially supported by the Scientific Fund of the Ministry of Education, Science and Culture.

References

[1] S.Grossmann and H.Fujisaka: Phys.Rev.A26,1779(1982)
[2] H.Fujisaka and S.Grossmann: Z.Physik B48,261(1982)
[3] T.Geisel and J.Nierwetberg: Phys.Rev.Lett.48,7(1982)
[4] M.Schell,S.Fraser and R.Kapral: Phy.Rev. A26,504(1982)
[5] H.Fujisaka: Prog.Theor.Phys.(to be published)
[6] H.Fujisaka: to be submitted to Prog.Theor.Phys.
[7] M.Inoue and H.Koga: Prog.Theor.Phys.68,2184(1982)
H.Koga,H.Fujisaka and M.Inoue: Phys.Rev. October(1983)
[8] S.Grossmann and S.Thomae: to appear in Phys.Letters
[9] H.Fujisaka and M.Inoue: in preparation
[10] A.S.Monin and A.M.Yaglom: Statistical Fluid Mechanics (The MIT Press,Cambridge, Massachusetts and London,England,1981)
[11] U.Frisch,P.-L.Sulem and M.Nelkin: J.Fluid Mech.87,719(1978)
[12] H.Fujisaka and M.Inoue: in preparation
[13] Y.Takahashi and Y.Oono: preprint

Analytic Study of Power Spectra of Intermittent Chaos

B.C. So, H. Okamoto, and H. Mori

Department of Physics, Kyushu University 33
Fukuoka 812, Japan

1. Introduction

One-dimensional maps provide us with useful models for the onset and the growth of turbulence [1]. In fact, the period doubling, the windows, the band splitting, and the intermittency can all be described by one-dimensional maps. Therefore one-dimensional maps are useful for studying the onset of turbulence. The power spectrum exhibits a measure-theoretical structure of orbits in phase space, and hence the power spectrum is indispensable to studying the structure and the growth of chaos.

Recently SHOBU et al.[2] clarified the shape of the power spectrum of the intermittent chaos due to a tangent bifurcation [3] from a phenomenological point of view and also by calculating the power spectrum of a piecewise linear map analytically. We also studied the power spectrum of an intermittent chaos for a different piecewise linear map [4]. In this talk we shall discuss important features of the above two maps and propose a new mechanism for the onset of intermittent turbulence.

2. Two types of intermittency

There may exist several types of the onset of intermittent chaos. We shall discuss two types. One type is an abrupt excitation of an infinite number of unstable periodic orbits and is called Type A. The other is a collapse of a pair of stable and unstable periodic orbits and is called Type B. Simple maps which exhibit these two types are the maps f_A and f_B shown in Fig.1 and Fig.2 respectively. If a map is piecewise linear and satisfies the finite Markov condition, then the power spectrum of the map can be calculated analytically with the use of the Perron-Frobenius operator. This is the technical reason why we take piecewise linear maps.

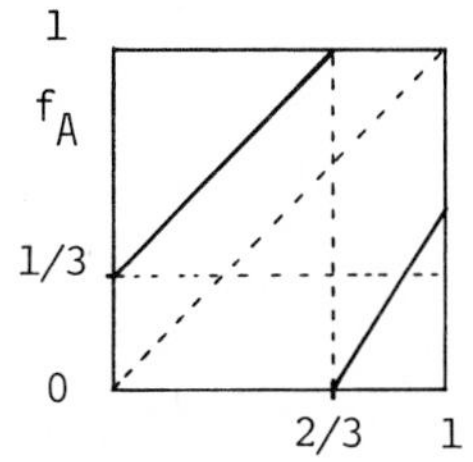

Fig.1 $f_A(x)$

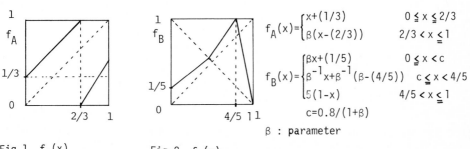

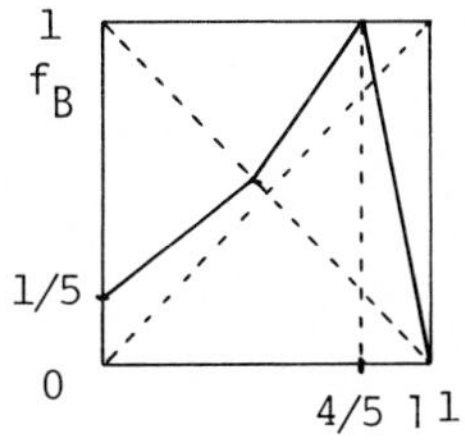

Fig.2 $f_B(x)$

$$f_A(x)=\begin{cases} x+(1/3) & 0 \leqq x \leqq 2/3 \\ \beta(x-(2/3)) & 2/3 < x \leqq 1 \end{cases}$$

$$f_B(x)=\begin{cases} \beta x+(1/5) & 0 \leqq x < c \\ \beta^{-1}x+\beta^{-1}(\beta-(4/5)) & c \leqq x < 4/5 \\ 5(1-x) & 4/5 < x \leqq 1 \end{cases}$$

$$c=0.8/(1+\beta)$$

β : parameter

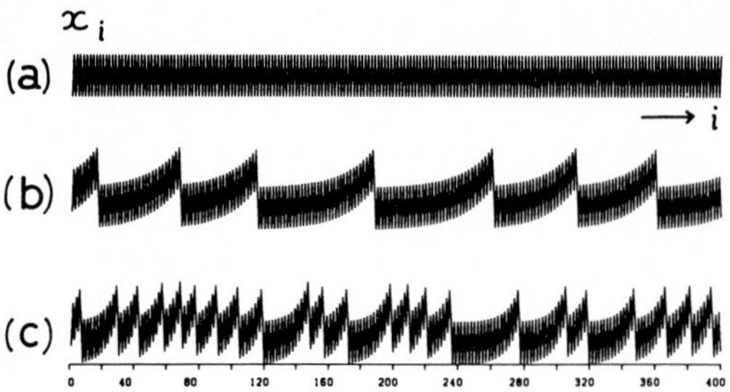

Fig.3 Orbit $\{x_i\}$ of the map f_A: (a) $\beta < 1$ (b) $\beta = 1.114465$ (c) $\beta = 1.324718$

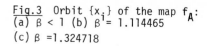

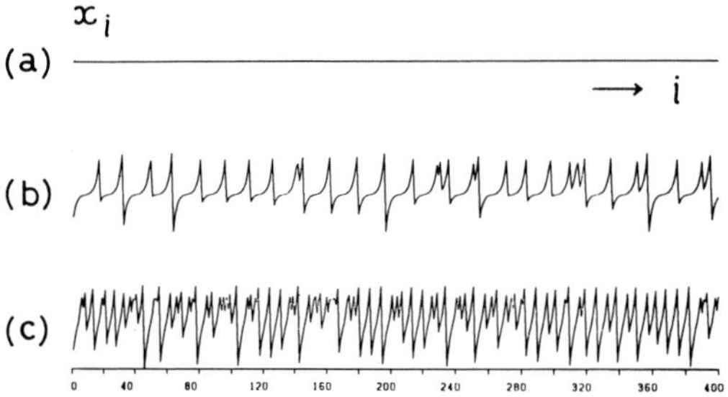

Fig.4 Orbit $\{x_i\}$ of the map f_B: (a) $\beta < 0.6$ (b) $\beta=0.605$ (c) $\beta= 0.75$

The fundamental property which characterizes the map is the sequence of periodic orbits for their coexistence. Since $f_B(x)$ is a continuous function of x, Sarkovskii's ordering holds, i.e.,

$$3 \vdash 5 \vdash 7 \vdash . . . \vdash 2^n \cdot 3 \vdash 2^n \cdot 5 \vdash 2^n \cdot 7 \vdash . . . \vdash 2^n \vdash 2^{n-1} \vdash . . . \vdash 2 \vdash 1 \quad (2.1)$$

for $f_B(x)$, where $j \vdash k$ indicates that if a periodic orbit of period j exists, then a periodic orbit of period k also exists. On the other hand, we found

$$1 \vdash 6 \vdash 4 \vdash 2 \vdash 9 \vdash 7 \vdash 12 \vdash 10 \vdash 15 \vdash . . . \vdash 3n+2 \vdash 3(3n+2)+3 \vdash 2(3n+2)+3$$
$$\vdash 3(3n+2)+3\cdot 2 \vdash 2(3n+2)+3\cdot 2 \vdash 3(3n+2)+3\cdot 3 \vdash . . . \vdash 3 \quad (2.2)$$

for $f_A(x)$. The most outstanding difference between (2.1) and (2.2) is that the subsequence 2^n of (2.1) indicates period-doubling bifurcations leading to a chaotic state, whereas (2.2) indicates an abrupt transition from a periodic state of period three to a chaotic state. Figures 3 and 4 show orbits of these maps at several excitation parameter values.

3. Analytic method for calculating the power spectrum $S(\omega)$

Let us consider an orbit $\{x_i\}$ generated by a map f(x) ;

$$x_i = f(x_{i-1}) = f^i(x_0) , \quad (3.1)$$

where $f^i(x) = f(f^{i-1}(x))$. Let us assume that f(x) is ergodic in the interval I = [0,1]. Then, for almost all initial points x_0, the orbit $\{x_i\}$ is nonperiodic and covers I densely so that the long-time average of a function g(x) can be replaced by the space average as

$$\lim_{N\to\infty} \frac{1}{N} \sum_{i=0}^{N-1} g(x_i) = \int_0^1 dx\, P^*(x) g(x) \equiv \langle g(x) \rangle , \quad (3.2)$$

where $P^*(x)$ is the invariant density of the map f. The invariant density satisfies $\{HP^*\}(x)=P^*(x)$ in terms of the Perron-Frobenius operator H

$$\{Hg\}(x) \equiv \int_0^1 dy\ \delta(x-f(y))g(y) \quad = \quad \sum_i g(y_i)/|f'(y_i)| \ . \tag{3.3}$$

The time-correlation function of orbits C_t can be written as [5]

$$C_t = \int_0^1 dx\ P^*(x)f^t(x)\delta x \quad = \int_0^1 dx\ x\ H^t\{P^*(x)\delta x\} \ , \quad (\ \delta x = x-\langle x\rangle). \tag{3.4}$$

The power spectrum $S(\omega)$ is given by

$$S(\omega) = \sum_{t=-\infty}^{\infty} C_t\ e^{-i\omega t} \ , \quad (\ C_{-t} \equiv C_t\ ). \tag{3.5}$$

If we can expand $P^*(x)\delta x$ in terms of the eigenfunctions $\{\Psi_j\}$ of H, $H\Psi_j=\nu_j\Psi_j$, as

$$P^*(x)\delta x \quad = \quad \sum_j\ b_j\Psi_j(x) \ , \tag{3.6}$$

then we obtain

$$C_t \quad = \quad \sum_j B_j\nu_j \ , \quad B_j = b_j\int_0^1 dx\ x\ \Psi_j(x) \quad , \tag{3.7}$$

$$S(\omega) = \quad \sum_j B_j\ /(1-\nu_j e^{-i\omega}) \quad + \text{c.c.} \quad -1\ . \tag{3.8}$$

Therefore, the power spectrum consists of Lorentzian spectral lines at frequencies $\omega = \omega_j$ with widths $\gamma_j = -\ln|\nu_j|$, where $\nu_j = |\nu_j|\exp(i\omega_j)=\exp(-\gamma_j+i\omega_j)$.

Thus, if one can find the eigenfunction expansion (3.6), then one can study the time-correlation function C_t and the power spectrum $S(\omega)$ analytically. This method was initiated and developed for various piecewise linear maps by our group [5,2,4].

Before the onset of intermittent chaos the power spectrum consists of a single line with zero width, representing a stable periodic state. After the onset, the power spectrum consists of a number of Lorentzian lines and the shape of the power spectrum critically depends on whether the separation $\Delta\ \omega_j = \omega_{j+1}-\omega_j$ is larger than the width γ_j or not. Which case occurs depends on the mechanism of chaos, the excitation parameter and the external noise [2].

4. Comparison between Type A and Type B

Figures 5 and 6 show the invariant density of the maps f_A and f_B, respectively. The invariant density of f_B has a single peak at the center of a narrow channel [2], whereas the invariant density of f_A has three peaks at the periodic points of period three [4].

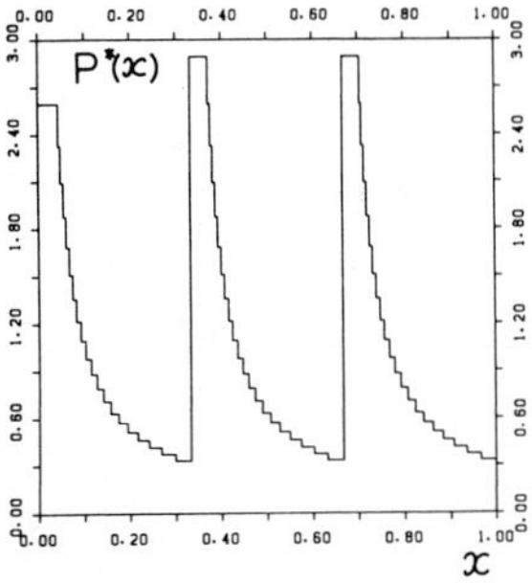

Fig.5 Invariant density of the map f_A at β=1.114465

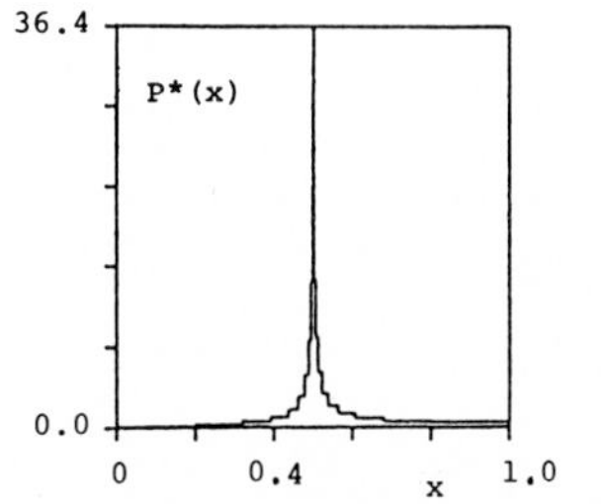

Fig.6 Invariant density of the map f_B at β= 0.602003

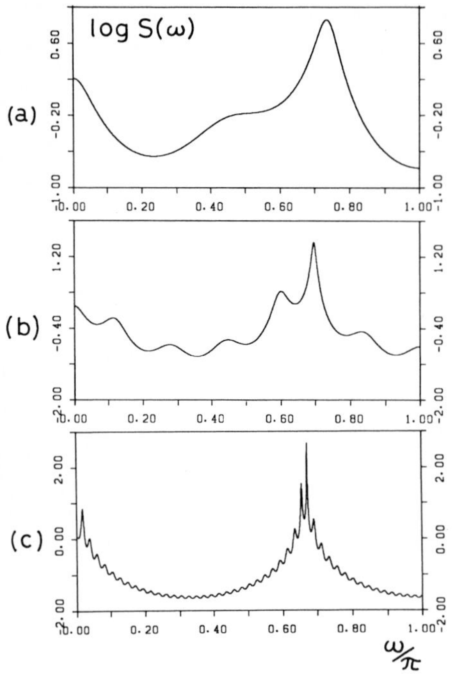

Fig.7 Power spectrum of the map f_A:
(a) β = 1.618034 (b) β = 1.324718
(c) β = 1.085450
The onset point is β = 1.

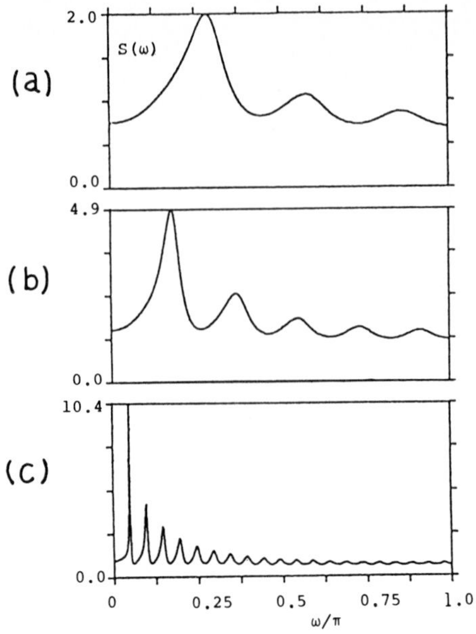

Fig.8 Power spectrum of the map f_B:
(a) β = 0.742959 (b) β = 0.633268
(c) β = 0.600012
The onset point is β = 0.6.

Figures 7 and 8 show the power spectra of the maps f_A and f_B, respectively. Analyzing the exact expression for the power spectra, we obtain the following results.

(1) The power spectra of the maps f_A and f_B consist of a large number of Lorentzian lines as shown in Figs. 7 and 8. The number of lines blows up in both maps, as the excitation parameter decreases to the onset point.

(2) In the vicinity of the onset point, the power spectrum has the highest Lorentzian line around frequency ω_0, where ω_0 is the frequency of the periodic state before the onset of chaos. Indeed Fig.7(c) for the map f_A has the highest line around $\omega_0 = 2\pi/3$, corresponding to a periodic state of period 3, and Fig. 8(c) for f_B has the highest line around $\omega_0 = 0$, corresponding to a fixed point before the onset.

(3) Figure 7(c) for the map f_A has two dominant peaks; the highest peak locates around $\omega_0 = 2\pi/3$ with an envelope of the power law $1/|\omega-(2\pi/3)|^4$, and the second-highest peak locates around $\omega = 0$ with an envelope of $1/|\omega|^2$. On the other hand,

Fig. 8(c) for f_B has one dominant peak around $\omega = 0$ and its envelope obeys the power law $1/|\omega|^2$. It should be noted that the power laws for the envelope are the asymptotic behavior in the vicinity of the onset point [4]. Their exponents, 4 and 2, can be shown to be universal irrespective of details of the models [2].

(4) The highest Lorentzian line around frequency ω_0 determines the most important feature of the time-correlation function C_t, which exhibits an interesting behavior when $\omega_0 \neq 0$. We consider the case $\omega_0 \neq 0$. Let $\omega_L = \omega_0(1+\hat{q})$ and γ_L be the frequency and the width of this highest line, respectively, where $\hat{q}$ represents the misfit of the frequency. Figure 7(c) for the map f_A has $\omega_0 = 2\pi/3$ and $\hat{q} = 0.008019$, $\gamma_L = 0.0013637$. Figure 9 shows the time-correlation function C_t which corresponds to the power spectrum Fig.7(c).The time-correlation function consists of a rapid osillation with period 3 and a slow oscillation of the amplitude with period about 125. The period 3 represents the frequency ω_0 , whereas the long period 125 arises from the misfit $\hat{q}$. Indeed $T_{amp} = 1/\hat{q} = 124.7$. The damping of Fig.9 is given by γ_L. Figure 9 shows an asymmetry in positive and negative directions as time proceeds. This asymmetry can be obtained by taking into account the second-highest Lorentzian line [4]. Thus the main features of the time-correlation function can be reproduced by only a few highest Lorentzian lines.

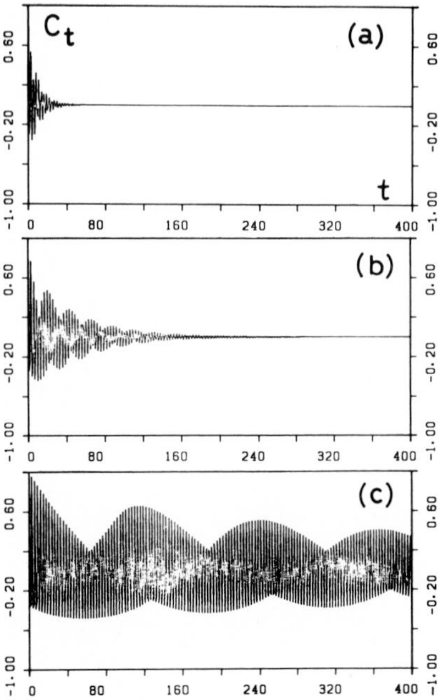

Fig.9 Time-correlation function C_t of the map f_A : (a) $\beta = 1.618034$ (b) $\beta = 1.324718$ (c) $\beta = 1.085450$

5. Concluding remarks

We have considered two simple maps for intermittent chaos; Type B is a typical map for the tangent bifurcation and Type A is a typical map for the abrupt excitation of an infinite number of unstable periodic orbits. We have compared main features of the two maps and found important differences. One of them is about the dominant peak of the power spectrum around ω_0, where ω_0 is the frequency of the periodic state before the onset of chaos. Namely, the envelope of the peak for Type B obeys the power law $1/|\omega - \omega_0|^2$, whereas that for Type A obeys the power law $1/|\omega - \omega_0|^4$.

Thus it may be concluded that there exists the onset mechanism for intermittent chaos which is different from the tangent bifuraction. This viewpoint is under further study with the hope that the intermittent chaos found by Bergé et al. in the Bênard convection of silicone oil [6] could be one candidate for Type A.

References

1. J.-P. Eckmann: Rev. Mod. Phys. 53 , 643 (1981)
   H.L. Swinney: Physica D , (1983)
2. K. Shobu, T. Ose and H. Mori: "Shapes of the Power Spectrum of Intermittent Chaos near its Onset Point" J. Phys. Soc. Japan (to be submitted)
3. Y. Pomeau and P. Manneville: Commun. Math. Phys. 74 , 189 (1980)
4. B.C. So, N. Yoshitake, H. Okamoto and H. Mori: "Correlations and Spectra of an Intermittent Chaos near its Onset Point" J. Stat. Phys. (to be submitted)
5. H. Mori, B.C. So and T. Ose: Prog. Theor. Phys. 66 , 1266 (1981)
   H. Shigematsu, H. Mori, T. Yoshida and H. Okamoto: J. Stat. Phys. 30 , 649 (1983)
   T. Yoshida, H. Mori and H. Shigematsu: J. Stat. Phys. 31 , 279 (1983)
6. P. Bergé, M. Dubois, P. Manneville and Y. Pomeau: J. Phys. Lett. 41 L341 (1980)

Part V

Bifurcations and Normal Forms

Versal Deformation of Singularities and Its Applications to Strange Attractors

S. Ushiki

Institute of Mathematics, Yoshida College, Kyoto University
Kyoto 606, Japan

1. Introduction

The aim of this article is to present a systematic mathematical method for the study of global bifurcation diagrams of families of autonomous ordinary differential equations. We employ the normal form and versal family theory developed in USHIKI[22], which is an improved version of classical normal form theory for singularities of vector fields. The classical theory for normal forms is known and has been employed in many authors to study the bifurcation of vector fields. See POINCARE[16][17][18], BIRKHOFF[6], ARNOLD[2][3][4][5], TAKENS[21] and BROER[8] for the classical theory and its modern version. See ARNOLD[4][5], BOGDANOV[7], LANGFORD[13], LANGFORD-IOOSS[14], GUCKENHEIMER[9], HOLMES[10][11], and ARNEODO-COULLET-SPIEGEL-TRESSER[1] for some applications of the normal forms theory to bifurcation problems.

The normal forms theory, started by H.Poincaré, aims to construct a "universal" family of vector fields. Suppose the following situation: we are given a system of ODEs on the Euclidean space $\mathbb{R}^n$; we are to find the simplest form of the given system which can be obtained by a change of coordinates; if we had a list of normal forms of vector fields and we could tell to which simplest form a system can be tranformed, then the above problem is easy to solve. The theory of Jordan normal forms for matrices can be regarded as the first-order part of such normal forms theory of singularities of vector fields.

The author[22] improved the normal form theory by computing the coordinate transformations considered as a Lie group action acting over the Lie algebra of jets in an exhaustive manner, so that the obtained normal forms is unique, i.e., there corresponds only one normal form for any given system in the list of normal forms of vector fields up to certain higher order terms. The list also provides the versal deformation for these singularities.

This theory is applied to the Lorenz family of ordinary differential equations [15] and the Rössler's family[19]. These families are embedded in the family of normal forms by suitable smooth changes of coordinates. By rescaling the time and by renormalizing the family to the normal form, we shall obtain a one-parameter family of families of systems, which connects the original family having quite complicated dynamics studied by many people, to a very simple, but degenerate, "integrable" system. Here, "integrable" means that it has explicit formula for the solutions. The result can be interpreted as follows. There exists "Lorenz attractor" in the neighborhood of the origin in some perturbation of the system $\dddot{x} = -x^3$, $\dot{z} = x^2$. Similar results are obtained for the Rössler's family of ODEs. See USHIKI-OKA-KOKUBU[23] for the detail. In the case of Rössler's attractor, the rescaled attractor does not shrink in all directions. It becomes thin and finally almost one-dimensional in the limit.

2. Jordan normal form and Poincaré's normal form

Consider a system of ordinary differential equations :

$$\dot{x}_i = f_i(x_1,\dots,x_n) \qquad i=1,\dots,n \tag{2.1}$$

and suppose that the origin is singular, i.e., $f_i(0) = 0$, $i=1,\dots,n$. Let D_i denote the partial differential operator $\partial/\partial x_i$ for $i=1,\dots n$. We regard these operators as the basis of tangent vectors in the Euclidean space $\mathbb{R}^n$. We write (2.1) in the vector form :

$$\dot{x} = f_1(x)D_1 + \dots + f_n(x)D_n. \quad (x = {}^t(x_1,\dots,x_n)). \tag{2.2}$$

We write it also in the form :

$$\dot{x} = F(x) = F_1(x) + F_2(x) + \dots, \tag{2.3}$$

where $F(x)$ denotes the right-hand side of (2.2) and $F_k(x)$ denotes the k-th order homogeneous part of $F(x)$. Let A be the $n \times n$ matrix corresponding to the linear part $F_1(x)$, i.e., $Ax = F(x)$.

Let us see how a linear change of coordinates transforms (2.1). Let P be a nonsingular $n \times n$ matrix. Define a linear change of coordinates by $y = Px$. Then in the y coordinates, (2.3) can be written in the form :

$$\dot{y} = P\dot{x} = PF(P^{-1}y) = PF_1(P^{-1}y) + PF_2(P^{-1}y) + \dots. \tag{2.4}$$

Linear transformation of coordinates P transforms the matrix A into PAP^{-1}. Hence (2.1) can always be transformed into the form whose linear part is of Jordan normal form. Any system (2.1) which is singular at the origin can be transformed into the Jordan normal form by a coordinate change up to terms of order two.

Now let us recall the Poincaré's normal forms theory. Suppose $F(z)$ is a complex vector of power series in complex vector $z = (z_1,\dots,z_n) \in \mathbb{C}^n$, which vanishes at the origin. Let w be the vector of eigenvalues of the linear part of F. Let $m = (m_1,\dots,m_n)$ denote an n-tuple of nonnegative integers. We call m a multi-index. We set

$$z^m = z_1^{m_1} \cdot z_2^{m_2} \cdot \dots \cdot z_n^{m_n}, \quad |m| = m_1+\dots+m_n, \tag{2.5}$$

and $(m,w) = m_1w_1+m_2+\dots+m_nw_n$. A monomial vector field z^mD_i is said to be resonant if $w_i = (m,w)$ holds and the length $|m| \geq 2$. H. Poincaré proved the following theorem.

Theorem 2.1 (Poincaré) System of ordinary differential equations $\dot{z} = F(z)$ defined by a vector of formal power series can be transformed by a formal transformation $z = y + \dots$ into the form

$$\dot{y} = F_1 + W(y), \qquad (2.6)$$

where all monomial terms in the power series $W(y)$ are resonant.

We call (2.6) the Poincaré's normal form. Essentially the same result is found in much more elegant form in TAKENS[21]. He and ARNOLD[5] computed the explicit forms of normal forms instead of the abstract statement in the above theorem and they studied in detail the global phase portraits of these singularities. The normal forms computed by Arnold and Takens are, however, unsatisfactory for the following reason. The family of normal forms can be considered as the set of representatives for each equivalence class of vector fields under the coordinate transformations. If a certain family is a family of normal forms, the family derived from this family by applying a coordinate transformation simultaneously to all systems of the family is also a family of normal forms. We cannot claim the family to be unique. We can expect, however, that the normal form (2.6) should be determined uniquely from the original system $\dot{z} = F(z)$. Unfortunately, Poincaré-Arnold-Takens theory does not give the uniqueness of the normal form in the above sense.

3. Versal family of vector fields

Consider a family of systems :

$$\dot{x} = F_r(x),\ x \in \mathbb{R}^n,\ r \in \mathbb{R}^p \qquad (3.1)$$

which has a singular point at the origin for all values of parameter r. How can we transform the family into a normal form by a family of coordinate transformations smoothly depending upon the parameter r ? A family of systems $\dot{y} = G_s(y)$ on $\mathbb{R}^n$ with parameter $s \in \mathbb{R}^q$ is said to be k-versal if for any family (3.1) such that $F_0(x) = G_0(x)$, there exists a family of coordinate transformations $y = T_r^0(x)$, smoothly depending on the parameter r and a smooth mapping of parameters $s = t(r)$, such that the system $\dot{x} = F_r(x)$ transformed by $y = T_r(x)$ and the system $\dot{y} = G_{t(r)}(y)$ coincides up to terms of order higher than k. ARNOLD[3] gave the solution for this problem in the case of matrices. He gave the versal family of matrices, which can be considered as the normal forms for families of matrices. The versal family is also the versal family of first order for ordinary differential equations with a singular point at the origin.

It is known that most of degenerate singularities must have an infinite number of unfolding parameters for the unfolding family to be versal, if we don't neglect any higher order terms (see ICHIKAWA [12]).

4. Normal forms for some degenerate singularities

In this section, we reproduce the normal forms obtained in [22] for the systems of ODEs on $\mathbb{R}^2$ and $\mathbb{R}^3$, whose linear part includes zero or pure imaginary eigenvalues. These results for truncated normal forms are obtained by computing, up to higher order terms, the effect of all the coordinate transformations exhaustively. See [22] for the detail. In the following we use the notations D_x, D_y, and D_z as basis of tangent vectors $\partial/\partial x$, $\partial/\partial y$, and $\partial/\partial z$ in $\mathbb{R}^2$ or $\mathbb{R}^3$. We use also polar coordinates (r,θ) and the notations $D_\theta = -yD_x + xD_y$ and $rD_r = xD_x + yD_y$.

Let X be a smooth vector field on $\mathbb{R}^2$ (or $\mathbb{C}^2$). Assume that X vanishes at the origin and that the linear part of X is yD_x. Then X can be transformed by a transformation of coordinates into one of the following forms up to terms of order 5.

$$\text{(a)} \quad yD_x \pm x^2D_y + uxyD_y + wx^3D_y + qx^3D_y,$$

$$\text{(b)} \quad yD_x \pm xyD_y + v_1x^3D_y + v_2x^2yD_y + qx^3D_y + sx^4D_y,$$

$$\text{or (c)} \quad yD_x + w_1x^3D_y + w_2x^2yD_y + qx^3D_y + sx^4D_y, \text{ with } w_1^2+w_2^2=1.$$

Parameters u, v_1, v_2, w, q, s, w_1, w_2 are uniquely determined from the 4-th order Taylor expansion of X at the origin.

Versal families for these singularities are obtained by adding the supplementary parameters. For instance, a versal family of fourth order for a system of type (a) above, say X_* for $u = u_0$, $w = w_0$ and $q = q_0$, is given by adding the Arnold's versal family for linear part to the family of normal forms :

$$\text{(a')} \quad X_* + (p_1x + p_2y)D_y + uxyD_y + wx^3D_y + qx^3D_y.$$

The following is a list of simplest normal forms with nonhyperbolic linear parts in $\mathbb{R}^3$ (or $\mathbb{C}^3$). Nongeneric cases are omitted. Normal form (4.1) is of 5-th order and (4.2) and (4.3) are of third order. All parameters are uniquely determined from the original system.

$$D_\theta+azrD_r+bzD_\theta\pm r^2D_z\pm z^2D_z+cz^3D_z+dz^2rD_r+ez^2D_\theta+fr^5D_r+gr^4D_\theta, \quad (4.1)$$

$$yD_x\pm x^2D_y+axyD_y+bxzD_z\pm z^2D_y+cyxD_y+dz^2D_z$$
$$+ex^3D_y+fx^3D_z+gz^3D_z+hxyzD_y+iyz^2D_y, \quad (4.2)$$

$$yD_x+zD_y\pm x^2D_z+axyD_z+bxzD_z+cx^2yD_z+dxz^2D_z+ex^3D_z. \quad (4.3)$$

5. Normalization and rescaling of Lorenz system

In this section we construct a one-parameter family of systems of ODEs, which connects the Lorenz system to an "integrable" system. The Lorenz system is :

$$\dot{x} = sy - sx, \ \dot{y} = -xz + rx - y, \quad \dot{z} = xy - bz. \quad (5.1)$$

This system has a symmetry: the system takes exactly the same form when it is transformed by coordinate transformation $(x,y,z) \rightarrow (-x,-y,z)$. The normal form theory works also for such equivariant systems taking only equivariant systems and equivariant coordinate transformations. The third-order versal family of vector field yD_x under the symmetry above is given by

$$\dot{x} = y,\ \dot{y} = Ax+By+Eyz+Fxz+Hx^2y+Iyz^2+Jx^3,\ \dot{z} = Cz+x^2+Gz^2+Kz^3,$$

where F is a nonzero constant and A, B, C, E, G, H, I, J, K are unfolding parameters. For a degenerate case

$$\dot{x} = y,\ \dot{y} = -x^3,\ \dot{z} = x^2, \tag{5.2}$$

the versal family for this system is computed as :

$$\dot{x} = y,\ \dot{y} = Ax+By+Eyz+Fxz+Hx^2y+Iyz^2+Jx^3+Lxz^2,$$

$$\dot{z} = Cz+x^2+Gz^2+Kz^3; \tag{5.3}$$

here all parameters are unfolding parameters.

The linear part of (5.1) at the origin has triple zero eigenvalues when $(s,r,b) = (-1,1,0)$. Assume $s \neq 0$ and $b \neq 2s$ and execute the change of variables for (5.1): $X = 2x$, $Y = 2s(y-x)$, $Z = Q(2sz-x^2)$, where $Q = 1 - b/2s$. Put $A = s(r-1)$, $B = -(1+s)$, $C = -b$, and $S = 2Qs$. We obtain :

$$\dot{x} = y,\ \dot{y} = Ax+By-Sxz-x^3,\ \dot{z} = Cx+x^2, \tag{5.4}$$

which is a subfamily of (5.3). Observe that (5.4) has four parameters whereas (5.1) contains only three parameters.

Now change the time scale from t to $t' = t/p$ and renormalize the obtained system by coordinate change $X = px$, $Y = p^2y$, $Z = pz$. The renormalized system is of the form

$$\dot{x} = y,\ \dot{y} = p^2Ax+pBy-pSxz-x^3,\ \dot{z} = pCx+x^2, \tag{5.5}$$

which shows that there exists a "miniature" Lorenz attractor in the neighborhood of the origin for small parameter p, since for any positive value p, the system is analytically conjugate to the Lorenz system.

REFERENCES

[1] Arneodo,A., Coullet,P.H., Spiegel,E.A. and Tresser,C.: Asymptotic chaos, to appear in PHYSICA D.

[2] Arnold,V.I.: On matrices depending on a parameter, Russ. Math. Surveys 26 (1971) 29-43.

[3] Arnold,V.I.: Lectures on Bifurcation in versal families, Russ. Math. Surveys 27 (1972) 54-123.

[4] Arnold,V.I.: Loss of stability of self oscillations close to resonance and versal deformations of equivariant vector fields, Funct. Anal. Appl., 11 (1977) 85-92.

[5] Arnold,V.I.: Chapitres supplémentaires de la théorie des équations différentielles ordinaires, Editions Mir, Moscow, 1980.

[6] Birkhoff,G.D.: Dynamical systems, Amer. Math. Soc. Colloquium Publications, New York (1927).

[7] Bogdanov,R.I.: Versal deformation of a singular point of a vector field on a plane in the case of zero eigenvalues, Proceedings of I.G.Petrovskii Seminar, 2 (1976) 37-65.

[8] Boer,H: Formal normal form theorems for vector fields and some consequences for bifurcations in the volume preserving case, Lecture Notes in Math. 898, Dynamical Systems and Turbulence, Warwick 1980, Springer, (1981) 54-74.
[9] Guckenheimer,J: On a codimension two bifurcation, Lecture Notes in Math. 898, Dynamical Systems and Turbulence, Warwick 1980, Springer, (1981) 99-142.
[10] Holmes,P.J.: Center manifolds, normal forms and bifurcation of vector fields, Physica 2D (1981) 449-481.
[11] Holmes,P.J.: A strange family of three-dimensional vector fields near a degenerate singularity, J. Diff. Eq. 37 (1980) 382-403.
[12] Ichikawa,F: Finitely determined singularities of formal vector fields, Invent. Math. 66 (1982) 199-214.
[13] Langford,W: Periodic and steady-state mode interactions lead to tori, SIAM J. Appl. Math., 37 (1979) 22-48.
[14] Langford,W., Iooss,G.: Interactions of Hopf and pitchfork bifurcations, Workshop on Bifurcation Problems, Birkhäuser Lecture Notes, (1980).
[15] Lorenz,E.N.: Deterministic nonperiodic flows, J. Atmospheric Sci., 20, (1963), 130-141.
[16] Poincaré,H.: Thèse (1879), Oeuvre I, Gauthier-Villars (1928) 69-129.
[17] Poincaré,H.: Mémoire sur les courbes définies par une équation différentielle, I, II, III, and IV, J.Math. Pures Appl. (3) 7 (1881) 375-422, (3) 8 (1882) 251-286, (4) 1 (1885) 167-244, (4) 2 (1886) 151-217.
[18] Poincaré,H.: Les méthodes nouvelles de la mécanique céleste, I (1892), III (1899).
[19] Rössler,O.E.: Continuous chaos - four prototype equations, Ann. New York Acad. Sci., 316, (1979), 376-392.
[20] Rössler,O.E.: Different types of chaos in two simple differential equations, Z. Naturforsch, 31a, (1976), 1664-1670.
[21] Takens,F.: Singularities of vector fields, Publ. Math. IHES, 43 (1973) 47-100.
[22] Ushiki,S.: Normal forms for singularities of vector fields, preprint.
[23] Ushiki,S., Oka,H., and Kokubu,H.: Attracteurs étranges engendré par une singularité des systèmes intégrables, preprint.

Some Codimension-Two Bifurcations for Maps, Leading to Chaos

G. Iooss

Math. Département, Université de Nice, Parc Valrose
F-06034 Nice, France

I. INTRODUCTION

Consider a family of mappings F_μ in a Banach space, depending on a real parameter μ, and assume that 0 is a fixed point of F_0. The derivative at this point being noted T_0, an elementary bifurcation occurs in each of these cases :

i) 1 is a simple eigenvalue of T_0, the remaining part of its spectrum being of modulus less than 1. This is the "saddle-node" bifurcation where, while μ crosses 0, two fixed points (a saddle point and a node) meet together and disappear. By some extra nonlocal phenomenon this can produce "intermittency" of a simple kind [1].

ii) -1 is a simple eigenvalue of T_0, the remaining part of its spectrum being of modulus less than 1. This is the "flip" bifurcation where, while μ crosses 0, the fixed point is changed from a node to a saddle, and two periodic points appear of period 2 for μ on one side of 0, they are node (resp.saddle) if the fixed point is a saddle (resp.node). A succession of such bifurcations seems to be frequent in physics (for instance in hydrodynamics [2]).

iii) λ_0 and $\overline{\lambda}_0$ are simple eigenvalues of T_0 on the unit circle, $\lambda_0^n \neq 1$ for n=1,2,3,4, the remaining part of the spectrum being of modulus less than 1. This is the "Hopf bifurcation" for maps which lead , while μ crosses 0, to the creation of an invariant circle under F_μ, growing from 0, attracting (resp. repelling) if it appears on the side where the fixed point is repelling (resp. attracting). The dynamic on the invariant circle depends on μ and is related to the rotation number of the diffeomorphism of the circle (the restriction of F_μ to the invariant circle) [3], [4].

The object of our work is now to consider a two-parameter family of maps F_μ, i.e., $\mu=(\mu_1,\mu_2)\in \mathbb{R}^2$. We assume that for $\mu=0$ the dimension of the invariant subspace for T_0, belonging to the eigenvalues on the unit circle, is $\geqslant 2$, and we are interested in studying the dynamical behavior of F_μ for μ close to 0. A problem of this type was studied by A.Chenciner[5]. There are a lot of different cases of interest, but we shall concentrate on two of them. The first reduces to two dimensions and is related to a now classical work on vector fields, the second reduces to three dimensions and shows us possibility of new types of chaotic regimes.

II. FIRST EXAMPLE

We assume in what follows that T_o has a double eigenvalue 1 which is not semi-simple, the remaining of its spectrum being strictly inside the unit disc. A first reduction leads to a map on a two-dimensional center manifold[3] containing 0 and tangent to the invariant subspace belonging to the eigenvalue 1 for $\mu=(0,0)$. The manifold depends smoothly on μ. The second work is to put the map into a normal form[6] by choosing an appropriate system of coordinates. Here, a suitable choice leads to the map $(x,y)\longmapsto(X,Y)$ with

$$(1)\qquad \begin{cases} X = x+y+\mu_1 x + b x^2+O(|\mu|+\|x\|)^3 \\ Y = \mu_2 + y + x^2 + O(|\mu|+\|x\|)^3 \end{cases}$$

where $|\mu|=|\mu_1|+|\mu_2|$ and $\|x\|=|x|+|y|$.

We specify our study in a part of the parameter plane (μ_1,μ_2) described by a new scaling :

$$(2)\qquad x=\tau^2\bar{x}\ ,\quad y=\tau^3\bar{y}\ ,\quad \mu_1=\tau^2\bar{\mu}\ ,\quad \mu_2=-\tau^4$$

i.e., $\mu_2<0$ and μ_1 of order $\sqrt{-\mu_2}$. The map F_μ is now written $\bar{F}_{\tau\bar{\mu}}:(\bar{x},\bar{y})\longmapsto(\bar{X},\bar{Y})$ with

$$(3)\qquad \begin{cases} \bar{X} = \bar{x}+\tau\bar{y}+\tau^2(\bar{\mu}\bar{x}+b\bar{x}^2)\ O(\tau^3), \\ \bar{Y} = \bar{y}+\tau(\bar{x}^2-1)+O(\tau^3). \end{cases}$$

We can introduce the auxiliary differential equation

$$(4)\qquad \begin{cases} \dfrac{dx}{dt} = y+\tau[(b-\tfrac{1}{2})x^2+\tfrac{1}{2}+\bar{\mu}x] \\ \dfrac{dx}{dt} = x^2-1-\tau x y \end{cases}$$

which is such that if the "time τ map" is noted $\Phi_{\tau\bar{\mu}}$, then we have

$$(5)\qquad \bar{F}_{\tau\bar{\mu}}=\Phi_{\tau\bar{\mu}}+O(\tau^3).$$

The interest of the system (4) is that it is an order τ perturbation of the Hamiltonian vector field $x+\frac{y^2}{2}-\frac{x^3}{3}$.

On the auxiliary system (4) we can see easily the following facts : there is a Hopf bifurcation to a periodic solution starting at $\bar{\mu}_C=2(b-1)$, this cycle grows while $\bar{\mu}$ varies from $\bar{\mu}_C$ to $\bar{\mu}_h$ which corresponds to the occurrence of an homoclinic orbit, starting and ending at the hyperbolic fixed point close to (1,0). For $\bar{\mu}$ at the exterior of the interval $(\bar{\mu}_C,\bar{\mu}_h)$ there is no closed orbit and the dynamic is dominated by the behavior at the neighborhood of the two fixed points, the focus and the saddle. The study of phase portraits was done by Bodganov[7], see also Arnold[6].

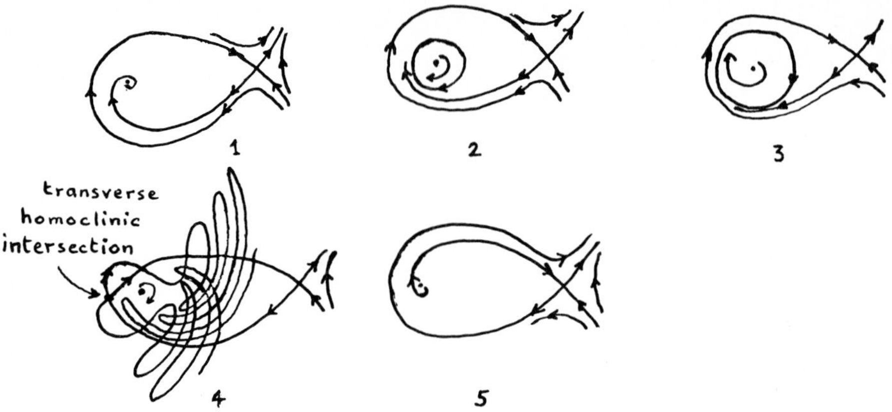

Fig.1 Phase portraits for the map (3), $\bar{\mu}$ varying

Now, what can be proved on the map (3)? In fact, the persistence of a Hopf bifurcation for μ close to $\bar{\mu}_C$ for the map can be proved even though the pair of eigenvalues crossing the unit circle is very close to the strong resonance 1 [8]. The persistence of growing invariant circles can also be proved, using the fact that the circles occurring for the vector field(3) are hyperbolic (i.e.,attracting or repelling), and that the strongness of the perturbation (5) of $\Phi_{\tau\bar{\mu}}$ is not sufficient to destroy the existence of such an invariant circle[8].
The previous proof does not work when we get close to $\bar{\mu}_h$ and it would be interesting to study up to what point the invariant circle exists, in the neighborhood of $\bar{\mu}_h$. Now, for $\mu=\bar{\mu}_h$ we know that for the map $\Phi_{\tau\bar{\mu}}$ we have a homoclinic orbit, and we perturb this map by higher order terms to obtain the real map $F_{\tau\bar{\mu}}$.
What happens here is that we have a transverse intersection of the stable and unstable manifolds of the saddle point (close to (1,0)), for $\bar{\mu}$ in a little interval close to $\bar{\mu}_h$ [9]. This situation is known to create a chaotic behavior of horse-shoe type[10].

III. SECOND EXAMPLE

Let us assume in what follows that T_0 has a simple eigenvalue in -1 and a pair of simple eigenvalues $\lambda_0, \bar{\lambda}_0$ on the unit circle such that $\lambda_0^n \neq 1$ for n=1,2,3,4,5,6 and $\lambda_0^n \neq -1$ for n=4 and 5. The remaining of the spectrum of T_0 lies inside the unit disc. A first reduction leads to a map on a three-dimensional center manifold[3], depending smoothly on μ. Now we make a nonlinear change of coordinates to find a normal form in $\mathbb{R}\times\mathbb{C}$ for F_μ: $(x,z) \longmapsto (X,Z)$ with

$$(6)\qquad \begin{cases} X = -x - \mu_1 x + a_1 x^3 + a_2 x|z|^2 + O(|x|+|z|)^5, \\ Z = \lambda_0[z + (\mu_2 + i\nu)z + b_1 x^2 z + b_2 z|z|^2] + O(|x|+|z|)^5. \end{cases}$$

Let us put (6) in polar coordinates : $z = y e^{i\theta}$, then the new form of F_μ is :

$(x,y,\theta) \longmapsto (X,Y,\Theta)$ in $\mathbb{R}^2 \times T^1$:

$$(7) \quad \begin{cases} X = -x - \mu_1 x + a_1 x^3 + a_2 x y^2 + \text{h.o.t.} \\ Y = y + \mu_2 y + \beta_1 x^2 y + \beta_2 y^3 + \text{h.o.t} \\ \Theta = \theta + \arg \lambda_\circ + \nu + \gamma_1 x^2 + \gamma_2 y^2 + \text{h.o.t.} \end{cases}$$

where h.o.t. means "higher order terms" whose exact form has no interest here except that at order 5 in X and Y we have no θ, thanks to the nonresonant conditions on $\lambda_\circ$. The coupling between θ and x and y, in X and Y comes at order of at least 6 in (x,y).

III.1 Occurence of chaos of the previous type

We specify our study in a part of the parameter plane (μ_1, μ_2) by the choice of new scales :

$$(8) \qquad x = \tau \bar{x} \ , \ y = \tau \bar{y} \ , \ \mu_1 = \tau^2 \bar{\mu}_1 \ , \ \mu_2 = \tau^2 \bar{\mu} ,$$

where $\bar{\mu}_1 = \pm 1$. The map is now written $\bar{F}_{\tau\bar{\mu}}$ ($\bar{\mu}_1 = +1$ in the following), and we have

$$(9) \quad \begin{cases} \bar{X} = -\bar{x} + \tau^2 [-\bar{x} + a_1 \bar{x}^3 + a_2 \bar{x} \bar{y}^2] + O(\tau^4) \\ \bar{Y} = \bar{y} + \tau^2 [\bar{\mu} \bar{y} + \beta_1 \bar{x}^2 \bar{y} + \beta_2 \bar{y}^3] + O(\tau^4) \\ \Theta = \theta + \arg \lambda_\circ + \nu + \tau^2 (\gamma_1 \bar{x}^2 + \gamma_2 \bar{y}^2) + O(\tau^4) \end{cases} ,$$

where $\bar{\mu}$ is of order 1 now and τ is small. The idea is to use the same type of argument as in example 1. For this aim, consider the iterated map $\bar{F}^2_{\tau\bar{\mu}}$. It can also be put into a form looking like (5).

Let us note $\bar{F}^2_{\tau\bar{\mu}}(\bar{x},\bar{y},\theta) = (\bar{X}^{(2)}, \bar{Y}^{(2)}, \Theta^{(2)})$, then we may write

$$(10) \quad \begin{cases} (\bar{X}^{(2)}, \bar{Y}^{(2)}) = \Phi_{\tau\bar{\mu}} (\bar{x},\bar{y}) + O(\tau^6) \\ \Theta^{(2)} = \theta + 2 \arg \lambda_\circ + 2\nu + 2\tau^2 (\gamma_1 \bar{x}^2 + \gamma_2 \bar{y}^2) + O(\tau^4) , \end{cases}$$

where $\Phi_{\tau\mu}$ is the "time-$2\tau^2$ map" of a differential equation

$$(11) \quad \begin{cases} \dfrac{dx}{dt} = x - a_1 x^3 - a_2 x y^2 + O(\tau^2) \\ \dfrac{dy}{dt} = \bar{\mu} y + \beta_1 x^2 y + \beta_2 y^3 + O(\tau^2) , \end{cases}$$

where the $O(\tau^2)$ terms can be computed, knowing the $O(\tau^4)$ terms in $\bar{X}$ and $\bar{Y}$ in (9). By change of scales we can assume that a_1 and β_2 are ± 1. The interesting cases are when they both have the same sign. So we consider in what follows the case $a_1 = \beta_2 = +1$.

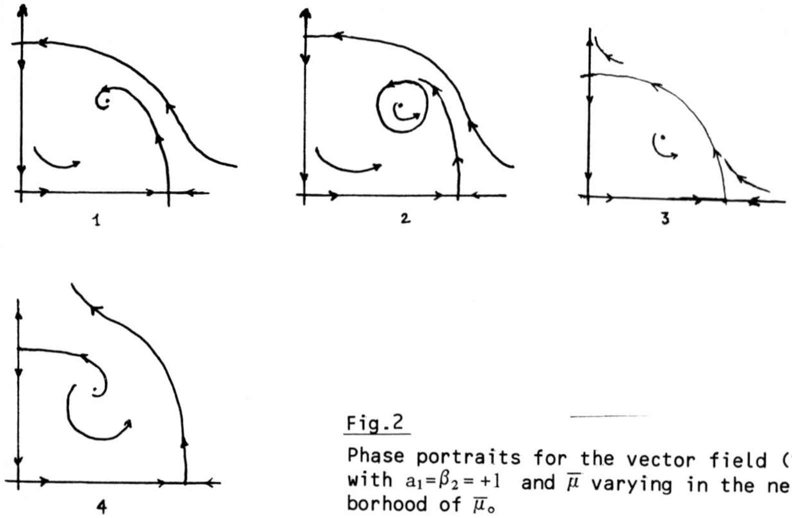

Fig.2

Phase portraits for the vector field (11), with $a_1=\beta_2=+1$ and $\bar{\mu}$ varying in the neighborhood of $\bar{\mu}_o$

The interest of (11) is that if we suppress the $O(\tau^2)$ terms, then it is integrable for a value $\bar{\mu}=\bar{\mu}_o$ with the following first integral :

(12) $$H = \frac{x^p y^q}{q}\left(1-x^2+\frac{1}{\bar{\mu}_o}y^2\right),$$

with $\bar{\mu}_o = -\frac{1+\beta_1}{1+a_2}$, $p = -2\frac{1+\beta_1}{1-a_2\beta_1}$, $q = -2\frac{1+a_2}{1-a_2\beta_1}$.

One of the most interesting cases occurs when $a_2\beta_1>1$, $\beta_1>0$, $a_2>0$.
On the differential equation (11) there is a value $\bar{\mu}_c$ corresponding to a Hopf bifurcation : $\bar{\mu}_c = \bar{\mu}_o+\tau^2\mu_c'$. The 4 symmetric closed orbits grow while $\bar{\mu}$ varies. Then there is a value $\bar{\mu}_h$ where a heteroclinisation occurs (see figure 2, next page). In fact, the study of the phase portraits of system (11) is yet an open problem, and there are only reasonable conjectures on the number of possible closed orbits between $\bar{\mu}_c$ and $\bar{\mu}_h$. This probably depends on the other parameters (those which appear in order τ^4 terms in (9)).

Now, for the map (9), what can be proved here, while $\bar{\mu}$ lies in a neighborhood of $\bar{\mu}_o$ of order τ^2? In fact a closed orbit for (11) would correspond to the existence of an invariant T^2 torus for the map (10), so two T^2 tori would be exchanged by the map (9) (which correspond to the 4 closed orbits symmetric with respect to x and y axis). Such an existence of T^2 tori invariant under $\bar{F}^2_{\tau\bar{\mu}}$, exchanged under $\bar{F}_{\tau\bar{\mu}}$, can be proved, provided we are not too close to the bifurcation point $\bar{\mu}_c$, and not too close to the heteroclinisation point $\bar{\mu}_h$. The way of proving it is very similar to the one used in the first example[8].

While $\overline{\mu}$ approaches $\overline{\mu}_h$, what happens is not quite clear, and knowing what it would be with only two dimensions (see the first example), it is hard to think about the same situation coupled with the angular variable ! i.e., the perturbation created by coupling terms depending on θ on the Cartesian product of the horse-shoe dynamic with the dynamic of a diffeomorphism of the circle. Nevertheless, let us say that this type of chaotic behavior is closely related to the chaotic dynamic shown in the first example.

III.2 New types of chaos

a) Now, at the exterior of a little neighborhood of $\overline{\mu}_c$ we know the existence of an invariant couple of circles exchanged under the map $\overline{F}_{\tau\overline{\mu}}$ (corresponding to the 4 foci of system (11) leading to Hopf bifurcation). Moreover we know the existence of the two T^2 tori exchanged under $\overline{F}_{\tau\overline{\mu}}$ (corresponding to the 4 closed orbits of (11) resulting from Hopf bifurcation). In fact it cannot be true that there is a bifurcation from the two invariant circles to the two invariant tori for $\overline{F}^2_{\tau\overline{\mu}}$, because this bifurcation is nongeneric and would require too many conditions to be realised[11]. The neighborhood of $\overline{\mu}_c$ where we know nothing is very small but it exists and it is expected that a very precise numerical experience should show a chaotic behavior here.

b) There is in fact a simpler situation here which corresponds to a chaotic behavior. When μ_2 changes its sign in (7), it is not hard to prove that for F_μ the fixed point at the origin bifurcates into an invariant circle, close to the circle $x = 0$, $y^2 = -\mu_2/\beta_2$.
Now, if we forget the dependency in θ, and reduce (7) to a map in two dimensions $(x,y) \longmapsto (X,Y)$, we have a pair of fixed points on the y axis. This pair bifurcates into two pairs of periodic points of period 2, as can be easily computed in looking at the fixed points of $\tilde{F}^2_\mu$ (the reduced F_μ in (x,y) plane). This occurs along a curve in μ_1, μ_2 plane close to the line $a_2\mu_2+\beta_2\mu_1 = 0$.

Now taking account of the coupling in θ, it is no longer possible to prove that we have a true bifurcation from an invariant circle into an invariant pair of circles for F_μ, even though we can prove the existence of both objects outside a neighborhood of the line $a_2\mu_2+\beta_2\mu_1 = 0$. The reason for this shortcoming has the same source as in **a)**. It is due to the lack of normal hyperbolicity of the invariant circle for the truncated map, close to this line. The idea of proof to show the persistence (for instance) of such a circle is based on this hyperbolicity, even a very weak one (Ruelle-Takens method[12]). This problem is, in its spirit, very similar to the one treated by A.Chenciner in[5], so we can hope in the future for some progress on the chaos produced here.

References

[1] Pomeau,Y., P.Manneville, *Intermittency and the Lorenz model*, Phys.Lett. 75 A, 1, 1979

[2] Giglio,M., S.Musazzi, U.Perini, *Transition to chaos via a well ordered sequence of period doubling bifurcations*, Phys.Rev.Lett. 47, 243, 1981

[3] Iooss,G., *Bifurcation of maps and applications*, North Holland Math.Studies, 36, 1979

[4] Marsden,J.E., M.Mc Cracken, *The Hopf bifurcation and its applications*, Applied Math. Sci. 19, Springer Verlag, 1976

[5] Chenciner,A., *Bifurcations de difféomorphismes de* $\mathbb{R}^2$*, au voisinage d'un point fixe elliptique. Chaotic behavior of deterministic system*, Les Houches Summer School 1981, G.IOOSS, R.HELLEMAN, R.STORA ed., North Holland, (to appear in 1983)

[6] Arnold,V., Chapitres supplémentaires de la Théorie des équations différentielles ordinaires, Mir, Moscou, 1980

[7] Bogdanov,R., Trudy Sem.Petrov, 2, 1976, 23

[8] Iooss,G., *Persistance d'un cercle invariant par une application voisine de "l'application temps* τ*" d'un champ de vecteurs intégrable.* Partie I: *En dehors de la bifurcation de Hopf*, Partie II: *Voisinage d'une bifurcation de Hopf. Exemples d'applications*, C.R.Acad.Sci.Paris, t.296,I, 27-30 et 113-116, 1983

[9] Gambaudo,J.M., *Perturbation de "l'application temps* τ*" d'un champ de vecteurs intégrable de* $\mathbb{R}^2$, C.R.Acad.Sci. Paris,I, (to appear)

[10] Newhouse,S.E., *The creation of non-trivial recurrence in the dynamics of diffeomorphisms. Chaotic behavior of deterministic systems*, Les Houches Summer School 1981, G.IOOSS, R.HELLEMAN, R.STORA ed., North Holland, (to appear in 1983)

[11] Chenciner,A., G.Iooss, *Bifurcations de tores invariants*, Arch. Rat. Mech. Anal. 69, 2, 109-198, and 71, 301-306, 1979

[12] Ruelle,D., F.Takens, *On the nature of turbulence*, Com. Math. Phys. 20, 167-192 and 23, 343-344, 1971.

Bifurcations in Doubly Diffusive Convection

E. Knobloch

Department of Physics, University of California
Berkeley, CA 94720, USA

0. INTRODUCTION

In this lecture I would like to describe some recent ideas and techniques that I believe to be of great potential usefulness in studying a variety of nonlinear phenomena. The emphasis will be on bifurcation phenomena, since the properties of a nonlinear system can rarely be elucidated systematically at parameter values substantially far from their bifurcation values. I would also like to emphasize the usefulness of these techniques in studying systems described by partial differential equations. It is for this reason that I have chosen a specific fluid dynamical system to illustrate the ideas that I shall describe. Apart from my familiarity with doubly diffusive systems, I think it is generally helpful in this field to be as specific as possible.

I. DOUBLY DIFFUSIVE CONVECTION

For the purposes of this lecture I shall include under the term doubly diffusive convection several different phenomena that have one thing in common: in all cases the convection occurs in a layer of fluid heated from below in the presence of a restraint [1]. The restraint can take the form of a stabilizing solute gradient (thermosolutal convection[2]), an imposed magnetic field (magnetoconvection[3]) or rotation. With a simple transformation the problem of convection in binary fluid mixtures driven by the Soret and Dufour effects also reduces to this general form[4]. In each case the destabilizing thermal buoyancy is opposed by a stabilizing effect: negative buoyancy due to the solute content, the Lorentz force from the perturbed magnetic field, or the Coriolis force. Because of this competition, a much richer variety of dynamical behavior can occur at small amplitudes than in ordinary Rayleigh-Bénard convection, and this makes doubly diffusive systems ideal objects for study from a mathematical point of view. My own interest comes primarily from the ubiquity with which these systems arise in astrophysics and geophysics.

The competition between the destabilizing and stabilizing effects manifests itself already in linear theory, where not surprisingly, the motion-free conductive state remains stable for larger values of the Rayleigh number than in ordinary convection. But there are two essential differences. The first makes an appearance already in linear theory, and is the possibility that the conduction state loses stability not to exponentially growing disturbances, but to overstable disturbances. This is an example of a Hopf bifuraction. Physically, the reason for this can be seen by considering a fluid element in a fluid layer heated from below and containing a stabilizing solute gradient. If the element is displaced upwards it rapidly gives up its heat to the ambient medium while maintaining its solute content. This makes it overdense relative to the ambient medium, and the fluid element descends faster than it moved upwards, overshooting on its downward excursion. In this way exponentially growing oscillations result. From this simple discussion, it is readily seen that the effect can occur only because heat diffuses faster than the solute (or the other restraining agents), and then only if the restoring force

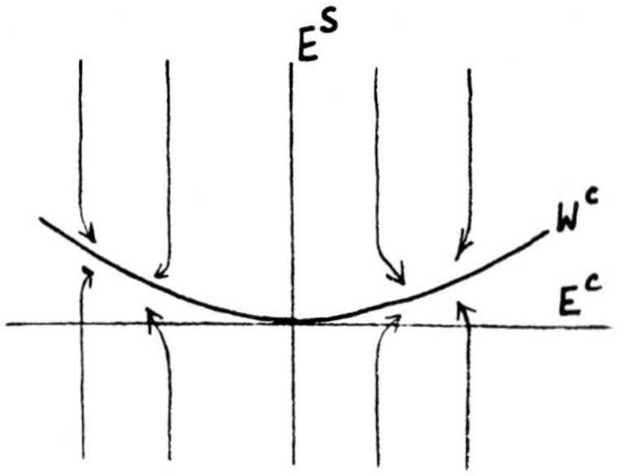

Fig. 1 The center manifold W^c

(e.g.,the stabilizing solute gradient) is sufficiently strong to overcome viscous dissipation. The second difference appears in nonlinear theory. When convection does set in, the motion typically reduces the effectiveness of the restraint. For example, the overturning of the fluid mixes the fluid reducing the stabilizing solute gradient through the bulk of the fluid and confining the stabilizing gradients to boundary layers at the top and bottom. This means that finite amplitude convection can occur at Rayleigh numbers less than the critical value predicted by linear theory. Such convection is called subcritical, and evidently can occur when the restraining agent does not diffuse so fast that it cannot be "expelled" from the body of the fluid. Both steady and oscillatory convection can be subcritical.

In the following I therefore assume that the Prandtl number (Lewis number) characterizing the diffusion of the restraining agent (e.g., the solute) satisfies

$$\tau \equiv \kappa_S/\kappa_T < 1, \tag{1.1}$$

where κ_T is the thermal diffusivity, and that the magnitude of the stabilizing effect is described by a dimensionless number R_S. I next make a crucial assumption, partly for mathematical convenience and partly because it is suggested by convection experiments, and suppose that the convection pattern that is established is spatially periodic. This periodicity may be in the form of two-dimensional rolls, triangles, squares, rectangles or hexagons. In all cases the assumption of spatial periodicity enables us to solve the linear stability problem: it makes the eigenvalues discrete and of finite multiplicity. Under these conditions the center manifold theorem states that at a bifurcation point a reduction of the original (here infinite-dimensional) system to a finite (and generally small) dimensional system is possible. This provides an essential simplification of the analysis of the system. Specifically[5] Theorem: Let Z be a smooth Banach space and let F_t be a C^0 semiflow defined on a neighborhood of $0 \in Z$ for $0 < t < \tau$. Assume $F_t(0) = 0$, and for $t > 0$, $F_t(x)$ is C^{k+1} jointly in t and x. Assume that the spectrum of the linear semigroup $DF_t(0): Z \to Z$ is of the form $\exp[t(\sigma_1 \cup \sigma_2)]$ where σ_1 lies on the imaginary axis and σ_2 lies in the half plane $\mathrm{Re}\, \sigma_2 < -\sigma < 0$. Let E^c be the generalized eigenspace corresponding to the part of the spectrum on the unit circle. Assume $\dim E^c = d < \infty$. Then there exists a neighborhood V of 0 in Z and a C^k submanifold $W^c \subset V$ of dimension d passing through 0 and tangent to E^c at 0 such that (a) if $x \in W^c$, $t > 0$ and $F_t(x) \in V$, then $F_t(x) \in W^c$, (b) if $t > 0$ and $F_t^{\,n}(x)$ remains defined and in V for all $n = 0, 1, 2, \ldots$, then $F_t^{\,n}(x) \to W^c$ as $n \to \infty$.

In Fig. 1 I show a typical situation in which trajectories in the "stable" directions contract on a rapid time scale (obtained from linear theory) towards the plane E^c, but because of nonlinear terms that become important near E^c they settle on the curved surface W^c. Thus after an initial transient, the

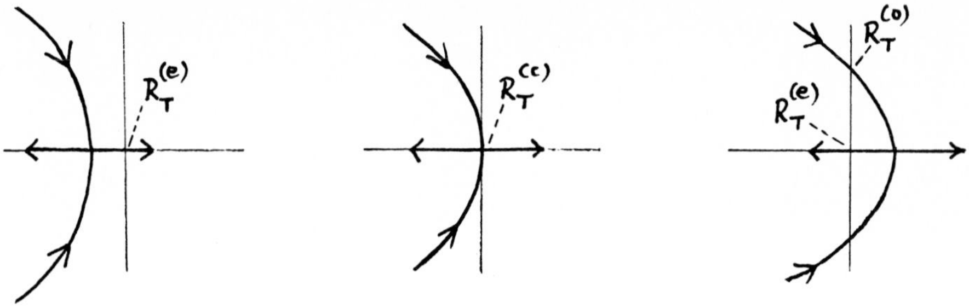

(a) $R_S<R_S^{(c)}$ (b) $R_S=R_S^{(c)}$ (c) $R_S>R_S^{(c)}$

Fig. 2 The eigenvalues in the complex plane as a function of R_T

phase space trajectories are confined to the vicinity of W^c, and asymptotically the dynamics takes place on W^c and on a *long* time scale, since on W^c the evolution is governed by the small nonlinear terms. It is the possibility of a rigorous reduction to a small-dimensional system that makes the study of bifurcations of such great value. I illustrate this with a discussion of the case where convection occurs in the form of parallel rolls, and shall return to the other spatial patterns in § IV.

In all doubly diffusive problems in two dimensions, the pure conduction state can lose stability in one of three ways (Fig. 2).

In Fig. 2(a), valid for $R_S<R_S^{(c)}$, as the thermal Rayleigh number R_T increases a *single* real eigenvalue becomes positive when R_T exceeds $R_T^{(e)}$. The nonlinear terms halt the growth of the instability, and motion equilibrates at a small time-independent amplitude a. Because the spatial periodicity (translation through half a wavelength) implies a symmetry between clockwise and counterclockwise motions, the one-dimensional dynamics *on* the center manifold (i.e., at $R_T^{(e)}$) is of the form

$$\dot{a} \equiv g(a) = -R_2^{(e)} a^3 + 0(a^5), \tag{1.2}$$

where the *nondegeneracy* condition $R_2^{(e)}(R_T^{(e)}, R_S) \neq 0$ is assumed to hold. For R_T *near* $R_T^{(e)}$ one obtains the familiar Landau equation describing the pitchfork bifurcation:

$$\dot{a} = (R_T-R_T^{(e)})\, a - R_2^{(e)} a^3 + 0(a^5,(R_T-R_T^{(e)})\, a^3). \tag{1.3}$$

Note that if $R_2^{(e)} \neq 0$, it may be evaluated at $R_T^{(e)}$: The dynamics described by the truncated form of (1.3) are *structurally stable*, i.e., they are not affected qualitatively by the presence of the error terms. The associated motion will be stable if the pitchfork is supercritical ($R_2^{(e)} > 0$), and unstable if not ($R_2^{(e)} < 0$) as shown in Fig. 3.

Fig. 3(b) is unsatisfactory as a bifurcation diagram: in a physical system the trajectories cannot escape to infinity, and the system equilibrates at a large amplitude, i.e., one expects the subcritical solution to "turn around."

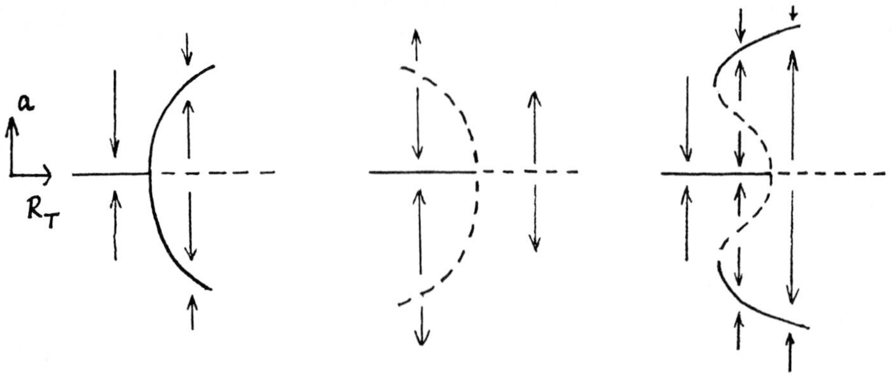

(a) $R_2^{(e)} > 0$ (b) $R_2^{(e)} < 0$ (c) $R_2^{(e)} \lesssim 0,\ R_4^{(e)} > 0$

Fig. 3 Bifurcation diagrams for the pitchfork

To make this behavior accessible to analysis, it must be made to occur at small enough amplitude. One can do this by looking at values of R_S, say, close to *degeneracy:* $R_2^{(e)}(R_{Sc}) = 0$. At $R_T = R_T^{(e)}$, $R_S = R_{Sc}$, the bifurcation is degenerate: the dominant nonlinear term is quintic,

$$\dot{a} = - R_4^{(e)} (R_T^{(e)}, R_{Sc})\, a^5 + O(a^7), \tag{1.4}$$

if $R_4^{(e)} \neq 0$. Near this bifurcation (i.e., R_T, R_S near $R_T^{(e)}$, R_{Sc}), we have

$$\dot{a} = \alpha a + \beta a^3 - R_4^{(e)} a^5, \tag{1.5}$$

where $R_4^{(e)}$ is evaluated at the degenerate bifurcation ($R_4^{(e)} \neq 0$), and α, β are linear combinations of $R_T - R_T^{(e)}$ and $R_S - R_{Sc}$. This is an example of a codimension 2 bifurcation: two parameters α, β are required to unfold it. A typical bifurcation diagram for fixed $R_S - R_{Sc}$ is shown in Fig. 3(c), and captures the essence of the finite amplitude subcritical instability. We speak of $(R_T^{(e)}, R_{Sc})$ as the *organizing center*. Note that the amplitude a is given *naturally* in terms of the distance from the bifurcation by the requirement that the terms on the right of (1.5) are all of the same order.

The above results hold while the Z_2 symmetry of the problem remains unchanged. However, imperfections in the boundaries may allow heat to flow out sideways (no container is infinite!). To study such symmetry-breaking effects in a general framework, one seeks to embed the pitchfork in a structurally stable family of "nearby" vector fields $g(a, R_T; \underline{\alpha})$ that do not have the Z_2 symmetry. Such an embedding is called a *universal unfolding,* and for the pitchfork is

$$\dot{a} = \alpha + (R_T - R_T^{(e)})a + \beta a^2 - R_2^{(e)} a^3 . \tag{1.6}$$

The breakup of the pitchfork for $R_2^{(e)} > 0$ is shown in Fig. 4 as a function of the unfolding parameters α, β.

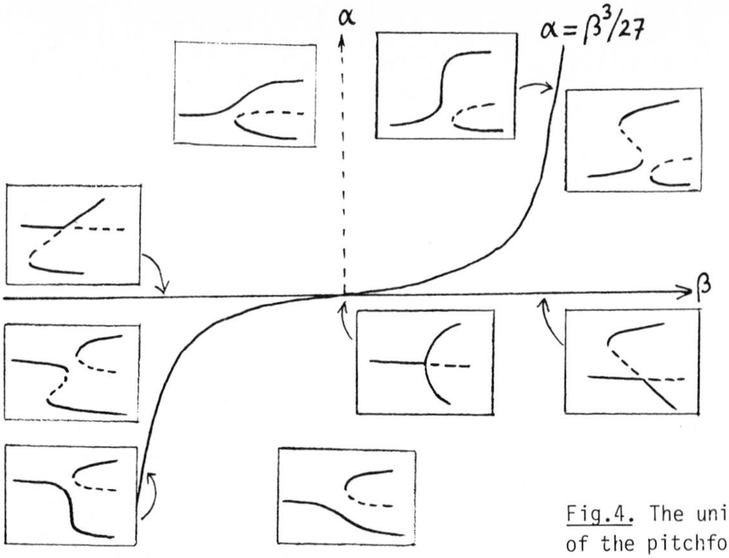

Fig.4. The universal unfolding of the pitchfork

The cases shown in Fig. 4 are *all* the possible ways in which a small symmetry-breaking effect (as measured by the parameters α, β) can destroy the pitchfork. We see that some ways are more likely than others. In fact unless one varied α and β together in a very special way, the thin "wedges" in the first and third quadrant would be unobserved. In the light of this general theory, the results of explicit and tedious calculations for Rayleigh-Bénard convection [6, 7] come as no surprise at all.

Returning now to Fig. 2 observe that when R_S exceeds $R_S^{(c)}$ stability is now lost at a Hopf bifurcation that occurs when the dominant pair of complex eigenvalues crosses into the right half-plane. Since a pair of eigenvalues is involved the dynamics on the center manifold is two-dimensional. Near the bifurcation the phase of the oscillations decouples and their amplitude a obeys

$$\dot{a} = (R_T - R_T^{(o)})\, a - R_2^{(o)} a^3, \tag{1.7}$$

provided $R_2^{(o)} \neq 0$ (cf. 1.3). As R_T increases further the unstable pair of eigenvalues in Fig. 2(c) coalesces on the positive real axis (when $R_T = R_T(i)$), and turns into two real eigenvalues, one of which continues to increase, but the other decreases and crosses back into the left half plane at $R_T^{(e)}$. Thus as R_T increases there are two bifurcations off the trivial solution, the first producing finite amplitude oscillations, and the second finite amplitude steady convection. Two immediate questions suggest themselves: how do these solution branches interact at finite amplitude, and how do oscillations disappear, since at large enough R_T convection is overturning.

The key to answering questions of this type, and indeed the key to the dynamics of these problems is provided by the case shown in Fig. 2b. This occurs when $R_S = R_S^{(c)}$, and there are then two zero eigenvalues. One's immediate reaction is to doubt the usefulness of studying the problem at one special parameter value but I will show that because this parameter value divides the two possibilities shown in Fig. 2(a,c), it plays a very special role, and the dynamics for parameter values nearby are not only accessible analytically, but also capture the essential aspects of the dynamics for other parameter values. The idea I wish to emphasize is that degenerate (i.e., multiple) bifurcations are the things to study, and the more degenerate the better.

I will now illustrate this with a specific example.

II. TWO-DIMENSIONAL THERMOSOLUTAL CONVECTION

The equations describing convection in a horizontal layer of thickness d in the presence of a dissolved solute of concentration S are, in the Boussinesq approximation,

$$\rho_0(\partial_t \underline{u} + \underline{u} \cdot \nabla \underline{u}) = -\nabla(p-p_0) + \underline{g}(\rho - \rho_0) + \rho\, \nu \nabla^2 \underline{u} \tag{2.1}$$

$$\partial_t T + \underline{u} \cdot \nabla T = \kappa_T \nabla^2 T \tag{2.2}$$

$$\partial_t S + \underline{u} \cdot \nabla S = \kappa_S \nabla^2 S \tag{2.3}$$

$$\nabla \cdot \underline{u} = 0 \tag{2.4}$$

together with the Boussinesq equation of state

$$\rho = \rho_0(1-\alpha(T-T_0) + \beta(S-S_0)), \tag{2.5}$$

where the subscript zero denotes a reference value. Thus the buoyancy force in (2.1) that drives the motion arises from changes in temperature and solute concentration; these in turn are advected by the motion, and reduced by diffusion. In studying the two-dimensional problem it is convenient to introduce the stream function ψ defined by

$$\underline{u} = (-\partial_z \psi, 0, \partial_x \psi) \tag{2.6}$$

and to nondimensionalize the equations by measuring distances in terms of d and time in terms of the thermal conduction time d^2/κ_T across the thickness d. In the absence of motion the temperature T and solute concentration S are both linear functions of the vertical coordinate z. I denote by θ and Σ the nondimensional departures of T and S from their conduction form. If ΔT and ΔS are the temperature and solute differences imposed across the layer,

$$T-T_0 = \Delta T(1-z+\theta(x,z,t)), \qquad S-S_0 = \Delta S(1-z+\Sigma(x,z,t)). \tag{2.7}$$

Eliminating the pressure from (2.1) by cross-differentiation, one obtains the basic equations:

$$\sigma^{-1}[\partial_t \nabla^2\psi + J(\psi,\nabla^2\psi)] = R_T\partial_x\theta - R_S\partial_x\Sigma + \nabla^4\psi \tag{2.8a}$$

$$\partial_t\theta + J(\psi,\theta) = \partial_x\psi + \nabla^2\theta \tag{2.8b}$$

$$\partial_t\Sigma + J(\psi,\Sigma) = \partial_x\psi + \tau\nabla^2\Sigma\,, \tag{2.8c}$$

where J denotes the Jacobian, and the four dimensionless numbers are

$$R_T = \frac{g\alpha\Delta T d^3}{\kappa_T \nu} . \quad R_S = \frac{g\beta\Delta S d^3}{\kappa_T \nu} . \quad \sigma = \nu/\kappa_T . \quad \tau = \kappa_S/\kappa_T .$$

The boundary conditions that I shall choose for illustration purposes correspond to stress-free perfectly conducting boundaries at the top and bottom, and to periodicity in the horizontal in cells of (nondimensional) half width λ:

$$\psi = \partial_z^2\psi = \theta = \Sigma = 0 \text{ on } z = 0,1 \tag{2.9a}$$

$$\psi = \partial_x^2\psi = \partial_x\theta = \partial_x\Sigma = 0 \text{ on } x = o,\lambda . \tag{2.9b}$$

Note that with these boundary conditions Equations (2.8) are invariant under the translations $x \to x + n\lambda$, with n an integer. A consequence of this symmetry is that the pure conduction solution $\psi = \theta = \Sigma = 0$ is a solution for all values of the parameters. We shall see that this symmetry is responsible for the character of the nonlinear solutions as well.

With the boundary conditions chosen the eigenfunctions of the linearized problem are trigonometric functions. It is convenient therefore to expand ψ, θ and Σ in these functions:

$$\psi = \frac{2}{\pi}(2p)^{1/2}\lambda \sin \pi x/\lambda \sin \pi z \; a(pt) + \ldots \tag{2.10a}$$

$$\theta = 2\left(\frac{2}{p}\right)^{1/2} \cos \pi x/\lambda \sin \pi z \; b(pt) - \frac{1}{\pi} \sin 2\pi z \; c(pt) + \ldots \tag{2.10b}$$

$$\Sigma = 2\left(\frac{2}{p}\right)^{1/2} \cos \pi x/\lambda \sin \pi z \; d(pt) - \frac{1}{\pi} \sin 2\pi z \; e(pt) + \ldots . \tag{2.10c}$$

Here the first terms are the linear eigenfunctions, and the second terms arise at second order in an amplitude expansion. They represent the concentration of the temperature and solute gradients into boundary layers at the top and bottom of the cell. Substitution of (2.10) into (2.8) yields the following equations for the modal amplitudes[8]

$$a' = \sigma[-a + r_T b - r_S d] + \ldots \tag{2.11a}$$

$$b' = -b + a(1-c) + \ldots \tag{2.11b}$$

$$c' = \varpi[-c + ab] + \ldots \tag{2.11c}$$

$$d' = -\tau d + a(1-e) + \ldots \tag{2.11d}$$

$$e' = \varpi[-\tau e + ad] + \ldots . \tag{2.11e}$$

Here $\varpi = 4\pi^2/p$, $(r_T, r_S) = (\pi^2/\lambda^2 p^3)(R_T, R_S)$ and $p = \pi^2(1+1/\lambda^2)$. Observe that the symmetry $x \to x+n\lambda$ of the partial differential equations induces the symmetry $(a,b,d) \to (-a, -b, -d)$. Equations (2.11) are sufficient to describe the local behavior near both bifurcation points: the terms omitted do not contribute.

Following the idea of the preceding section, I focus on the codimension two bifurcation of Fig. 2(b). Unlike ordinary perturbation calculations, the calculation I shall describe does not select a distinguished parameter as a function of which the solutions are studied. It is a two-parameter calculation, and the results of any one-parameter calculation can be found by following the appropriate line through parameter space. The analysis proceeds in four stages that are routine to implement (cf. [9]):

(1) the linear part of the equations is transformed into Jordan normal form,
(2) the center manifold is calculated up to terms of the required degree,
(3) coordinate changes are found that put the equations into normal (i.e., simplest form) on the center manifold,
(4) the dynamical analysis of the normal form fixes the degree which is necessary, so that steps (2) and (3) would need to be repeated to higher degree if the initial calculations did not yield a system for which all of the desired dynamical information is insensitive to the addition of higher degree terms.

For step (3) there is a general theory which tells us which terms can be removed by such coordinate changes, and which cannot [e.g.,10]. It turns out that this depends only on three pieces of information, all of which are known a priori:

(1) the nature and multiplicity of the eigenvalues that are crossing the imaginary axis, and specifically the form of the associated Jordan block. In the present problem this is

$$\begin{pmatrix} 0 & 1 \\ 0 & 0 \end{pmatrix}. \tag{2.12}$$

(2) Whether the trivial solution is a solution for all parameter values, or not. In the present problem it is.

(3) Any symmetry properties of the system. In the present problem there is an induced Z_2 symmetry.

The theory shows that in this case the dynamics near the bifurcation have to be of the form [11]

$$\begin{aligned} q' &= p \\ p' &= \alpha q + \beta p + Aq^3 + Bq^2 p , \end{aligned} \tag{2.13}$$

where α and β are unfolding parameters (they vanish at the bifurcation), and A and B are coefficients that have to be explicitly computed. The resulting dynamics depends in an essential way on whether $A > 0$ or $A < 0$, but the theory allows a classification of the possible dynamics and an investigation of their structural stability even in the absence of a calculation of the coefficients.

The linearization of (2.11) about the trivial solution a=b=c=d=e given by

$$L = \begin{pmatrix} -\sigma & \sigma r_T & 0 & -\sigma r_S & 0 \\ 1 & -1 & 0 & 0 & 0 \\ 0 & 0 & -\varpi & 0 & 0 \\ 1 & 0 & 0 & -\tau & 0 \\ 0 & 0 & 0 & 0 & -\tau\varpi \end{pmatrix} \tag{2.14}$$

has two zero eigenvalues when

$$r_T = r_T^{(c)} \equiv \frac{\sigma + \tau}{\sigma(1-\tau)}, \quad r_S = r_S^{(c)} \equiv \frac{\tau^2(1+\sigma)}{\sigma(1-\tau)}. \tag{2.15}$$

For these parameter values L is brought into Jordan form by the matrix

$$P = \begin{pmatrix} 1 & 0 & 0 & 1 & 0 \\ 1 & 1 & 0 & -(\sigma+\tau)^{-1} & 0 \\ 0 & 0 & 1 & 0 & 0 \\ \tau^{-1} & \tau^{-2} & 0 & -(1+\sigma)^{-1} & 0 \\ 0 & 0 & 0 & 0 & 1 \end{pmatrix} \tag{2.16}$$

so that

$$P^{-1}LP = \begin{pmatrix} 0 & -1 & 0 & 0 & 0 \\ 0 & 0 & 0 & 0 & 0 \\ 0 & 0 & \varpi & 0 & 0 \\ 0 & 0 & 0 & -\Delta & 0 \\ 0 & 0 & 0 & 0 & -\varpi\tau \end{pmatrix} \qquad \Delta \equiv 1+\sigma+\tau. \tag{2.17}$$

In terms of new coordinates (u, v, w) defined by $(u,v,c,w,e)^T = P^{-1}(a,b,c,d,e)^T$ the system (2.11) becomes

$$\begin{pmatrix} u \\ v \\ c \\ w \\ e \end{pmatrix}' = \begin{pmatrix} -v \\ 0 \\ -\varpi c \\ -\Delta w \\ -\varpi\tau e \end{pmatrix} + \frac{u+w}{D} \begin{pmatrix} -\frac{c}{\tau^2} + e \\ \frac{\Delta c}{\tau(1+\sigma)} - \frac{\Delta e}{\sigma+\tau} \\ D\varpi(u+v - \frac{w}{\sigma+\tau}) \\ \frac{c}{\tau^2} - e \\ D\varpi(\frac{u}{\tau} + \frac{v}{\tau^2} - \frac{w}{1+\sigma}) \end{pmatrix} \tag{2.18}$$

where $D \equiv \det P = \Delta^2(1-\tau)/\tau^2(1+\sigma)(\sigma+\tau)$. (2.19)

To reduce the system (2.18) to a two-dimensional system on the center manifold, one has to eliminate the damped modes (c,w,e). These modes contract rapidly towards the center manifold, and one has to determine its form to sufficient accuracy. Since the center manifold is tangent to the (u,v) plane, we introduce a near-identity change of coordinates

$$x = c + Q_1(u,v), \qquad y = e + Q_2(u,v) \tag{2.20}$$

where

$$Q_i(u,v) = \alpha_i u^2 + \beta_i uv + \gamma_i v^2, \qquad i = 1,2. \tag{2.21}$$

The coefficients are chosen in such a way that the expressions for x',w',y' contain no quadratic terms in u,v. Then we may set x=w=y=0 as an approximation to the center manifold. The dynamics on the center manifold is therefore of the form

$$\begin{aligned} u' &= -v + A_1u^3 + B_1u^2v + C_1uv^2 + O(5) \\ v' &= A_2u^3 + B_2u^2v + C_2uv^2 + O(5)\,. \end{aligned} \tag{2.22}$$

These equations can be put by a further coordinate transformation of the form

$$q = u + P_1(u,v), \qquad p = -v + P_2(u,v)$$
$$\text{where } P_i(u,v) = \alpha_i u^3 + \beta_i u^2 v, \qquad i = 1,2, \tag{2.23}$$

into the normal form

$$\begin{aligned} q' &= p + O(5) \\ p' &= Aq^3 + Bq^2 p + O(5), \end{aligned} \tag{2.24}$$

with $A = -A_2$ and $B = 3A_1 + B_2$.

These equations are valid at the multiple bifurcation. The next step includes the contribution of small departures of the parameters (r_T, r_S) from $(r_T^{(c)}, r_S^{(c)})$. Writing $(\tilde{r}_T, \tilde{r}_S)$ for $r_T - r_T^{(c)}, r_S - r_S^{(c)}$, one readily sees that in this case a linear contribution arising from the matrix

$$L = \begin{pmatrix} 0 & \sigma\tilde{r}_T & 0 & -\sigma\tilde{r}_S & 0 \\ 0 & 0 & 0 & 0 & 0 \\ 0 & 0 & 0 & 0 & 0 \\ 0 & 0 & 0 & 0 & 0 \\ 0 & 0 & 0 & 0 & 0 \end{pmatrix} \tag{2.25}$$

has to be added to the normal form. In the normal form this contributes P^{-1} LP, with the result

$$\begin{pmatrix} q' \\ p' \end{pmatrix} = R\begin{pmatrix} q \\ p \end{pmatrix} + \begin{pmatrix} 0 \\ Aq^3 + Bq^2 p \end{pmatrix} + O(|u|^5, \tilde{r}_T|u|^3, \dots), \tag{2.26}$$

where

$$R = \begin{bmatrix} (1+\sigma+\tau\Delta)(\sigma/\Delta^2)(\tilde{r}_T - \tilde{r}_S/\tau) & -1 \\ -\sigma\tau(\tilde{r}_T - \tilde{r}_S/\tau)/\Delta & -\sigma\tau(\tilde{r}_T - \tilde{r}_S/\tau^2)/\Delta \end{bmatrix}. \tag{2.27}$$

A small rotation of the (p,q) variables finally puts (2.23) into Arnol'd's normal form

$$\begin{pmatrix} q' \\ p' \end{pmatrix} = \begin{pmatrix} p \\ \alpha q + \beta p + Aq^3 + Bq^2 p \end{pmatrix} \tag{2.28}$$

$$\begin{aligned} \text{where } \alpha &= -\det R = (\tau\tilde{r}_T - \tilde{r}_S)\sigma/\Delta \\ \beta &= \operatorname{tr} R = \sigma[(1+\sigma)\tilde{r}_T - (\sigma+\tau)\tilde{r}_S]/\Delta^2, \end{aligned} \tag{2.29}$$

and A and B can be evaluated at $(r_T^{(c)}, r_S^{(c)})$ provided the nondegeneracy conditions

$$A \neq 0, \qquad B \neq 0 \tag{2.30}$$

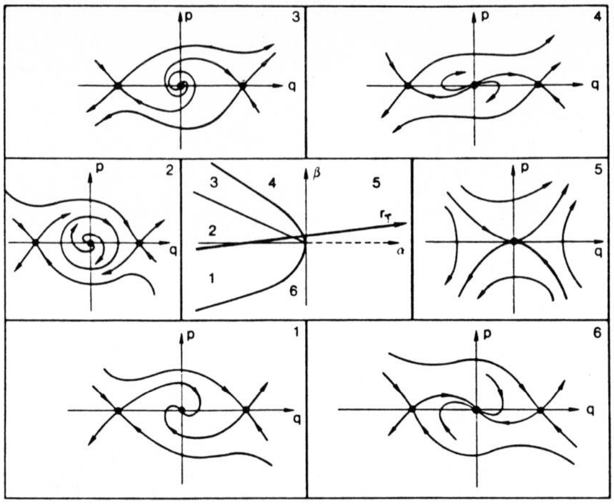

(a)

Fig.5. The dynamics of the normal form (2.28) in (a) the α-β plane, and (b) the amplitude-Rayleigh number diagram

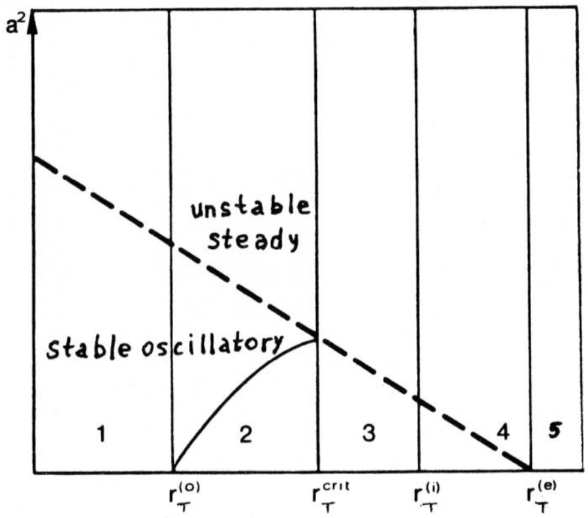

(b)

are satisfied. The result is

$$A = 1, \quad B = -(1 + \frac{2}{\varpi})(1 + \frac{1+\sigma}{\tau\Delta}) . \tag{2.31}$$

The above result was obtained by Knobloch and Proctor [12] using a perturbation method. The dynamics described by the system (2.28) is shown in Fig. 5, both in the α-β plane, and also as the more conventional bifurcation diagram obtaining for fixed $r_S (> r_S^{(c)})$ and increasing r_T. The latter is found by describing the line

$$(1+\sigma)\alpha - \Delta\tau\beta + \sigma(1-\tau)(r_S - r_S^{(c)}) = 0 \tag{2.32}$$

in the α-β diagram (cf. 2.29).

The next stage of the analysis requires a consideration of the structural stability of the phase portraits. Here a celebrated "theorem"[11] states that the phase portraits shown persist for sufficiently small but finite $(\tilde{r}_T, \tilde{r}_S)$. In fact numerical calculations using both the fifth-order model [8] and the partial differential equations [13] show that $(\tilde{r}_T, \tilde{r}_S)$ can be $O(1)$, and the bifurcation analysis still describes qualitatively the dynamics.

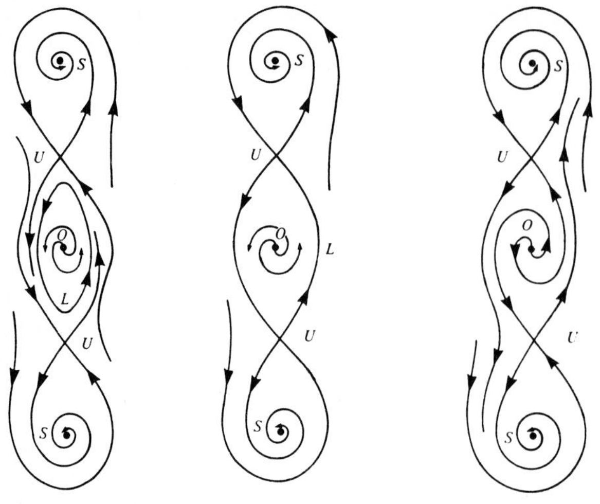

Figure 6. Sketches of the solution trajectories projected onto the a′-a plane for (a) $r_T^{(0)} < r_T < r_T^{(c)}$, (b) $r_T = r_T^{(c)}$ and (c) $r_T^{(c)} < r_T < r_T^{(c)}$. 0 denotes the static solution (0,0) the unstable steady solutions and L the limit cycle. cf. ref. [8]

At this stage one notices that the bifurcation diagram of Fig. 5(b) is subject to the same criticism that was applied to the subcritical pitchfork (Fig. 2(b)). In order to get the branch of steady solutions to "turn around" it is necessary to consider a more degenerate bifurcation: a codimension three bifurcation in which a third parameter λ, say, is chosen such that $A(\lambda_c)=0$, i.e., the steady solutions bifurcate vertically. In this case the normal form cannot be truncated at third degree and will include higher order terms:

$$q' = p, \quad p' = \alpha q+\beta p+\gamma q^3+Bq^2p+Cq^5+Dq^4p+\dots \tag{2.33}$$

Here α,β,γ are unfolding parameters (linear combinations of $\tilde{r}_T, \tilde{r}_S$ and $\tilde{\lambda}\equiv\lambda-\lambda_c$) and $B,C,D \neq 0$. By a rescaling of the variables

$$t \to \bar{t}/\varepsilon^2, \qquad q \to \varepsilon\bar{q}, \tag{2.34}$$

the equations become

$$\bar{q}'' = \bar{\alpha}\,\bar{q} + \bar{\beta}\,\bar{q}' + \bar{\gamma}\,\bar{q}^3 + B\,\bar{q}^2\,\bar{q}' + C\,\bar{q}^5 + O(\varepsilon^2), \tag{2.35}$$

where the prime now denotes the barred time. Thus close enough to the bifurcation where the amplitude is sufficiently small (i.e., $\varepsilon \to 0$), it is sufficient to analyze the dynamics described by

$$q'' = \alpha q + \beta q' + \gamma q^3 + Bq^2q' + Cq^5, \quad B \neq 0, \quad C \neq 0, \tag{2.36}$$

as a function of the three parameters α,β,γ. While this is lengthy [13], when $C < 0$, all the solutions do indeed remain bounded. Note, however, that the calculation of the coefficients B and C now requires the inclusion of additional modes in the system (2.11). One may conclude that near this particular multiple bifurcation a satisfactory description of the dynamics characteristic of doubly diffusive convection has been found. In fact, numerical calculations using both the fifth-order model [8] and the partial differential equations [14,15] show that the departures of r_T, r_S and λ from their critical values can all be $O(1)$ and the dynamics qualitatively unchanged from those described here. Eventually, however, the dynamics ceases to be near two-dimensional, and new phenomena appear. This is briefly described in the next section.

III. APPEARANCE OF CHAOS

An important feature of the phase portraits shown in Fig. 5(a) is the presence of the infinite period heteroclinic orbit connecting the two saddle points. This orbit lies on the line dividing regions 2 and 3 (cf. Fig. 6). Since it is well known that homoclinic and heteroclinic connections are often associated with chaos, it is worthwhile studying the fate of this heteroclinic orbit as the parameters are varied away from the codimension three bifurcation. The fifth-order system (2.11) is very useful in this respect, and for reasons discussed in Ref. 8 it gives qualitatively correct results. Analogous calculations can be done for the partial differential equations, though with greater effort [14,15]. For fixed values of r_S, λ that are not too large, one finds that as r_T increases the period of the oscillations along the oscillatory branch increases and tends to infinity very rapidly [8] as r_T approaches $r_T^{crit}(r_S,\lambda)$, the value of r_T at which the heteroclinic orbit is present (Fig. 7). For larger r_S,λ the saddle points on the branch of unstable steady solutions become saddle foci, the orbit spiralling into the fixed point before escaping. For still larger values of r_S,λ one finds a bifurcation to asymmetrical oscillations, that is followed as r_T increases by a Feigenbaum cascade of period-doubling bifurcations leading to a semiperiodic (aperiodic) regime. In this region semiperiodic bands merge in pairs as r_T increases and periodic windows are present,

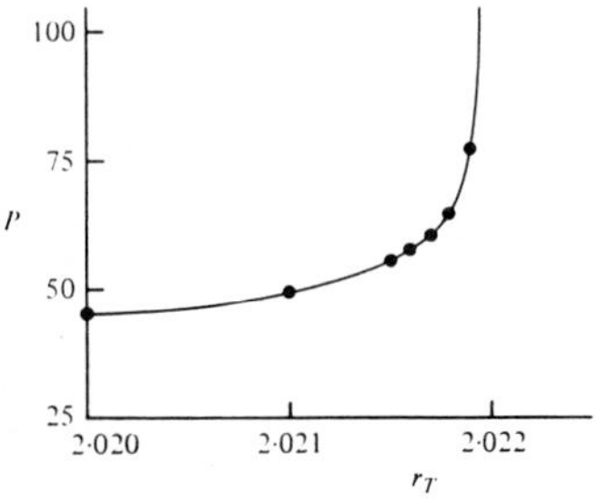

Fig.7. The period P as a function of r_T near r_T^{crit}, for $\sigma = 10$, $\tau = 0.4$, $\varpi = 8/3$ and $r_S = 0.5$

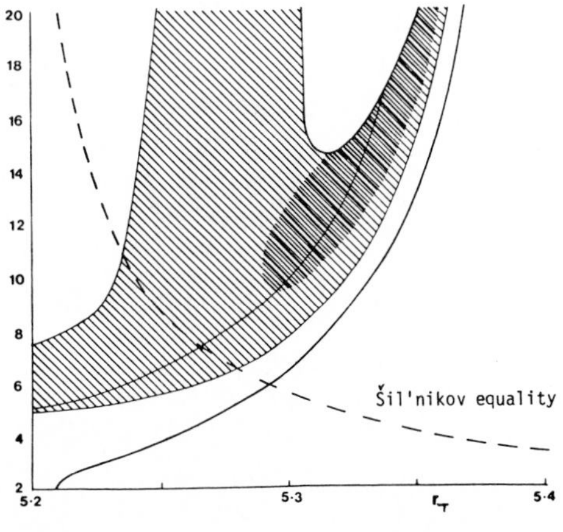

Fig.8. The σ-r_T plane for a model of magnetoconvection [17]. The hatched region indicates bifurcation to asymmetry followed by period-doubling. Solid lines denote heteroclinic or homoclinic orbits

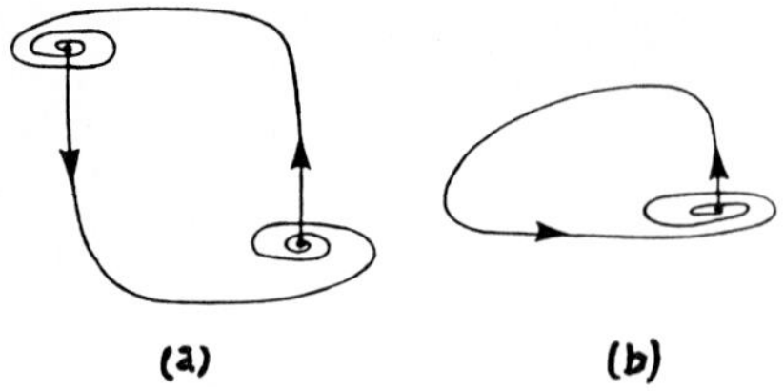

Fig. 9. A sketch of (a) a heteroclinic orbit, and (b) a homoclinic orbit

but as r_T increases further the whole sequence of bifurcations occurs in reverse with the original symmetric orbit becoming stable again, thereby forming a complete bifurcation bubble [16,17]. Typically this is followed by a hysteresis bifurcation to another branch of stable oscillations on which a similar bifurcation bubble is found. The bubbles may burst in the middle: no stable oscillatory solutions may exist for certain parameter intervals [17]. Such behavior appears to be associated with the appearance of secondary heteroclinic and homoclinic orbits [17]. A typical set of results obtained in a detailed study of a fifth-order model of magnetoconvection [17] is shown in Figure 8.

The key to the understanding of the complicated story just described is provided by the eigenvalues S_i (i=1,2...) at the saddle points when $r_T = r_T^{crit}(r_S,\lambda)$ Near the codimension three bifurcation the eigenvalues may be ordered as follows: $S_1 = O(1) > 0$, $S_2 = O(\varepsilon) < 0$, and $S_3,\ldots = O(1) < 0$, where r_T, r_S, $\lambda = O(\varepsilon^2)$. As ε increases, S_2 and S_3 become equal and thereafter become a complex conjugate pair with $\mathrm{Re}\,S < 0$ and $|\mathrm{Re}\,S| > S_1$. The heteroclinic orbit now connects two saddle foci and is approximately three-dimensional (Fig. 9). As ε increases further S_1 may increase faster than $\mathrm{Re}\,S$ decreases so that the inequality

$$S_1 > |\mathrm{Re}S| > 0, \qquad \mathrm{Im}S \neq 0 \tag{3.1}$$

becomes satisfied at $r_T^{crit}(r_S)$. For this case Šil'nikov [18] proved that the return map associated with the heteroclinic orbit will contain a countable number of horseshoes. For the case of a homoclinic orbit γ this is illustrated in Fig. 10. The map T_1 is described entirely by the linearization near the saddle P,

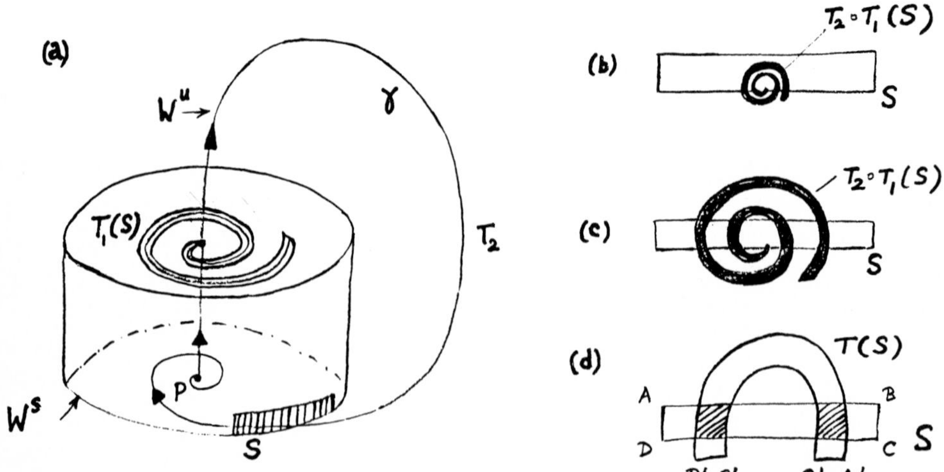

Fig. 10. The Šil'nikov mechanism

and maps a strip S on one face of a small 'cube' into a logarithmic spiral $T_1(S)$ on the top face that is perpendicular to the unstable direction. Since one end of the spiral lies on the unstable manifold and the rest is close to it, the spiral gets carried along γ, and because the orbit is homoclinic it is laid across the original strip S. If this mapping is T_2, we see that $S \cap T_2T_1(S)$ can have one of two forms (Fig. 10). If $S_1 < |\mathrm{Re}S|$ the spiral $T_1(S)$ that is formed is too small and T_2 maps it into S; successive applications of T_2T_1 then show that the return map T_2T_1 has only one fixed point, corresponding to the original orbit γ. If, however, the Šil'nikov inequality (3.1) is satisfied the situation depicted in Fig. 10(c) occurs and since the spiral gets stretched to infinite length, a countable number of horseshoe intersections will be present in the return map. The dynamics of a horseshoe return map (Fig. 10(d)) under forward and backward iterations can be analyzed in great detail. The analysis shows that the dynamical system contains an uncountable number of bounded aperiodic orbits, as well as all periodic orbits of arbitrarily high periods. These orbits are all, however, nonstable since in the neighborhood of each there is a nonrecurrent orbit, and we would not expect to see them except perhaps as transients. But we have to ask where have all these orbits come from? The answer is that they are created in a complicated sequence of bifurcations. Yorke and Alligood [19] showed rigorously that the creation of a horseshoe requires an infinite number of cascades of period-doubling bifurcations. The work of refs. [17,20] suggests strongly that the bifurcation bubbles and multiple oscillatory branches are characteristic of the approach to a heteroclinic or homoclinic orbit. It is likely that there are typically both parameter intervals in which no stable oscillations exist, as well as intervals in which infinitely many stable oscillations coexist. It is this behavior that is to some extent observable numerically and in experiments. It is important to realize, however, that such complex behavior has basically a simple explanation: a homoclinic or a heteroclinic orbit is present in the system with the inequality (3.1) being satisfied.

IV. THE PATTERN SELECTION PROBLEM

In studying convection in three dimensions one faces the question: what is the pattern of convection that the convective instability will evolve to? Even under the simplifying assumption that the resulting pattern is doubly periodic in the plane, a number of solutions is possible. The linear problem predicts only the critical wave number of the instability, but not the direction. Any superposition of rolls with different orientation is a solution, provided their wave vectors lie on the critical circle. Only in the nonlinear regime can their relative stability be studied. In this section I outline the results of two recent studies of this problem [21,22].

For the doubly periodic lattice one can choose the hexagonal lattice. There are then 6 wave vectors that are simultaneously unstable: $\pm \underline{k}_1$, $\pm \underline{k}_2$, $\pm \underline{k}_3$, where $\underline{k}_1 + \underline{k}_2 + \underline{k}_3 = 0$. Four planforms of particular interest fit on the hexagonal lattice: rolls (R), hexagons (H), triangles (T) and rectangles (RA). In restricting the problem to the hexagonal lattice one can still study the competition between these possible planforms in the nonlinear regime.

It is convenient to write quantities of interest in terms of complex amplitudes $z_\alpha(t)$ as follows:

$$f(x_1, x_2, y, t) = \sum_{\alpha=1}^{3} z_\alpha(t)\, e^{i\underline{k}_\alpha \cdot \underline{x}} f(y) . \qquad (4.1)$$

Then three complex modes are simultaneously unstable at $R_T = R_T^{(c)}$, i.e., there are 6 simultaneous zero eigenvalues. Note that this linear problem is defined by the lattice choice, and that the center manifold theorem applies. It follows that the problem can be reduced to three complex evolution equations for the amplitudes $z_\alpha(t)$. The form of these equations depends entirely on the imposed symmetry. The analysis thus applies equally to Rayleigh-Bénard convection, doubly diffusive convection or any other system with the same symmetry. Apart from the symmetry with respect to the hexagonal lattice, one may consider the situation in which the system possesses reflectional symmetry about the mid-place of the layer, or one in which it does not. This symmetry is present in the Boussinesq problem with identical boundary conditions at the top and bottom. The symmetry may be broken by a slight change in the boundary conditions, or by allowing for small non-Boussinesq effects (e.g., a temperature-dependent viscosity). This symmetry breaking may leave the linear problem self-adjoint, or it may make it nonself-adjoint. The purpose of the study [22]is an understanding of the possible stable states in the symmetric case, and the effects of breaking the symmetry in these two ways. The calculations are done using singularity theory. This is a powerful technique for studying steady-state bifurcations, particularly highly degenerate problems in the presence of a symmetry group.

In the case with no reflectional symmetry, two different kinds of hexagons are solutions: hexagons with upward flow in the center and downward flow along the sides (H^+), and the opposite (H^-). The possible steady-state solutions are described by the normal form $g_\alpha(z,\lambda) = 0$ (α=1,2,3), whose first component is

$$z_1(-\lambda + u_1) + \bar{z}_2\bar{z}_3 \;=\; 0, \tag{4.2}$$

where λ is the bifurcation parameter, and $u_\alpha = z_\alpha \bar{z}_\alpha$. The bifurcation diagram of $\sigma \equiv u_1+u_2+u_3$ as a function of λ is shown in Fig. 11(a). Observe that there are no stable solutions near the bifurcation. On the other hand,with the reflectional symmetry, H^+ and H^- are equally possible, and which is realized depends on the initial conditions. The two hexagon solutions are related by the symmetry, and I call them H. The problem is described by the normal form

$$z_1(-\lambda + a\sigma + u_1 + d\sigma^2) + cq\bar{z}_2\bar{z}_3 \;=\; 0, \qquad |c| = 1, \tag{4.3}$$

where $q = \mathrm{Re}\; z_1z_2z_3$, and a and d are coefficients that cannot be scaled away. Depending on the value of a, hexagons (H), regular triangles (RT) and rolls (R) can all be stable. A typical bifurcation diagram is shown in Fig. 11(b). Observe that the bifurcation diagrams are entirely different. In particular there

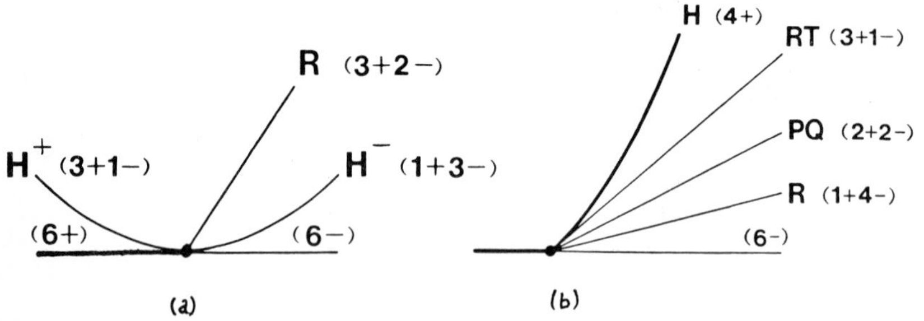

Fig. 11 Bifurcation diagrams for (a) the nonsymmetric problem, and (b) the symmetric problem with $-1/3 < a$, and $c=-1$. The patchwork quilt (PQ) is a regular rectangle (RA) solution. Heavy lines denote stable solutions

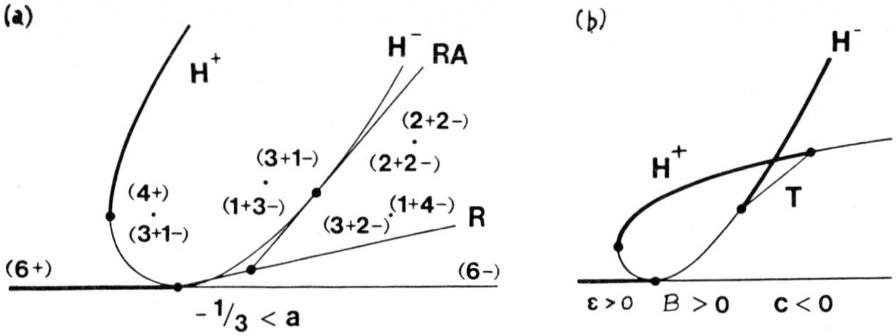

Fig. 12. Typical bifurcation diagrams for (a) the normal form (4.3), and (b) the normal form (4.4). Fig. (b) shows the "larger" amplitude behavior of the equal amplitude solutions (H,T)

are no triangle (T) or rectangle (RA) solutions in Fig. 11(a). How is the second transformed into the first as the symmetry is broken? This question can be answered by adopting the point of view expressed in § I, and studying the unfolding of a degenerate case in which the symmetry-breaking even order terms are small. The normal form for the nonsymmetric self-adjoint problem contains no quadratic terms. If the perturbation breaks the self-adjointness the unfolded normal form is

$$z_1(-\lambda + a\sigma + u_1 + d\sigma^2) + \bar{z}_2\bar{z}_3\,(-\varepsilon + b\sigma + u_1 + cq) = 0, \tag{4.4}$$

where ε is the unfolding parameter and all other coefficients are $O(1)$. A typical bifurcation diagram is shown in Fig. 12(a). Since the triangle solutions are still absent, it is necessary to consider a yet more degenerate bifurcation, in which the quadratic and quartic terms vanish simultaneously, and its universal unfolding. Then secondary bifurcations involving the triangles finally appear. For the equal amplitude solutions (H,T) the universal unfolding is

$$z[-\lambda + (3a+1)u + Aq] + \bar{z}^2(-\varepsilon + 3Bu + cq) = 0, \quad z=z_\alpha . \tag{4.5}$$

There are now three unfolding parameters ε, B and A that describe the effects of symmetry breaking. A typical bifurcation diagram is shown in Fig. 12(b).

Figure 12 shows not only the subcritical instability to H^+, but also how the two solutions RA, T that are absent in fig. 11(a) approach the origin as the system becomes closer and closer to being symmetric. This example illustrates well the motivation for and the utility of studying multiple bifurcation phenomena.

Acknowledgement: I am grateful to Marty Golubitsky and John Guckenheimer for their patience while teaching me bifurcation theory, and to Betty Duncan for her careful typing of the manuscript. This work was supported in part by the Alfred P. Sloan Foundation and the California Space Institute.

REFERENCES

1. N.O.Weiss: Phil. Trans. Roy. Soc. A256, 99(1964).
2. H.E.Huppert and J.S.Turner: J.Fluid Mech. 106, 299(1981).
3. M.R.E.Proctor and N.O.Weiss: Rep. Prog. Phys. 45, 1317(1982).
4. E.Knobloch: Phys. Fluids 23, 1918(1980).
5. J.Marsden and M.McCracken: The Hopf Bifurcation and Its Applications, Springer Verlag (1976).
6. P.G.Daniels: Proc. Roy. Soc.A358, 173(1977).
7. P.Hall and I.C.Walton: Proc. Roy. Soc. A358, 199(1977).
8. L.N.Da Costa, E.Knobloch and N.O.Weiss: J.Fluid Mech. 109, 25(1981).
9. J.Guckenheimer and E.Knobloch: Geophys. Astrophys. Fluid Dyn. 23, 247(1983).
10. J.Guckenheimer and P.Holmes: Nonlinear Oscillations, Dynamical Systems and Bifurcations of Vector Fields, Springer Verlag(1983).
11. V.I.Arnol'd: Functional Anal. and Applics. 11, 85(1977).
12. E.Knobloch and M.R.E.Proctor: J.Fluid Mech. 108, 291(1981).
13. J.Guckenheimer and E.Knobloch, in preparation.
14. D.Moore, J.Toomre, E.Knobloch and N.O.Weiss: Nature 303, 663(1983).
15. E.Knobloch, D.Moore, J.Toomre and N.O.Weiss, in preparation.
16. E.Knobloch and N.O.Weiss: Phys. Lett. 85A, 127(1981).
17. E.Knobloch and N.O.Weiss: Physica D, in press.
18. L.P.Shil'nikov: Soviet Math. Doklady 6, 163(1965).
19. J.Yorke and K.Alligood: Bull. Amer. Math. Soc., in press.
20. P.Glendinning and C.Sparrow: preprint.
21. E.Buzano and M.Golubitsky: Phil. Trans. Roy. Soc.A308, 617(1983).
22. M.Golubitsky, J.W.Swift and E.Knobloch: Physica D, in press.

Strange Attractors in a System Described by Nonlinear Differential-Difference Equation

Y. Ueda and H. Ohta

Department of Electrical Engineering, Kyoto University
Kyoto 606, Japan

1 Introduction

This report deals with strange attractors which occur in a system described by the following differential-difference equation:

$$\frac{d\theta(t)}{dt} + \sin\theta(t - L) = \delta. \tag{1}$$

This equation is a mathematical model of phase-locked loops (PLL) with time delay. Synchronized states of the PLL are represented by the equilibrium points of the equation. The pull-in region, i.e., the parameter region in which all initial conditions lead to quiescent steady states, was already reported with some regions correlated with asynchronized steady states [1]. This report surveys various types of steady states, especially chaotic steady states, in computer-simulated systems of Eq. (1).

2 Experimental Results

In this report, steady states will mean asymptotically stable equilibrium points, periodic solutions, and so forth of Eq. (1). In the following the experimental results will be described. Before giving the details, we present a brief review concerned with the steady solutions of Eq. (1).

2.1 Classification of steady states

The trajectories of the steady states of Eq. (1) can be represented on the cylindrical phase plane (θ, $\dot{\theta}$). The types of steady states in computer-simulated systems can be roughly classified as follows:

(a) Equilibrium points: $\theta = \mathrm{Sin}^{-1}\,\delta$, $\theta = \pi - \mathrm{Sin}^{-1}\,\delta$ (2)

(b) Limit cycle of the first kind: $\theta(t) = \phi(t)$, $\phi(t + T) = \phi(t)$ (3)

(c) Limit cycle of the second kind: $\theta(t) = \phi(t) \pm 2\pi t/T$, $\phi(t + T) = \phi(t)$ (4)

(c') N-kai limit cycle of the second kind: $\theta(t) = \phi_N(t) \pm 2N\pi t/T_N$, $\phi_N(t + T_N) = \phi_N(t)$ (5)

(d) Chaotic steady states: Similar to ordinary differential equations (ODE), a chaotic steady state is considered to be represented by a bundle of orbits where the bundle is asymptotically orbitally stable but contains infinitely many unstable periodic orbits [2].

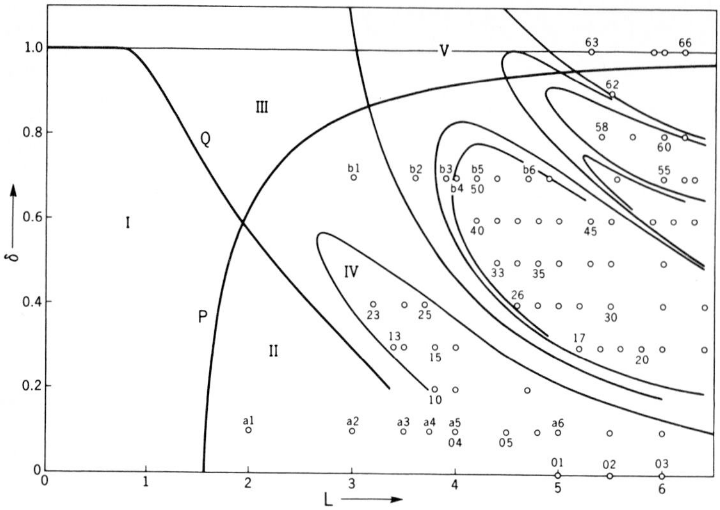

Fig. 1: Regions of different steady states for the system represented by Eq. (1)

2.2 Regions for different types of steady states

Figure 1 shows the regions on the (L, δ) plane in which different steady states are observed in the computer-simulated system described by Eq. (1). A stable equilibrium point exists between the curve P and the straight line $\delta = 1$ (Region I). This equilibrium point becomes unstable to the right of P (Region II) casting off a stable limit cycle of the first kind which surrounds it. Above the curve Q (Region III), there exists a limit cycle of the second kind. It disappears below Q. For large value of L, various steady states have been observed. Most of them are so complicated that it is almost impossible to locate their boundaries. Under such situations, only representative boundaries are depicted in Fig. 1. In the regions IV and V, there exist N(= 2, 4, 8,...)-kai limit cycles of the second kind. They will be seen in connection with the examples of steady states in the following section.

2.3 Representative examples of limit cycles and their bifurcations

In order to illustrate the whole aspect of the phenomenon, we show the steady states at twelve representative sets of parameters indicated by a1,..., a6, b1,..., b6 in Fig. 1. Figures (a1) to (a6) in Fig. 2 show an example of bifurcations for the limit cycle of the first kind in which the parameter L is increased with $\delta = 0.1$ in Eq. (1). Figures (b1) to (b6) show another example for the limit cycle of the second kind with $\delta = 0.7$. These figures show the trajectory doubling bifurcations explicitly and it seems that chaotic steady states occur through a cascade of the bifurcations. These situations closely resemble the case of ODE systems.

2.4 A collection of strange attractors

In order to survey an outline of chaotic steady states, we show the strange attractors at sixty six sets of parameters (L,δ) indicated by 01, 02,..., 66 in Fig. 1. A strange attractor is given by the set of points on the $(\dot{\theta}, \ddot{\theta})$ plane at the instants when a steady-state trajectory intersects $\theta = \mathrm{Sin}^{-1}\,\delta$, i.e.,

$$\{(\dot{\theta}(t_i),\ \ddot{\theta}(t_i)) \mid \theta(t_i) = \mathrm{Sin}^{-1}\,\delta\ (\mathrm{mod}\ 2\pi),\ i \in Z^{+}\}. \tag{6}$$

Figure 3 shows strange attractors thus obtained and each of them consists of 10,000 points. These figures demonstrate the great complexity of the phenomenon. The majority of strange attractors don't resemble the case of ODE systems and they are really strange and interesting.

3 Conclusion

Experimental studies of the steady states of the nonlinear differential-difference equation (1) have been carried out. From these results, various types of strange attractors have been clarified. The results reported here deserve further mathematical attention.

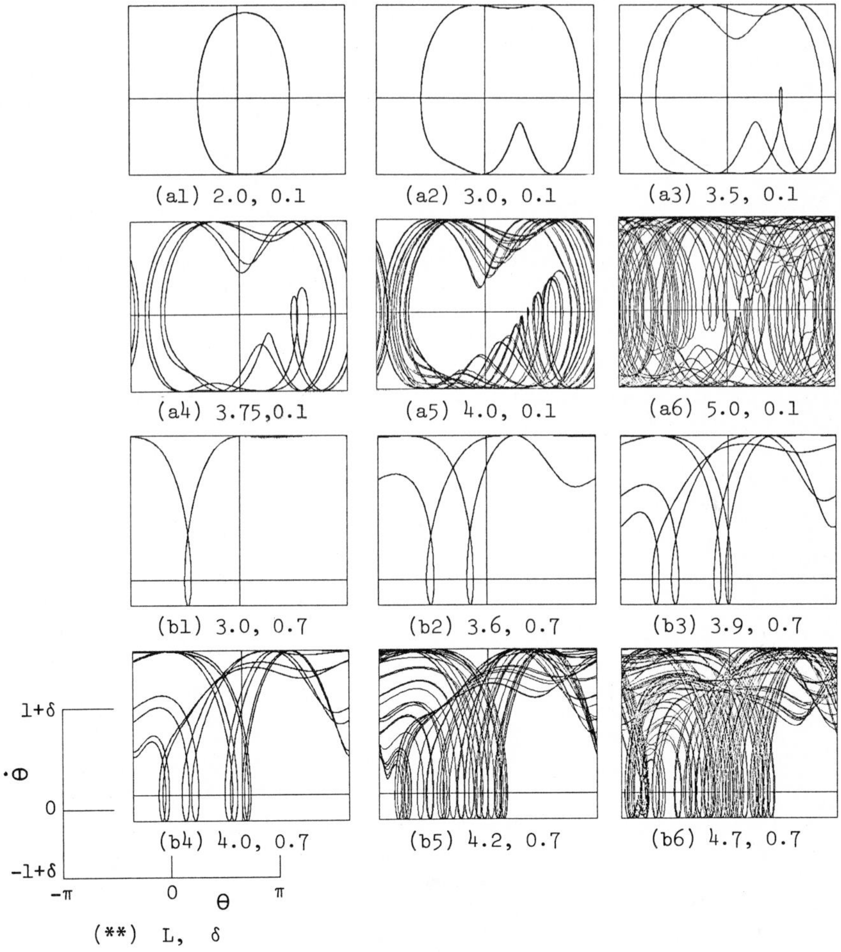

Fig. 2: Representative examples of limit cycles and their bifurcations

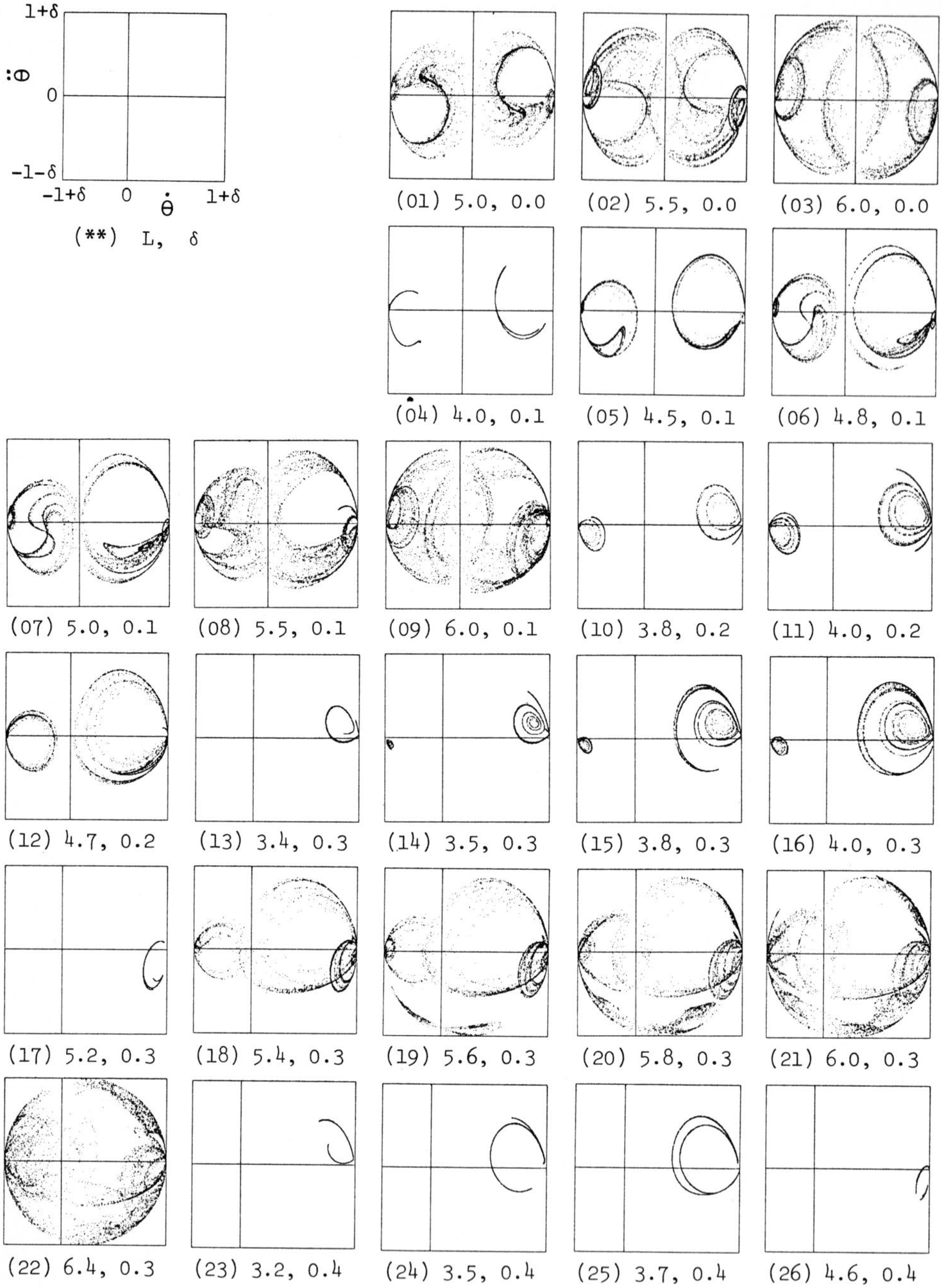

Fig. 3: Glossary of strange attractors

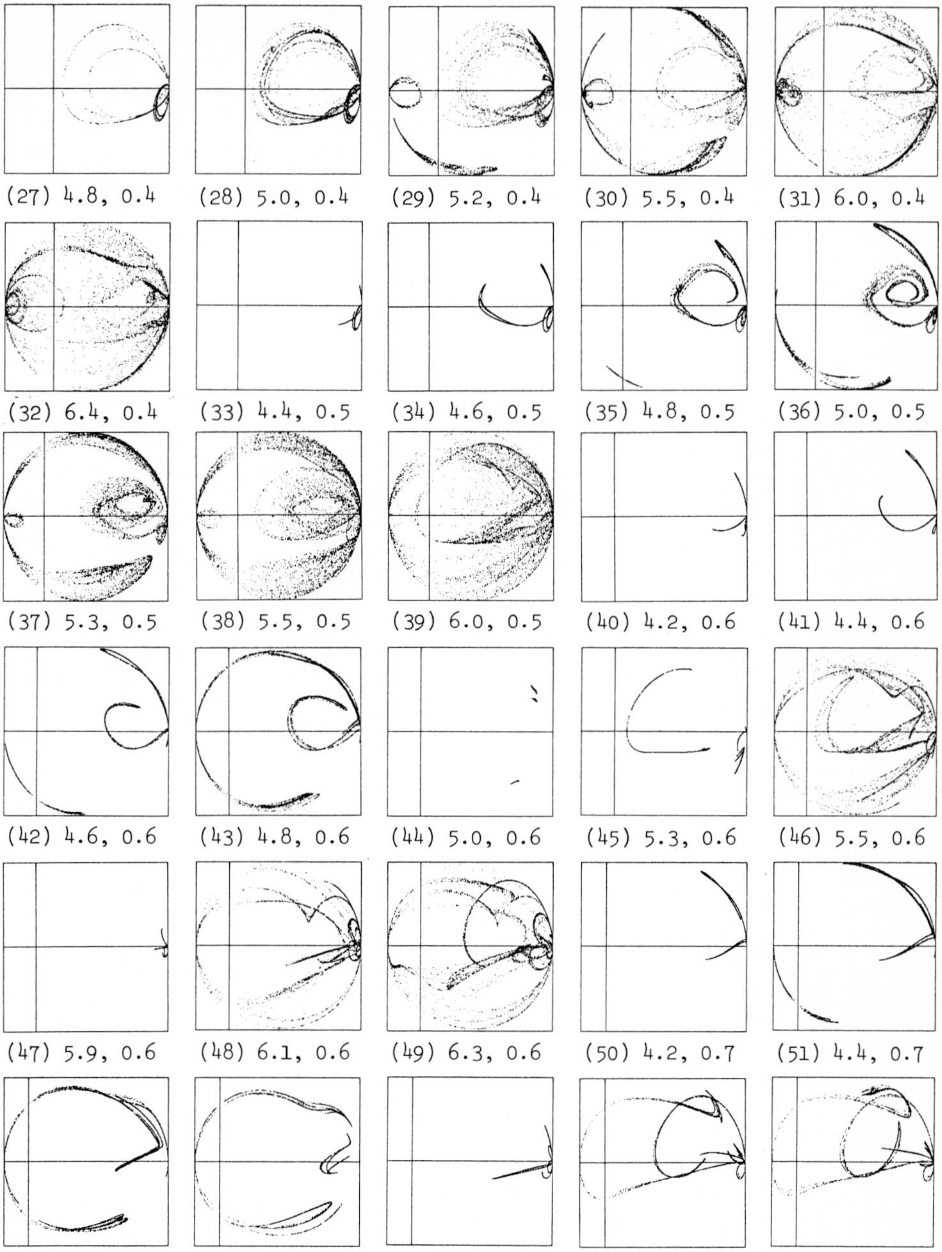

(27) 4.8, 0.4 (28) 5.0, 0.4 (29) 5.2, 0.4 (30) 5.5, 0.4 (31) 6.0, 0.4

(32) 6.4, 0.4 (33) 4.4, 0.5 (34) 4.6, 0.5 (35) 4.8, 0.5 (36) 5.0, 0.5

(37) 5.3, 0.5 (38) 5.5, 0.5 (39) 6.0, 0.5 (40) 4.2, 0.6 (41) 4.4, 0.6

(42) 4.6, 0.6 (43) 4.8, 0.6 (44) 5.0, 0.6 (45) 5.3, 0.6 (46) 5.5, 0.6

(47) 5.9, 0.6 (48) 6.1, 0.6 (49) 6.3, 0.6 (50) 4.2, 0.7 (51) 4.4, 0.7

(52) 4.7, 0.7 (53) 4.9, 0.7 (54) 5.55,0.7 (55) 6.0, 0.7 (56) 6.2, 0.7

Fig. 3: Continued

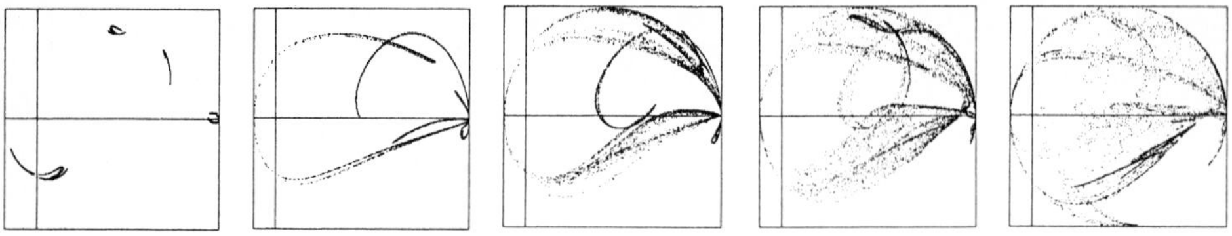

(57) 6.3, 0.7 (58) 5.4, 0.8 (59) 5.7, 0.8 (60) 6.0, 0.8 (61) 6.2, 0.8

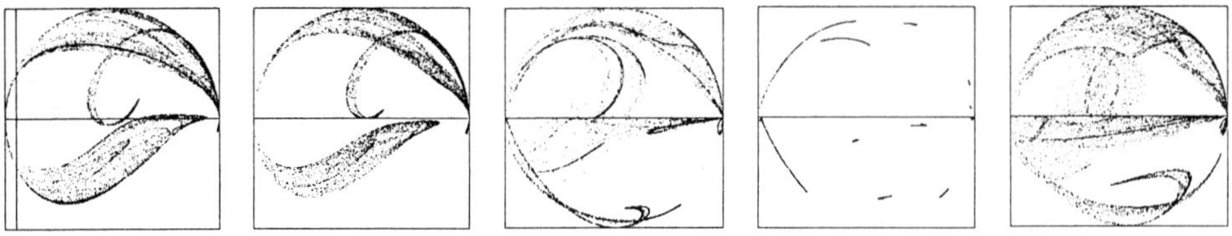

(62) 5.5, 0.9 (63) 5.3, 1.0 (64) 5.9, 1.0 (65) 6.0, 1.0 (66) 6.2, 1.0

Fig. 3: Continued

This work has been carried out in part under the Collaborating Research Program at the Institute of Plasma Physics, Nagoya University. The authors wish to express their sincere thanks to the staff of the Institute.

References

Y. Ueda: IEEE 24th Midwest Symposium on CAS Proc., pp. 549-553 (1981)
Y. Ueda: J. Statistical Physics, 20, 2, pp. 181-196 (1979)

Coupled Chaos

T. Yamada

Department of Physics, Kyushu Institute of Technology, Kitakyushu 804, Japan

H. Fujisaka

Department of Physics, Kagoshima University, Kagoshima 890, Japan

1. Introduction

In a coupled system consisting of N-equivalent nonlinear oscillators the uniform oscillation appears when the strength of couplings among the oscillators becomes sufficiently large, where N is the number of oscillators. The basic equations of the coupled system for N=2 can be written as:

$$\begin{aligned} \dot{\tilde{x}}(1) &= \tilde{f}(\tilde{x}(1);t)+D\{\tilde{x}(2)-\tilde{x}(1)\}, \\ \dot{\tilde{x}}(2) &= \tilde{f}(\tilde{x}(2);t)+D\{\tilde{x}(1)-\tilde{x}(2)\}, \end{aligned} \tag{1}$$

where each oscilltor is described by the state vector, $x(1)$ or $x(2)$,and D denotes the scalar coupling constant. Here the flow $\tilde{f}$ is assumed to be periodic in t:

$$\tilde{f}(\tilde{x}(i);t+T) = \tilde{f}(\tilde{x}(i);t), \text{ for } i=1,2. \tag{2}$$

The uniform oscillation can be realized in eq.(1) even if the temporal evolution of each oscillator behaves chaotically. (For simplicity, N=2 case will be considered, hereafter.) As the strength of the coupling constant is decreased, the uniform chaotic oscillation loses its stability and a nonuniform oscillation arises. The instability threshold of the uniform oscillation is given by $\lambda_L/2$, where λ_L is the largest Lyapunov exponent of the uncoupled system [1]:

$$\dot{\tilde{x}}(i) = \tilde{f}(\tilde{x}(i);t), \; i=1,2. \tag{3}$$

Various phases arise on the coupled nonlinear differential equations (1), depending on the strength of the coupling constant and the initial conditions.

2. Reduced Description

When the uncoupled system (3) has a mapping, a reduced description of the coupled system is possible. For the case where the uncoupled system has a one-dimensional mapping, the reduced equations can be written in the following form [2]:

$$x_{n+1}(1) = (1-\xi_D)g(x_n(1))+\xi_D g(x_n(2)),$$
$$x_{n+1}(2) = (1-\xi_D)g(x_n(2))+\xi_D g(x_n(1)), \qquad (4)$$

with $\xi_D = \{1-\exp(-2DT)\}/2$, where $x_n(i)$ (i=1 or 2) is the value of one of the state vector components at $t^n = nT$ $(n=0,1,2,\cdots)$ and g is the one-dimensional mapping function. The reduced equations (4) give the same threshold value for the instability point of the uniform oscillation.

3. Coupled Brussellator Model

The application of the present formulation is now made to the Brusselator model studied by Tomita and Kai [3]. For this model the flow term of eq.(3) is given by

$$\underset{\sim}{f}(\underset{\sim}{x};t) = \begin{pmatrix} x^2y-Bx+A-x+\tilde{a} \\ Bx-x^2y \end{pmatrix}, \quad \underset{\sim}{x} = \begin{pmatrix} x \\ y \end{pmatrix}, \qquad (5)$$

with $\tilde{a}=a\cos(\omega t)$. For the set of parameters, A=0.4, B=1.2, a=0.05 and $\omega=0.8$, the uncoupled system (3) shows a chaotic behavior. Moreover, by denoting the values of x at $t=t_n\equiv nT$ with $T=8\pi/\omega$ as x_n $(n=0,1,2,\cdots)$, the successive x_n form the one-dimensional map:

$$x_{n+1} = g(x_n), \qquad (6)$$

with the mapping function shown in Fig. 1. Numerical studies are done on eq. (1) and eq. (4) for the coupled Brusselator model. The bifurcation schemes are drawn in Figs. 2 and 3 with the iterated values of $x_n(1)$. There appear 3 chaotic phases, C_1, C_2 and C_3, a quasi-period phase Q.P and a periodic phase P. The C_3 is the uniform chaos. It can be clearly seen that the results in both figures give quite similar features though the transition points are slightly different. This fact shows the usefulness of the reduced description.

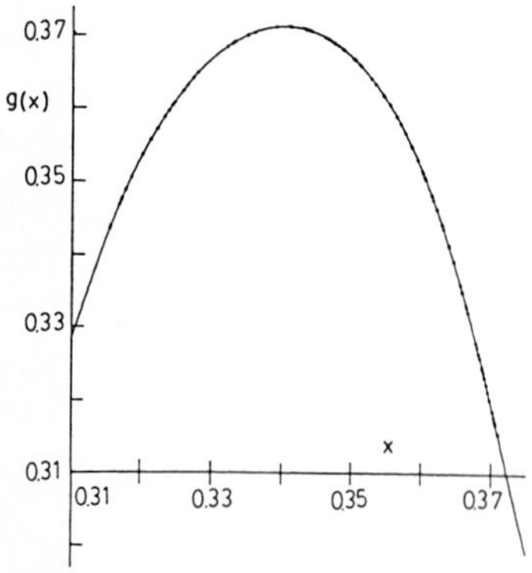

Fig. 1 The one-dimensional mapping function g for the Brusselator model (6). The dots are obtained by numerically calculating the uncoupled equation (3) and the solid line is drawn by using the method of least squares

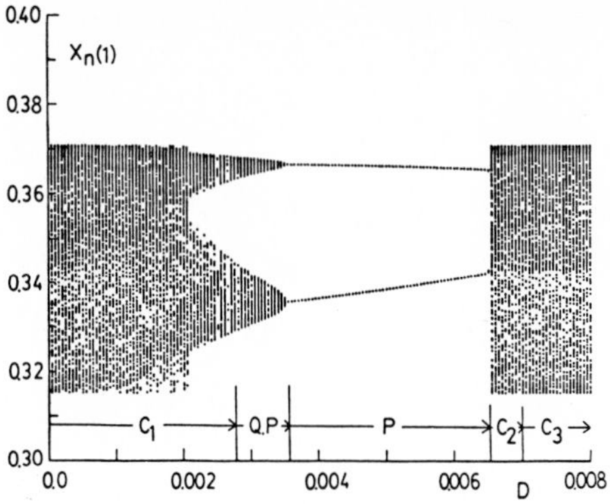

Fig. 2 The bifurcation scheme with the iterated values of $x_n(1)$ obtained by integrating eq. (1) for the coupled Brusselator model

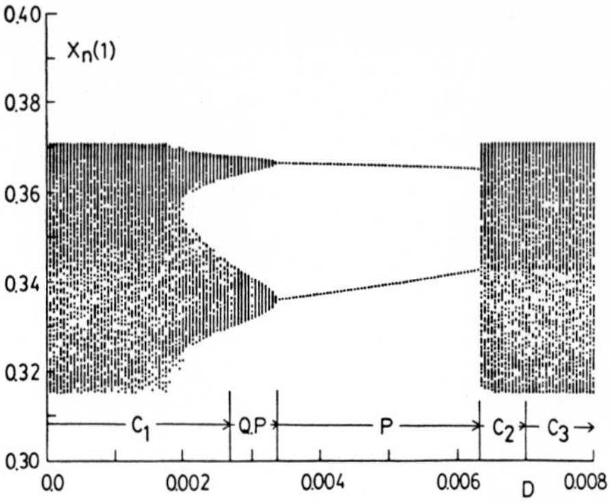

Fig. 3 The bifurcation scheme with the iterated values of $x_n(1)$ obtained by iterating eq. (4) for the same model as in Fig. 2

4. Summaries

The spatial variation of the state vectors becomes an important factor of study in consideration of the coupled chaos as well as the temporal variation. In the present study of the coupled system the appropriate mapping function can be constructed by retaining the essential properties of the original system described by the differential equations.

References

1. H. Fujisaka and T. Yamada, Progr. Theor. Phys. 69, 32 (1983)
2. T. Yamada and H. Fujisaka, Progr. Theor. Phys. , to appear
3. K. Tomita and T. Kai, Progr. Theor. Phys. Suppl. 61,1605 (1978)

Bifurcations in 2D Area-Preserving Mappings

K.-C. Lee and S.Y. Kim

Department of Physics, Seoul National University, Seoul 151, Korea

D.-I. Choi

Department of Physics, Korea Advanced Institute of Science and Technology
P.O. Box 150, Chonyangni, Seoul, Korea

1. Symmetric Structures of the Periodic Orbits

We consider a class of two-dimensional area-preserving mappings of

$$T : x_{n+1} = 2h(x_n) - y_n \; ; \; y_{n+1} = x_n . \qquad (1)$$

The area-preserving property of T is represented by det M = 1, where $M = \prod_{i=0}^{n-1}$ with

$$M_i = \begin{pmatrix} 2h'(x_i), & -1 \\ 1 \;\; , & 0 \end{pmatrix}$$

and this property is satisfied for any function h(x). The residue R is a convenient quantity in discussing the stability of an orbit [1] and it is given as

$$R = (2 - \mathrm{Tr}M)/4.$$

Also useful is the concept of reversibility which has greatly facilitated numerical studies of period doubling sequences. When T is reversible, the symmetry S of T exists and T = (TS)S with $S^2 = 1 = (TS)^2$ so that the inverse of T is given by T^{-1} = STS. The S and TS are complementary symmetries and they are

$$S \; : \; x_{t+1} = y_t \; ; \; y_{t+1} = x_t$$

$$TS \; : \; x_{t+1} = 2h(y_t) - x_t \; ; \; y_{t+1} = y_t .$$

The S reflects the point with respect to the symmetry line y = x while TS with respect to the symmetry curve x = h(y) keeping y constant. When h(x) is an antisymmetric function h(-x) = -h(x) another pair of symmetries that can not be derived from the previous ones exists. They are

$$S \; : \; x_{t+1} = -x_t \; ; \; y_{t+1} = y_t - 2h(x_t)$$

$$TS \; : \; x_{t+1} = -y_t \; ; \; y_{t+1} = -x_t$$

and the symmetry lines are now y axis (x = 0) and y = -x. These pairs are not unique because T^nS and TS T^{-n} for any integer n can be complementary pairs of T.

Symmetries are very useful in locating periodic orbits. An orbit on a symmetry curve with even numbered period can be found from the condition that the midpoint of the orbit is on the same symmetry curve. An orbit with odd numbered period on the S curve can be found from the condition that the (n+1)/2 way point lies on the TS curve. The structure of periodic orbits is obtained by operating T and T^{-1} on the midpoint for even and the (n+1)/2 way point for odd numbered periodic orbits. This operation shows

$$x_{\frac{n}{2}+\ell} = x_{\frac{n}{2}-\ell-1} \; ; \; y_{\frac{n}{2}+\ell+1} = y_{\frac{n}{2}-\ell}$$

$$(\ell = 0, 1, 2, \ldots, \tfrac{n}{2} - 1)$$

for the even numbered periodic orbit on S,

$$x_{\frac{n}{2}+\ell} = x_{\frac{n}{2}-\ell-2} \; ; \; y_{\frac{n}{2}+\ell+1} = y_{\frac{n}{2}-\ell-1}$$

$$(\ell = 0, 1, \ldots, \tfrac{n}{2} - 2)$$

for the even numbered periodic orbit on TS, while

$$x_{\frac{n+1}{2}+\ell-1} = x_{\frac{n+1}{2}-\ell-1} \; ; \; y_{\frac{n+1}{2}+\ell} \quad y_{\frac{n+1}{2}-\ell}$$

$$(\ell = 0, 1, \ldots, \tfrac{n-1}{2})$$

for the odd numbered periodic orbit with the (n+1)/2 way point on TS and the initial point on S.

2. Equivalence of Various 2D Area Preserving Mappings

It is shown in ref. 2 that various 2D area-preserving mappings with one parameter quadratic nonlinearity are equivalent. When we choose for h(x)

$$h_{HL}(x) = cx + x^2$$

the mapping of (1) becomes equivalent to Helleman's mapping while the quadratic mappings of DeVogelarere and Henon are obtained with

$$h_{DV}(x) = px - (1-p)x^2$$

$$h_{HN}(x) = \frac{1}{2}(1 - ax^2).$$

From these relations the equivalent parameters of one mapping can be obtained from those of the other mapping. For example, the Feigenbaum sequence of the bifurcation parameters p_n and the n-period fixed points $P_n^* = (x_n^*, y_n^*)$ of DeVogelarere mapping are obtained from those of Hellman's mapping $\{C_n\}$ and $\{P_n^{*HL}\}$

$$P_n = C_n$$

$$P_n^{*DV} = -\frac{1}{1-C_n} P_n^{*HL}.$$

Similarly Helleman's parameters are related to Henon's by

$$C_n^{\pm} = 1 \pm \sqrt{1 + a_n}$$

and

$$P_n^{*HL} = -\frac{a_n}{2} P_n^{*HN} - \frac{C_n}{2}.$$

One of the extensively studied mappings with nonlinearity beyond the quadratic form is the standard mapping of Chirikov given by

$$T_c : r_{n+1} = r_n - \mu \sin 2\pi\theta_n \;;\; \theta_{n+1} = \theta_n + r_{n+1}. \qquad (3)$$

The mapping T_c and T of (1) are equivalent if we choose for h(x)

$$h(x) = x - \frac{\mu}{2} \sin 2\pi x. \qquad (4)$$

The linear transformation of $T \to T_c$ is

$$Q_c : x = \theta \;;\; y = \theta - r \qquad (5)$$

and its inverse is given by

$$Q_c^{-1} : \theta = x : r = x - y. \qquad (6)$$

Any result obtained in one mapping can be translated into the other one by these transformations. For example, from the n-periodic orbit $(x_n{}^*, y_n{}^*)$ of T, the n-periodic orbit $(r_n{}^*, \theta_n{}^*)$ of T_c is easily found to be $(r_n{}^* = x_n{}^* - y_n{}^*)$. Two symmetries S and TS of (2) are given as

$$\begin{aligned} S_c &: r_{t+1} = -r_t \;;\; \theta_{t+1} = \theta_t - r_t \\ T_cS_c &: r_{t+1} = -r_t + \mu \sin 2\pi (r_t - \theta_t) \;;\; \theta_{t+1} = \theta_t + \mu \sin 2\pi (r_t - \theta_t) - 2r_t, \end{aligned} \qquad (7)$$

with $S_c{}^2 = 1 = (T_cS_c)^2$. Therefore the symmetry lines of y = x and x = h(y) in the Chirikov's coordinate become respectively

$$r = 0 \text{ and } r = \frac{\mu}{2} \sin 2\pi (r - \theta). \qquad (8)$$

The generalized Devogelarere mapping [1] of

$$T_D : X_D' = -Y_D + f(X_D) \;;\; Y_D' = X_D - f(X_D')$$

and T of (1) are shown to be equivalent by the transformations Q_D and $Q_D{}^{-1}$ of

$$\begin{aligned} Q_D &: x = X_D \;;\; y = Y_D + f(X_D) \\ Q_D{}^{-1} &: X_D = x \;;\; Y_D = y - f(x) \end{aligned}$$

with f(x) = h(x). GREENE et al. [1] show that two symmetries $S1_D$ and $S2_D$ of

$$\begin{aligned} S1_D &: X_D' = X_D \;;\; Y_D' = -Y_D \\ S2_D &: X_D' = Y_D + f(X_D) \;;\; Y_D' = X_D - X_D' \end{aligned}$$

are convenient because the (dominant) symmetry $S1_D$ is the X_D axis making the analysis of the period doubling bifurcation sequence rather simple.

3. Period-Doubling Bifurcations

In the following a sequence of period-doubling bifurcation is described. A mapping T with a parameter a has a stable interval. As the parameter a changes the stable orbit becomes hyperbolically unstable with reflection and new orbits with the doubled period are born. We consider first a case where a periodic orbit on TS symmetry line bifurcates on the same symmetry line. The initial point $P_0 = (x_0, y_0)$ of a 2n-period orbit can be found from the solution of the simultaneous equations $x_0 = h(y_0)$ and $x_n(p_0) = h(y_n(P_0))$. If we define $g_n(z) = y_n(P_0)$, with $x_0 = h(z)$ and $y_0 = z$, then some roots of the equation

$$\phi_n(z) \equiv g_n(z) - z = 0 \tag{9}$$

will give the initial point of the n-period orbit on the symmetry line. When we construct a function $\psi_{2n}(z)$ from $\phi_n(z)$ as

$$\psi_{2n}(z) = \phi_n(\phi_n(z) + z) + \phi_n(z) \tag{10}$$

then we can show [3] that some of the roots of the equation $\psi_{2n}(z) = 0$ give the initial point of the 2n period. We assume that the period-doubling bifurcation occurs at $a = a_n$. As a approaches a_n from below (assuming the stable interval to be $a < a_n$) both $\phi_n(y\ ;\ a < a_n)$ and $\psi_{2n}(y\ ;\ a < a_n)$ cross the y axis at $y_0 = y_n^*$ ($a < a_n$) of the initial point of the n-period orbit with the negative slopes, since an n-period orbit is also a 2n-period orbit. As a increases the negative slopes, $\phi'(y_n^*\ ;\ a < a_n)$, become steeper until $\phi'(y_n^*\ ;\ a = a_n) = -2$ while $|\psi'_{2n}(y_n^*\ ;\ a)| \to 0$ as $a \to a_n$. This can be seen by differentiating (9) as

$$\psi'_{2n}(z) = \phi'_n(\phi_n(z) + z)\ (\phi'_n(z) + 1) + \phi'_n(z)$$

and putting $z = y_n^*$ to get

$$\psi'_{2n}(y_n^*) = \phi'_n(y_n^*)(\phi'_n(y_n^*) + 2).$$

Thus the condition of the bifurcation of the n-period orbit at $a = a_n$ is that $\phi'_n(y_n^*\ ;\ a_n) + 2 = 0$ and $\phi'_n(y_n^*\ ;\ a \lessgtr a_n) + 2 \lessgtr 0$ where ϕ_n is defined in (9).

We can also show [3] that for the mapping T of (1) we have

$$\phi'_n(y_n^*\ ;\ a) = -2R. \tag{11}$$

From the observations of GREENE et al. [1], we see then that as R increases from $R < 1$ to $R > 1$ corresponding to $a < a_n$ and $a > a_n$ the stable elliptical orbit $(0 < R < 1)$ becomes unstable at $R = 1$ $(a = a_n)$ and it changes into a hyperbolic orbit with reflection $(R > 1)$ and thus a period-doubling bifurcation occurs. An odd periodic orbit usually does the period-doubling bifurcation on the S-symmetry line, in which case the above sequence of period-doubling bifurcation is reestablished, but replacing $\phi_n(z)$ in $\psi_{2n}(z)$ of (10) by $\xi_n(z)$ defined by

$$\xi_n(z) = \frac{1}{2}\,[\,x_n(z) + y_n(z)\,] - z$$

where $(x_n(z), y_n(z)) = T^n P(z,z)$, an n iterate of T, a point on the S-symmetry line P(z,z). Then we can show [4] that

$$\xi_n'(x_0\ :\ a) = -2R$$

similar to (11).

4. Universality of Bifurcations

Making use of (3)-(8) we have studied numerically [5] a sequence of period-doubling bifurcations of periodic orbits of Chirikov mapping. We followed one sequence on a symmetry line which leads to an accumulation point of the parameter value $\mu_c = 0.8259739 \ldots$. The sequence also confirms the universal scaling behavior and universal constants of $\delta = 8.721097 \ldots$, $\alpha = -4.0180767 \ldots$ and $\beta = 16.363896 \ldots$, where the Feigenbaum ratios are defined as

$$\delta = \lim_{k\to\infty} [(a_{k-2}-a_{k-1})/(a_{k-1}-a_k)]$$

$$\alpha = \lim_{k\to\infty} (d_{k-1}/d_k), \quad \beta = \lim_{k\to\infty} (e_{k-1}/e_k).$$

Here a_k denotes the parameter value at which 2^{k+1} periodic orbits are born and d_k and e_k are distances between bifurcated points on and off the symmetry line respectively. We are also studying the period-trebling cascades of $k \cdot 3^n$. Preliminary results show that the universal constant values are $\delta = 430$, $\alpha = -44$ and $\beta = -187$.

Acknowledgement

We wish to acknowledge the support of the Ministry of Education, Republic of Korea, through a grant to the Research Institute for Basic Sciences, Seoul National University.

References

1. J.M. Greene, R.S. Mackay, F. Vivaldi and J.J. Feigenbaum: Physica 3D, 468 (1981).
2. K.C. Lee: J. Phys. A16, L 137 (1983).
3. K.C. Lee and D.I. Choi: J. Phys. A16, L 55 (1983).
4. K.C. Lee: "Universal Behavior in the Onset of Chaotic Motion of a Hamiltonial Map" preprint (1983).
5. K.C. Lee, S.Y. Kim and D.I. Choi: "Period Doubling Bifurcations of Periodic orbits of Chirikov's Standard Mappings" preprint (1983).

Part VI

Soliton Systems

Chaotic Behavior Induced by Spatially Inhomogeneous Structures such as Solitons

M. Imada

Institute for Solid State Physics, University of Tokyo, Roppongi, Minato-ku
Tokyo 106, Japan

1. Introduction

One reason for the recent increasing interest in chaotic systems is due to the simplicity of the models: even a simple deterministic system with *a few* degrees of freedom shows chaotic behavior. Simple models with chaotic behavior are often used as models of systems with many degrees of freedom, when only small numbers of them are relevant to the evolution. This modeling is powerful for the investigation near the onset of the chaos. After one knows that very simple systems show chaotic behavior, it seems not so surprising if most of nonlinear systems with *infinite* degrees of freedom show chaotic behavior. Apparent exceptions are found in completely integrable systems. In those systems, time evolution is completely described by nonlinear normal modes including solitons. So far the onset of chaos has mainly been investigated by the reduction of system variables to a few relevant modes. Another approach is, however, possible by starting with the small deviation from the complete integrability when perturbation is added. The onset of chaos can be investigated without the reduction of variables in this case. The purpose of this paper is to show an aspect of the chaotic behavior in perturbed integrable systems. In this paper we restrict ourselves to perturbed sine-Gordon systems. We will show an example where spatial inhomogeneities like solitons have an important contribution to the chaotic behavior [1].

2. Damped Driven Sine-Gordon System

We start with the following equation of motion:

$$\phi_{tt} - c_0^2\phi_{xx} + \sin\phi = A\sin\omega t - \gamma\phi_t, \tag{1}$$

where the suffix denotes the derivative. The phase ϕ has dependences on spatial and time variables x and t. The time-dependent external field is characterized by its amplitude A and frequency ω. The dissipation is given by γ. If A and γ are both equal to zero, (1) reduces to the pure sine-Gordon equation. If we neglect the spatial dependence of the system, it is equivalent to the problem of a damped driven pendulum [2]. In the presence of the spatial dependence new features are expected because of the existence of nonlinear excitations [3,4]. In particular the numerical simulation of a finite-size system shows that the chaotic behavior is strongly influenced by the initial condition: the presence of solitons or breathers affects and complicates the chaotic behavior [4].

We, however, do not yet know the universal properties independent of specified initial conditions. Indeed it is desirable to find general properties of the system

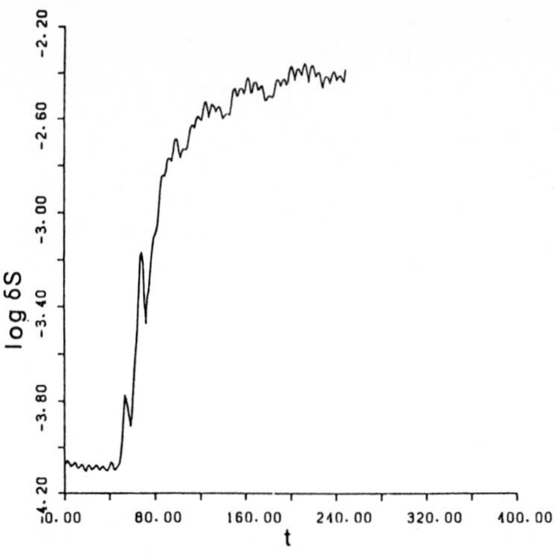

Fig.1 Logarithm of the total separation distance as a function of time in the driven damped sine-Gordon system. The parameters are chosen in (1) as follows: $A=0.1$, $\gamma=0.005$, $\omega=0.5$ and $c_0^2=30.0$

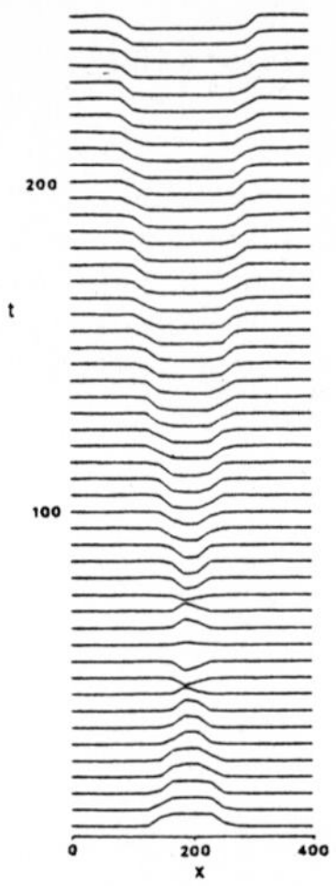

Fig.2 Time evolution of ϕ in case of Fig.1. We plot only every 20 lattice spacing for the schematic illustration of the displacement pattern, while the continuum approximation is expected to be satisfied in the actual simulation

of the infinite size when finite density of solitons and breathers is excited in the initial condition. To this purpose we will show an interesting aspect of the chaotic behavior in the driven damped sine-Gordon system. It is not yet known whether solitons and breathers continue to exist as long as one likes in the system. Actually in the case of $\gamma \neq 0$ and $A=0$, the total energy apparently approaches zero and any spatial structures are not possible in the limit $t\to\infty$. It is, however, likely that finite density of solitons and breathers continues to exist in the long time limit when the driving amplitude is sufficiently larger than the effect of dissipation. Therefore collisions and interaction of solitons are supposed to be a general feature in this parameter region. Using the numerical simulation we investigate the soliton-soliton collision as one of the elementary processes.

We define the local and total separation distance respectively by

$$\delta s(x,t) = \sqrt{[\delta\phi(x,t)]^2 + [\delta\phi_t(x,t)]^2} \tag{2}$$

and

$$\delta S(t) = \sqrt{\frac{1}{L}\int_0^L [\delta s(x,t)]^2 dx}, \tag{3}$$

where L is the system size and $\delta\phi$ is the difference of ϕ between two trajectories whose initial conditions have a slight difference given by $\delta\phi(x,0)$. In this paper we fix $\delta\phi(x,0)$ at 10^{-5} or 10^{-4}. Figure 1 shows the time dependence of $\log\delta S$ in the case shown in Fig.2, where the time evolution of ϕ is illustrated. We can clearly see that the rapid increase of logδS is caused by the soliton-soliton collision at $t \simeq 60$. If the soliton-soliton collisions occur constantly somewhere in the system of infinite size, logδS is expected to increase linearly with time. This means nothing but the positive Lyapunov number, that is the chaotic behavior. On the other hand when the damping becomes larger, it contributes to decrease δS. Therefore there is a competition between the interaction of nonlinear excitations and the damping. Of course when $\gamma=A=0$, the separation distance does not show any increases at the collision of solitons. Therefore the Lyapunov number seems to be zero in the pure sine-Gordon equation.

3. Sine-Gordon System under Energy Conserved Perturbation

By adding an energy-conserving perturbation term, we also analyze the orbital instability in a Hamiltonian system. As an example we add the following perturbation term H_I to the pure sine-Gordon Hamiltonian

$$H_I = (1 - R)^2 \frac{1 - \cos\phi}{1 + R^2 + 2R\cos\phi} - (1 - \cos\phi), \tag{4}$$

where R determines the deviation from the integrability because H_I becomes zero at R=0. We analyze the total separation distance when soliton-soliton collisions occur . Figure 3 shows the time dependence of the total separation distance for R= -0.05. The time evolution of ϕ is demonstrated in Fig.4 for the same case as Fig.3. The total separation distance rapidly increases when soliton-soliton collision takes place. Because of the periodic boundary condition, soliton-soliton collisions occur three times. Corresponding to the collisions, the total separation distance shows

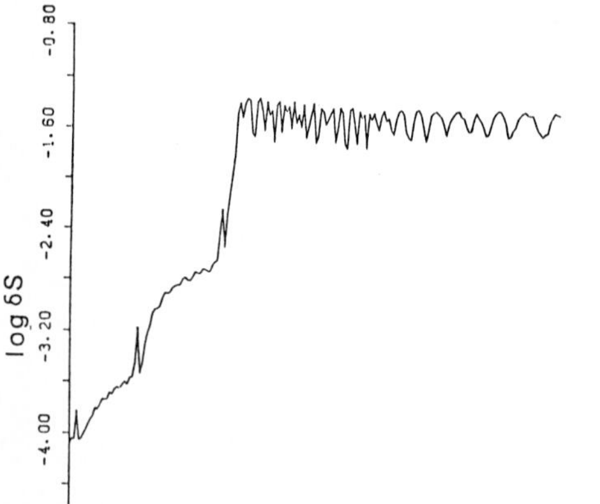

Fig.3 Time dependence of logδS in case of energy conserved perturbation. The parameter R is taken to be -0.05 and $c_0^2=60.0$

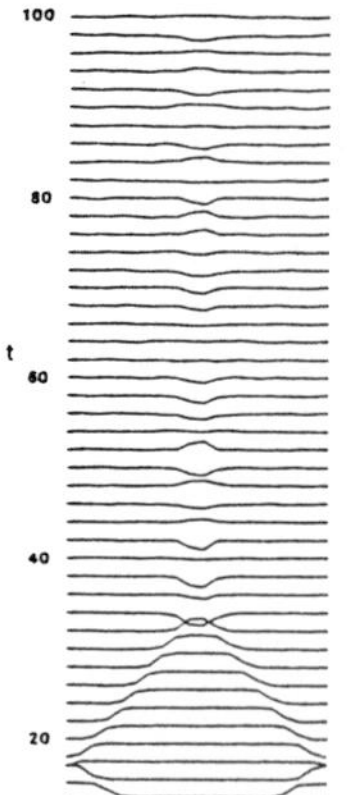

Fig.4 Time evolution of ϕ in case of Fig.3. The displacement pattern is plotted in the same way as Fig.2

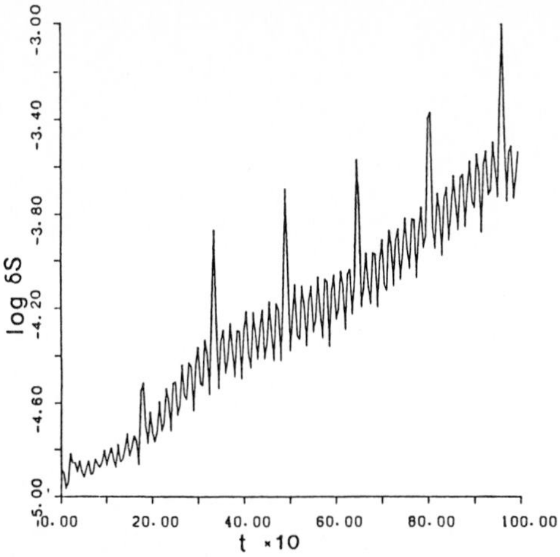

Fig.5 Time dependence of logδS in case of energy conserved perturbation. The parameter R is taken to be -0.01 and c_0^2=30.0

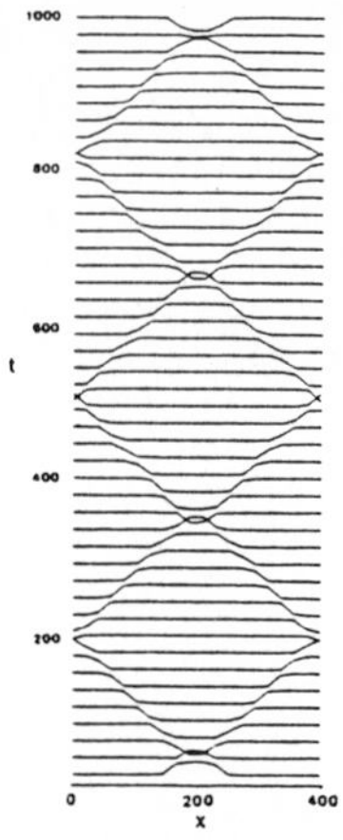

Fig.6 Time evolution of ϕ in case of Fig.5. The displacement pattern is plotted in the same way as Fig.2

the remarkable increase at t$\sim$3, t$\sim$16 and t$\sim$33. After the collision at t$\sim$33, two solitons decay to the small oscillation as seen in Fig.4. In this time region the total separation distance remains to be a constant value on an average until the end of the simulation. This interesting behavior can be interpreted that the orbital instability is dominantly generated by the soliton-soliton collision. Because of no dissipation, there is no tendency to decrease the total separation distance in this system. If the soliton-soliton collisions take place constantly, the exponential increase of the separation distance, that is the chaotic behavior, is expected. Indeed the linear increase of logδS is seen in Fig.5, where soliton-soliton collisions take place repeatedly during the simulation as seen in Fig.6. If we continued this simulation longer and longer, two solitons would sooner or later decay as in the case of Fig.3 and Fig.4. After that time the total separation distance may stop increasing.

It should, however, be noted that our purpose is to know the properties in the system of infinite size. When we assume finite energy density as the initial condition, are the soliton density and the collision rate finite at any time? This problem is closely related to the equipartition law. In the pure sine-Gordon equation the energy shared to a mode remains to be constant. Therefore the equipartition law is not satisfied. The problem is whether the equipartition law becomes satisfied by adding the perturbation term. This seems to be a renewal of the problem investigated by Fermi, Pasta and Ulam [5]. They have added a small perturbation term to the harmonic system to study the energy exchange among various modes. Our starting point is, however, different from theirs because our unperturbed system is a nonlinear integrable system. In addition our perturbation term is supposed to break the integrability actually. Indeed the energy exchange process among normal modes has been investigated numerically [1] by using the inverse scattering method. Energy switching processes from the soliton mode to others have always been observed in the presence of the perturbation. Therefore we expect the equipartition law in this case. This means that the soliton-soliton collisions take place at any time and they must be an important elementary process in the mechanism of the chaotic behavior.

4. Discussion

We have shown an approach to the onset of chaos in the system with infinite degrees of freedom. The chaotic behavior has been investigated in the region of small deviation from the complete integrability. Solitons, more specifically the soliton-soliton interaction, has turned out to make an important contribution to the chaotic behavior. If only the solitons are essential to the chaotic behavior, it may be possible to describe the chaotic behavior by reducing the degrees of freedom into variables of solitons only. This corresponds to the simplification to the system with a few degrees of freedom.

It should, however, be noted that we cannot neglect the phonon mode. If only the solitons make the orbital instability, the increase of the local separation distance (2) must be concentrated only near the position of solitons. This is not the case. As shown in [1], the orbital instability generated at the collision spot propagates independently of solitons. The local separation distance does not decrease after solitons go away. This clearly shows the importance of small oscillations, which make the problem complex. In addition the number of solitons is expected to be infinite in the infinite-size system. Then the reduction to a few degrees of freedom has not been successful so far.

In principle, by taking account of both the soliton and the phonon mode, we can apply the perturbation theory of the inverse scattering transform. The equation of motion for the scattering data is expected to show the chaotic behavior. The perturbation theory should, however, not be applied to the long time behavior, which is necessary to determine whether the system is chaotic or not. The future problem is to develop our approach to apply to fully developed chaos, where the reduction of variables is not successful.

References

1. M. Imada: J. Phys. Soc. Jpn. 52 1946 (1983)
2. B.A. Huberman, J.P. Crutchfield and N.H. Packard: Appl. Phys. Lett. 37 750 (1980)
3. J.C. Eilbeck, P.S. Lomdahl and A.C. Newell: Phys. Lett. 87A 1 (1981)
4. D. Bennett, A.R. Bishop and S.E. Trullinger: Z. Phys. B47 265 (1982). A.R. Bishop, K. Fesser, P.S. Lomdahl and S.E. Trullinger: preprint
5. E. Fermi: Collected Papers of Enrico Fermi Vol.II (Univ. of Chicago Press 1965) p.978

Chaotic Behaviour of Quasi Solitons in a Nonlinear Dispersive System

H. Nagashima

Faculty of Liberal Arts, Shizuoka University
Shizuoka 422, Japan

1. Nonlinear Wave Equation with Fifth-Order Dispersion

In real systems, when the third-order dispersion term becomes small, it is useful to consider the following equation instead of the Korteweg-de Vries (K-dV for short) equation;

$$u_t + uu_x - \gamma^2 u_{5x} = 0 \ , \tag{1}$$

where the suffix 5x represents the fifth-order partial derivative and γ^2 is a constant.

Here we summarize briefly the properties of (1) [1].

It has shown that (1) has only three conserved densities of polynomial types [2];

$$\begin{aligned} T_1 &= u \\ T_2 &= 1/2\ u^2 \\ T_3 &= 1/3\ u^3 - \gamma^2 (u_{2x})^2 . \end{aligned} \tag{2}$$

In terms of the third conserved quantity $I_3 = \int T_3 dx$, equation (1) is expressed as

$$u_t + \frac{1}{2}\left(\frac{\delta I_3}{\delta u}\right)_x = 0, \tag{3}$$

where $\delta I_3/\delta u$ is the functional derivative of I_3 with respect to u.

One solitary-wave solution is written using a function f as $u = \lambda \times f\{\lambda^{1/4}(x-\lambda t)\}$, where λ is a parameter, however the explicit form of the function f is not obtained yet and it has been obtained numerically [1,3]. According to the above expression of a solitary wave, the amplitude is proportional to the velocity and the width of the wave is inversely proportional to the fourth root of the velocity; these properties have been demonstrated by computer simulation [1]. Moreover, the simulation shows that one solitary wave propagates stably, and a two solitary-wave interaction is classified into two types (T and B) according to the ratio of two waves. The type T interaction takes place when the amplitude ratio of a smaller wave to the larger one is more than 0.94564, while the type B interaction occurs in the other cases.

Two solitary waves are stable in the type T interaction and almost stable in another interaction.

Multi-(more than three) solitary waves are generally unstable in their simultaneous collision, and after the collision their amplitudes vary, and they sometimes produce additional waves whose amplitudes are small compared with those of original waves.

According to the relative amplitudes and positions of solitary waves before interaction, a bound state of the two waves emerges after interaction.

The above-mentioned properties of solitary waves are due to the insufficient number of the conserved quantities of the system, and the solitary wave is called a quasi soliton.

2. Results of Numerical Solutions

We have carried out the computer simulation of (1) by using a conventional leap-flog scheme with double precision and set $\gamma^2=7\times10^{-7}$.

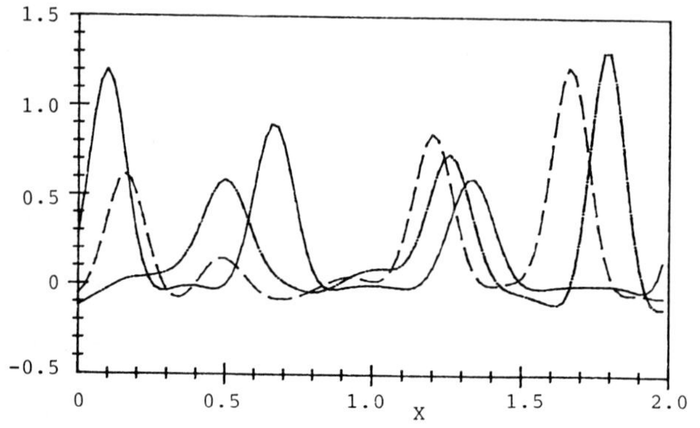

Fig. 1 Profiles of quasi solitons at t=0 ——, 64 —·—·- and 128 — — —

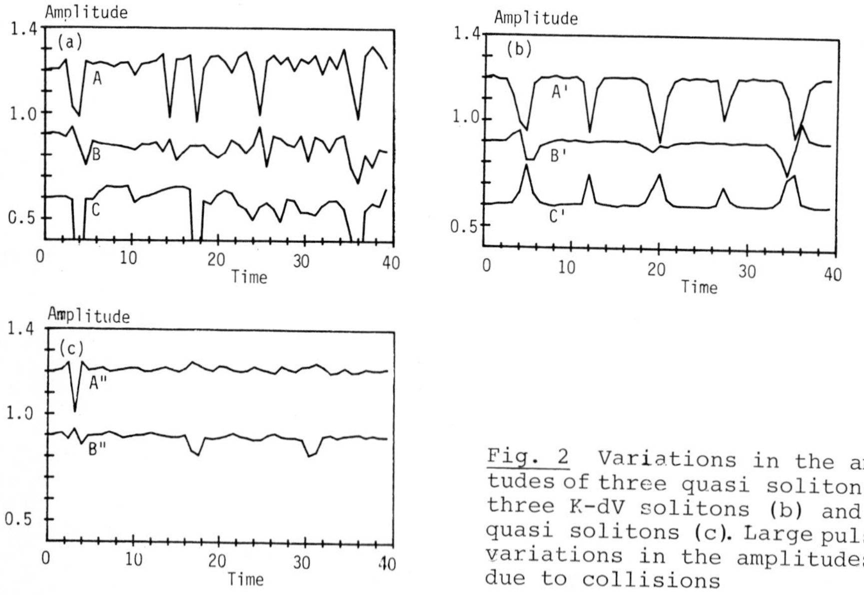

Fig. 2 Variations in the amplitudes of three quasi solitons (a), three K-dV solitons (b) and two quasi solitons (c). Large pulselike variations in the amplitudes are due to collisions

Figure 1 shows the development of the wave which is initially three separate quasi solitons with amplitudes 1.207, 0.907 and 0.602.

It should be noted that additional small waves emerge from the orignal waves, whose amplitudes vary through the mutual collisions.

Variations in three waves' amplitudes are shown in detail in Fig.2(a) from t=0 to 40. The three-wave interaction takes place at t=2~6 (see Fig. 2(a)).

After the interaction, the amplitudes of waves B and C slightly vary in the opposite sense to each other.

For t>20, the amplitudes of quasi solitons are scattered because two small waves with amplitudes about 0.1 emerge and coexist.

The imperfect soliton-like properties are evident in contrast to the properties of the K-dV solitons as in Fig. 2(b) (the K-dV equation computed is $u_t+uu_x+\delta^2=0$ and $\delta=0.022$, and the initial amplitudes of the solitons A', B' and C' are 1.2, 0.9 and 0.6, respectively).

On the other hand, a two-quasi-soliton solution is almost stable as shown in Fig. 2(c).

Any additional solitary waves do not emerge, however there exist very small ripples after the interaction at t=3.

Each quasi soliton's amplitude fluctuates around the initial value.

3. Characteristic Exponents

In order to examine the stability of the system described by (1), we estimate the largest characteristic exponent (Lyapunov number) [4].

Defining the distance d(t) between two waves u(x,t) and u'(x,t) by

$$d(t) = \sqrt{\int_0^h \{u(x,t)-u'(x,t)\}^2 dx} \quad , \tag{4}$$

we estimate the following quantity;

$$L_n(\tau,x,d) = 1/(n\tau)\sum_{i=1}^{n} \ln(d_i/d) \quad , \tag{5}$$

where h (=2) in (4) is the spatial period of the waves with conditions $u_{nx}(0,t)=u_{nx}(h,t)$ n=0,1,2,····. The notation of the symbols on the right-hand side of (5) is found in [5,6]. In the case of the K-dV equation, the distance d(t) of the two initially close orbits grows proportionally to time [4,7]. Then, L(t), an abbreviated expression of $L_n(\tau,x,d)$ and $t=\tau n$, varies as $L(t)=\ln(at+b)/t$, where a and b are constants. Further, it has been shown [4] that for small t, the dependence of L(t) of the quasi solitons described by (1) on time is similar to that of the solutions of the K-dV equation.

Therefore, we evaluate the difference $L^D(t)=L(t)-\ln(at+b)/t$ determining the values of a and b from L(t) of small t (usually t<30) by the least square method. Figure 3 shows $L^D(t)$ as a function of time for the cases of the K-dV equation and (1) for the same initial waves as in Fig. 2. The results indicate that for the three quasi solitons $L^D(t)$ is positive for sufficiently large t, while it converges to zero for the two quasi solitons and the K-dV solitons. Therefore, $L(\infty)$ is positive for the three quasi solitons, and as long-term behaviour, the distance between two initially close orbits grows as $d(t)\sim d(0)\exp(L(\infty)t)$.

Consequently, the correlation between the two whole waves vanishes for large t. This situation is called chaos.

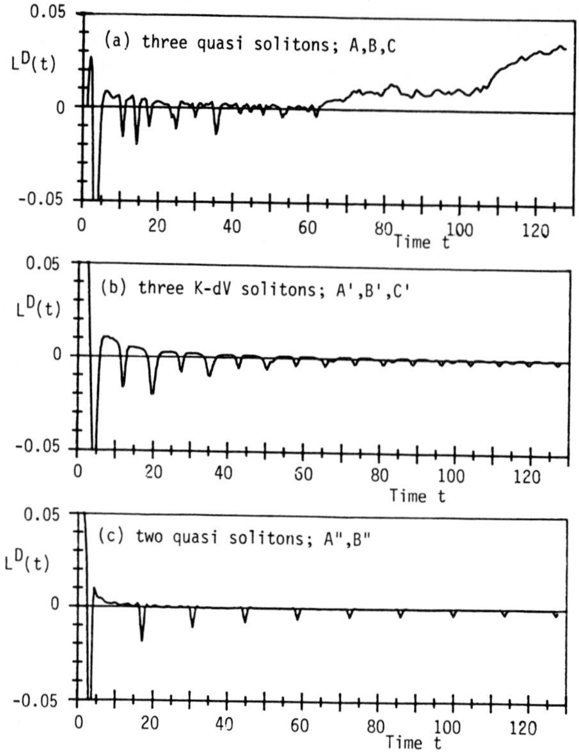

Fig. 3 $L^D(t)=L(t)-\ln(at+b)/t$ as a function of t, where a and b are determined by L(t) for t=0.8-30. Same initial waves as in Fig.2

On the other hand, the two quasi solitons are almost stable to the mutual collision and do not exhibit chaotic behaviour as shown in Fig. 3(c).

The chaotic behaviour of the system bearing quasi solitons results from the imperfect solitonlike properties of them. The quasi soliton is unstable in three-wave interaction. The amplitude and the position of each quasi soliton after the interaction sensitively depend on the whole waveform just before the interaction. The difference between two waves u and u' is enhanced by the interaction, and successive interactions due to the periodic boundary conditions make the distance grow exponentially [4,7].

References

1. H. Nagashima and K. Kuwahara: J. Phys. Soc. Jpn. 50, 3792 (1981)
2. G. Z. Tu and M. Z. Cheng: Computing Center of Academia Sinica (1979) preprint

3. T. Kawahara: J. Phys. Soc. Jpn. 33, 260 (1972)

4. H. Nagashima: To be published

5. G. Benettin, L. Galgani and J. Strelcyn: Phys. Rev. A14, 2338 (1976)

6. T. Nagashima and I. Shimada: Prog. theor. Phys. 58, 1318 (1977)

7. K. Yoshimura and S. Watanabe: J. Phys. Soc. Jpn. 51, 3028 (1982)

Part VII

Fluid Dynamics

Inviscid Singularity and Relative Diffusion in Intermittent Turbulence

H. Mori and K. Takayoshi

Department of Physics, Kyushu University 33
Fukuoka 812, Japan

1. Introduction

Fully developed turbulence exhibits outstanding asymptotic phenomena which obey power laws in length and time scale. In this talk we would like to summarize a scaling theory of the asymptotic phenomena and discuss a statistical variational principle which enables us to determine scaling exponents explicitly.

Let us consider turbulence whose small-scale fluctuations are statistically homogeneous and isotropic. When the Reynolds number R is sufficiently large, energy is transferred from large to small length scales and is distributed over a broad range from a large length scale L of the energy injection to a small length scale ℓ_d of the dissipation. This transfer of energy is due to the vortex stretching [1]. The vortex stretching leads to various interesting phenomena. First it generates an intermittency [2]. Then the energy spectrum takes the form

$$E(k) = Ck^{-\delta_0}(kL)^{-\mu\delta_1} \sim k^{-\delta}, \qquad (\delta = \delta_0+\mu\delta_1) \tag{1.1}$$

for wave numbers k satisfying $L^{-1} < k < \ell_d^{-1}$, where C is a positive constant independent of L and μ the intermittency exponent. For the three-dimensional energy cascade, we have $\delta_0=5/3$, $\delta_1=1/3$ [2,3] and experiments indicate $\mu = 0.25 \sim 0.5$ [4,5]. A variational principle of maximizing an information entropy of intermittency leads to $\mu \fallingdotseq 0.341$ in agreement with the experiments [6].

Secondly the vortex stretching brings about the relative diffusion of two particles [7]. Let $L_*^2(t)$ be the mean square of the relative distance of the two particles at time t. For steady turbulence, it takes the form

$$L_*^2(t) \sim t^{\psi} \tag{1.2}$$

for $L > L_*(t) >> L_*(0)$, where ψ is a positive exponent. For the 3d energy cascade, Richardson's 4/3 law leads to $\psi = 3$ [4]. However, recently Hentschel and Procaccia reanalyzed Richardson's data and obtained $\psi = 4.65$ [8]. A scaling theory leads to $\psi = 3/(1-\mu) \fallingdotseq 4.56$ in agreement with the experiments [7].

Thirdly the vortex stretching produces a singularity of vorticity field at a finite time t_* in the inviscid limit $R \to \infty$ so that an inviscid dissipation rate of a generalized enstrophy exists after t_*.

A general theory can be developed to determine the intermittency exponent μ, the diffusion exponent ψ, and the exponents of the inviscid singularities of generalized enstrophies explicitly for various normal cascades in a unified manner [9,10]. We shall discuss this theory by applying it to the 3d energy cascade, the 3d helicity cascade, the 2d enstrophy cascade and the temperature-fluctuation cascade in the Bénard convection.

A successful turbulence theory introduces a relevant reduction. For example, turbulence just after the onset is represented by a strange attractor with a finite dimensionality D. If $D \simeq 2$, then it is further reduced to a one-dimensional map. On the other hand, fully developed turbulence is described in terms of a self-similar cascade process in wave-number space or a self-similar vortex stretching. Such a reduction is useful for studying statistical quantities such as the time-correlation function of fluid velocity, the energy spectrum and transport properties.

In the present paper, we shall assume the self-similar vortex stretching. The self-similarity ensures power laws in length and time scale, and hence enables us to characterize turbulence by scaling exponents. The self-similarity also ensures the inviscid singularity at a finite time t_* as an accumulation point.

2. Inviscid dissipation rates

Basic quantities of turbulence are

$$Q_0 \equiv \langle u^2 \rangle /2, \quad \text{(mean energy)} \tag{2.1}$$

$$Q_1 \equiv \langle \underset{\sim}{u} \cdot \underset{\sim}{\omega} \rangle /2, \quad \text{(helicity)} \tag{2.2}$$

$$Q_2 \equiv \langle \omega^2 \rangle /2, \quad \text{(enstrophy)} \tag{2.3}$$

where $\underset{\sim}{u}$ is the fluid velocity and $\underset{\sim}{\omega} \equiv \text{rot}\ \underset{\sim}{u}$ the vorticity. Let us assume that turbulence is statistically homogeneous and isotropic. Then the average of quantities quadratic in $\underset{\sim}{u}$ and $\underset{\sim}{\omega}$, such as (2.1) $\sim$ (2.3), can be reduced to

$$Q_p = \int_0^\infty k^p\ E(k)\ dk, \tag{2.4}$$

where E(k) is the energy spectrum [2].

The inertial term of the Navier-Stokes equation conserves some of these quantities of lower orders, leading to

$$\eta_p \equiv -dQ_p/dt = 2\nu Q_{p+2} \tag{2.5}$$

for $p = 0, 1, \cdots, n$, where ν is the kinematic viscosity. The basic assumption of fully developed turbulence is that after an initial transient period, there exists a nonzero dissipation rate η_n in the inviscid limit $\nu \to 0$:

$$\infty > \eta_n \ (= \lim_{\nu \to 0} 2\nu Q_{n+2}) > 0. \tag{2.6}$$

Then cascade of Q_n is normal with the dissipation rate η_n, and dimensional analysis leads to $\delta_0 = (5+2n)/3$, $C \simeq \eta_n^{2/3}$ for (1.1). For the 3d nonhelical turbulence, we have n=0, $\delta_0 = 5/3$. For the 3d helical turbulence, n=1, $\delta_0=7/3$. For the 2d turbulence, n=2, $\delta_0=3$.

The existence of the inviscid dissipation rate η_n means that Q_p with $p \geq n+2$ diverges in the inviscid limit. This singularity arises from large wave numbers $k = \infty$ in the integral (2.4) and indicates that in an initial transient period, energy is transferred up to $k = \infty$ in a finite time t_*. This inviscid singularity ensures the existence of the inviscid dissipation rate. The building-up of the inviscid singularity and its critical exponents at t_* will be discussed later.

The dissipation rate η_p with $p < n$ vanishes in the inviscid limit $\nu \to 0$. Therefore the cascade of Q_p with $p < n$ is inverse [2].

In the Bénard convection, the energy Q_0 and the temperature fluctuation $Q_\theta \equiv \langle\theta^2\rangle/2$ are conserved without dissipation, and the quantities

$$Q_{\theta,p} = \int_0^\infty k^p \, E_\theta(k) \, dk, \qquad (Q_{\theta,0} \equiv Q_\theta) \tag{2.7}$$

are used instead of (2.4). Then a similar argument leads to $\delta_0 = 11/5$ [10,11]. The cascade of Q_θ is normal with a dissipation rate η_θ, whereas the cascade of the energy Q_0 is inverse.

3. Self-similar vortex stretching

A turbulent flow consists of a large number of eddies of various sizes [1,2]. Let us take one typical eddy of size L at time t=0 and consider its time evolution. The eddy is stretched and folded in time, and evolves into an assembly of smaller eddies of a mean size ℓ_t at time t. We assume that the time evolution is <u>self-similar.</u> The generation of intermittency means that the <u>volume V(t)</u> of the assembly of eddies decreases in time. Then the volume contraction ratio takes the form

$$\beta(t) \equiv V(t)/L^d = (\ell_t/L)^\mu \tag{3.1}$$

with a positive exponent μ, where d is the dimensionality of fluid space. It will turn out later that μ coincides with the intermittency exponent μ of (1.1).
Let $v(\ell)$ be the mean turnover velocity of eddies of size ℓ, and introduce the mean vorticity amplitude

$$\omega(\ell) \equiv v(\ell)/\ell \sim \ell^{-a}, \tag{3.2}$$

where $1 > a \geq 0$. The exponent a measures the degree of the enhancement of the enstrophy by the vortex stretching. The exponent δ of (1.1) is related to a by $\delta = 3+\mu-2a$ [9].

In order to know

1) what determines μ and a,
2) how the inviscid singularity is built up,

let us take a discrete cascade model. Namely, consider a hierarchy of eddies whose length scales are

$$\ell_i = L/b^i, \qquad (i = 0,1,2,\cdots) \tag{3.3}$$

with $b > 1$, and assume that eddies of size ℓ_i disintegrate into eddies of size ℓ_{i+1} with a mean lifetime τ_i which is of the order of the mean turnover time, i.e.,

$$\tau_i = 1/\omega(\ell_i) \sim \ell_i^{\,a}. \tag{3.4}$$

The condition (3.4) specifies the magnitude of b, and the wave-number doubling process of the dynamical interaction suggests $b \simeq 2$ [2,3].

The time $t=t_i$ at which $\ell_t=\ell_i$ is given by

$$t_i = \sum_{j=0}^{i-1} \tau_j = t_*(L)[1 - \Delta^{-i}], \tag{3.5}$$

where $\Delta \equiv \tau_i/\tau_{i+1}=b^a$ and

$$t_*(L) \equiv \tau_0/[1-\Delta^{-1}] = B/a\omega(L) \tag{3.6}$$

with $B \equiv a/[1-b^{-a}]$. Since $\Delta^{-i}=(\ell_t/L)^a$, (3.5) is inverted to give

$$r(t) \equiv \ell_t/L = [1 - (t/t_*)]^{1/a}. \tag{3.7}$$

It follows from (3.2) and (3.7) that the mean vorticity amplitude of the eddies of size ℓ_t is given by

$$\omega(t) \equiv \omega(\ell_t) = B/a[t_* - t]. \tag{3.8}$$

The eddy size ℓ_t reaches the dissipation length ℓ_d at a time

$$t_d \equiv t_*[1 - (\ell_d/L)^a]. \tag{3.9}$$

After this time the eddies are dissipated by viscosity. At the crossover time $\omega(t_d)=\omega(L)(L/\ell_d)^a$. Therefore, if $a > 0$, then the eddy size ℓ_t vanishes and $\omega(t)$ diverges at a finite time t_* in the inviscid limit $\ell_d \to 0$. If $a = 0$, on the other hand, $B = 1/\ln b$ and t_* becomes infinite so that (3.8) reduces to $\omega(t) = \omega(L)$. As will be shown later, we have $a > 0$ for the 3d turbulence and $a = 0$ for the 2d turbulence. Therefore (3.8) represents the inviscid singularity of the 3d turbulence. The singularity is the simple pole irrespective of a.

It follows from (3.5) that the divergence time t_* is the accumulation point of the sequence $\{t_i\}$. Therefore the inviscid singularity arises from the fact that the self-similar vortex stretching accumulates to a finite time t_* at a convergence ratio Δ^{-1} (< 1). The situation is similar to the period-doubling bifurcations for the onset of turbulence.

4. Inviscid singularities of generalized enstrophies

Let us take a small interval of wave number $\Delta k_i \equiv (k_i/\sqrt{b},\sqrt{b}k_i)$ around $k_i = 1/\ell_i$. Then the part of Q_p carried by the eddies of sizes in this interval is given by

$$\Delta Q_p(\ell_i) \equiv \int_{k_i/\sqrt{b}}^{\sqrt{b}k_i} k^p\, E(k)\, dk. \qquad (4.1)$$

It follows from (3.1) that the eddies of size ℓ_i occupy only a fraction $(\ell_i/L)^\mu$ of the space [2,3]. Hence

$$\Delta Q_p(\ell) \sim (\ell/L)^\mu\, v^2(\ell)/\ell^p \sim \ell^{\mu+2-2a-p}. \qquad (4.2)$$

The eddies of sizes in the interval Δk_i disintegrate with the mean lifetime τ_i. Therefore the mean transfer rate of Q_n from small to large wave numbers is given by

$$\eta_n = \Delta Q_n(\ell_i)/\tau_i \sim \ell_i^{\mu+2-3a-n}. \qquad (4.3)$$

Since the cascade of Q_n is normal, (4.3) must be independent of ℓ_i and agree with the inviscid dissipation rate. This leads to

$$a = (2-n+\mu)/3, \qquad (4.4)$$

whence $\delta=3+\mu-2a = (5+2n+\mu)/3$. Thus we obtain a and δ for various normal cascades as listed in Table 1.

It follows from (4.2) and (4.4) that

$$\Delta Q_p(\ell) \sim \ell^{a+n-p}. \qquad (4.5)$$

Table 1. Characteristic exponents of normal cascades+)

cascade	d	δ	a	μ	a	$\alpha\equiv a-\mu$
$\langle u^2\rangle/2$	3	$(5+\mu)/3$	$(2+\mu)/3$	0.341	0.780	0.439
				$\simeq$ 1/3	7/9	4/9
$\langle \underset{\sim}{u}\cdot\underset{\sim}{\omega}\rangle/2$	3	$(7+\mu)/3$	$(1+\mu)/3$	0.203	0.401	0.198
				$\simeq$ 1/5	2/5	1/5
$\langle \omega^2\rangle/2$	2	$(9+\mu)/3$	$\mu/3$	0	0	0
$\langle \theta^2\rangle/2$	3	$(11+3\mu)/5$	$(2+\mu)/5$	0.261	0.452	0.191
				$\simeq$ 1/4	9/20	1/5

+) $\underset{\sim}{\omega}$=rot $\underset{\sim}{u}$, and $\langle \underset{\sim}{u}\cdot\underset{\sim}{\omega}\rangle/2$ is the helicity, $\langle\omega^2\rangle/2$ the enstrophy, $\langle\theta^2\rangle/2$ the temperature fluctuation in the Bénard convection.

Let us consider the assembly of eddies evolved from eddies of size L in a time t. The Q_p of this assembly is given by

$$Q_p(t) = \int_{1/L}^{1/\ell_t} k^p\, E(k)\, dk \sim \ell_t^{-(p-n-a)} \qquad (4.6)$$

if $p > n+a$ and $\ell_t << L$. Inserting (3.7) into (4.6) leads to

$$Q_{n+2q}(t) \sim (t_* - t)^{-\gamma_q}, \qquad \gamma_q \equiv (2q-a)/a \qquad (4.7)$$

if $q > a/2$ and $(t_* - t) << t_*$. This represents the inviscid singularity of generalized enstrophies.

5. Variational principle for determining μ

Let M_i be the number of eddies of size ℓ_i per unit mass at time t. Following the β model [3], let us assume that each eddy of size ℓ_i disintegrates into N eddies of size ℓ_{i+1}. The offspring number N can be determined from (3.1) and (3.3) as

$$N = [V(t_{i+1})/\ \ell_{i+1}^d]/[V(t_i)/\ell_i^d] = b^D \qquad (5.1)$$

with $D \equiv d-\mu$. It is reasonable to assume that the disintegration of eddies obeys the exponential decay law with the mean lifetime τ_i [6]. Then

$$dM_i/dt = -(M_i/\tau_i) + b^D (M_{i-1}/\tau_{i-1}). \qquad (5.2)$$

In the steady state, (5.2) vanishes, leading to

$$M_i/M_0 = b^{(D-a)i} \sim \ell_i^{-D+a}. \qquad (5.3)$$

Using the sum of the volumes of eddies of sizes ℓ_i over $i \geq 1$,

$$\Omega(\mu) = \sum_{i=1}^{\infty} \ell_i^d\, M_i = L^d\, M_0/(b^{\mu+a} - 1), \qquad (5.4)$$

let us introduce the ratio

$$f(\mu) \equiv \Omega(\mu)/\Omega(\mu=0) = \frac{b^{a_0} - 1}{b^{\mu+a} - 1}, \qquad (5.5)$$

where $a_0 \equiv a(\mu=0)$. Then the information entropy of intermittency is given by [6]

$$H(\mu) \equiv -f \ln f - (1-f) \ln (1-f). \qquad (5.6)$$

This entropy represents the degree of the generation of intermittency. Let us assume that this becomes maximum for fully developed turbulence. Then, since $H(\mu)$ takes the maximum value $\ln 2$ at $f(\mu)=1/2$, (5.5) leads to

$$\mu + a = \frac{1}{\ln b} \ln (2b^{a_0} - 1). \qquad (5.7)$$

Let us put $a=a_0+\mu a_1$. Then $\mu \simeq a_0/(1+a_1)$ if $a_0 \ln b << 1$. Otherwise μ depends on b weakly. In fact $\mu \simeq \ln 2/(1+a_1)\ln b$ for $a_0 \ln b >> 1$.

We take $b = 2$ as suggested by the wave-number-doubling process of the dynamical interaction. Then we obtain values of μ and a listed in Table 1. The value of μ for the 3d energy cascade agrees with experiments on $\mu = 0.25 \sim 0.5$ [4,5]. The value for the 2d enstrophy cascade also agrees with numerical experiments $\mu = 0$ [12]. There exist no experiments on the other cascades.

The generalized enstrophy $Q_{n+2}(t)$, given by (4.7), and $Q_{\theta,2}(t)$, obtained from (2.7), have the inviscid singularity of the form $(t_*-t)^{-\gamma_1}$ with

$$\gamma_1 = \frac{2-a}{a} = \begin{cases} \frac{4-\mu}{2+\mu} = 1.56, & \text{(3d energy)} \\ \frac{5-\mu}{1+\mu} = 3.99, & \text{(3d helicity)} \\ \frac{8-\mu}{2+\mu} = 3.42, & \text{(Bénard).} \end{cases} \qquad (5.8)$$

For the 3d energy cascade, $\gamma_1=1.56$, $\gamma_2=4.16$, $\gamma_3=6.69$. A numerical study [13] leads to $\gamma_1 \simeq 0.8$, $\gamma_2 \simeq 4.2$, $\gamma_3 \simeq 9.9$. The agreement is fairly good but not satisfactory.

It follows from (5.3) and (3.3) that the number of eddies of sizes between ℓ and $\ell+d\ell$ is given by

$$M(\ell)\, d\ell = M_i di \sim \ell^{-D+a-1}\, d\ell. \qquad (5.9)$$

It follows from (3.6) that eddies of size ℓ produce the inviscid singularity at time $t_*(\ell) \sim \ell^a$. Therefore the inviscid singularities are distributed continuously on the time axis, whose distribution between t and t+dt is given by

$$N(t)\, dt = M(\ell)d\ell \sim t^{-D/a}\, dt. \qquad (5.10)$$

It is interesting that the distribution of the inviscid singularities is characterized by the fractal dimension D.

6. Relative diffusion in steady turbulence

It was shown [10] that the diffusion rate takes the form

$$\frac{d}{dt} L_*^2 = \begin{cases} \tilde{A}\, L_*^2\, \omega(L_*)\,(L_*/L)^{\mu} & \text{for } L > L_* \gg \ell_d \\ \tilde{A}\, L^2\, \omega(L) & \text{for } L_* > L, \end{cases} \tag{6.1}$$

where $\tilde{A} = 2\xi\phi(a+\mu)/2\mu B$, with ϕ being the folding exponent of the eddy elongation and ξ a positive number of order unity. If $\mu = 0$, then $\tilde{A} = 2\xi\phi/B$. Inserting (3.2) into $\omega(L_*)$ of (6.1) leads to

$$dL_*^2/dt = A\, L_*^{2-\alpha}, \qquad (\alpha \equiv a-\mu \geqq 0) \tag{6.2}$$

for $L > L_* \gg \ell_d$, where $A = \tilde{A}\, L^{\alpha}\omega(L)$. Exponent α takes different values as listed in Table 1.

For the 2d turbulence, $\alpha=0$ and (6.2) leads to $L_*^2(t) = L_*^2(0)\exp(At)$. This exponential growth agrees very well with a constant-level balloon experiment in the stratosphere on a large geophysical scale [14].

For $\alpha > 0$, (6.2) leads to (1.2) with $\psi = 2/\alpha$. For the 3d energy cascade, we obtain $2-\alpha = 1.56$, $\psi = 4.56$ in agreement with experiments [8]. We obtain $\psi = 10.1$ for the 3d helicity cascade, and $\psi = 10.5$ for the Bénard convection. There are, however, no experiments for these cases.

7. Decaying turbulence

Turbulence without energy injection decays in time. Then the Reynolds number R_t at time t decays as

$$R_t = R_0/(1+pt)^z, \qquad (1 \geqq z > 0) \tag{7.1}$$

with a decay constant p [15]. The decay occurs very slowly if $R_t \gg R_{crit}$, where R_{crit} is the critical Reynolds number which is of order 10^3. Indeed the lifetime of turbulence is estimated to be

$$T_{turb} \simeq 10^5 t_*(L) \tag{7.2}$$

for $R_0/R_{crit} = 10^2$, $z = 3/7$ [15]. Then formulas for the steady case can be extended to the decaying case by using the quasi-equilibrium assumption that the quantities local in time evolve in time only through R_t and

$$L_t = L_0(1+pt)^{(1-z)/2}, \tag{7.3}$$

which L_t represents the length scale of the largest eddies of the normal cascade range at time t.

Thus it turns out that (3.7) can be extended to

$$r(t) \equiv \ell_t/L = [1-(t'/t_*)]^{1/a} \tag{7.4}$$

Table 2. Diffusion exponent ψ of the relative diffusion

	d	$L_*^2(t) \sim t^{\psi}$	ψ	
steady ($p = 0$)	2	e^{At}, $(\alpha = 0)$	∞	$\langle \omega^2 \rangle /2$
$L_*(t) < L$	3	$t^{2/\alpha}$, $(\infty > \psi = \frac{2}{\alpha} > 2)$	10.5	$\langle \theta^2 \rangle /2$
			10.1	$\langle \mathbf{u} \cdot \boldsymbol{\omega} \rangle /2$
			4.56	$\langle u^2 \rangle /2$
$L_*(t) \gg L$		t	1.	purely random
decaying ($p \neq 0$)	2	$(1+pt)^{A/p}$, $(\alpha = 0)$		
$L_*(t) < L_t$	3	$[(1+pt)^{1-x}-1]^{2/\alpha}$, $(x \equiv 1-\alpha(1-z)/2)$		
		$[\ln(1+pt)]^{2/\alpha}$, $(x = 1)$		
$L_*(t) \gg L_t$		t^{1-z}, $(1 > \psi = 1-z > 0)$	0.80	$z = 1/5$ *)
			0.57	$z = 3/7$
		$\ln t$, $(z = 1)$	0	

*) z is related to the exponent σ of the large-scale structure of the initial energy spectrum $E(k,t=0) \sim k^{\sigma}$, $(kL \ll 1)$ by $z=(\sigma-1)/(\sigma+3)$. Hence $z=1/5$ for $\sigma=2$, $z=1/3$ for $\sigma=3$, $z=3/7$ for $\sigma=4$ and $z=1$ for $\sigma \to \infty$.

with a modified time

$$t' \equiv [(1+pt)^{1-y}-1]/(1-y)p, \tag{7.5}$$

where $y \equiv 1-a(1-z)/2$ [10]. This can be transformed into

$$r(t) \equiv \ell_t/L \sim [1-(t/t_*')]^{1/a} \tag{7.6}$$

in the vicinity of t_*', where

$$t_*'(L) \equiv \{[1+(1-y)pt_*(L)]^{1/(1-y)}-1\}/p. \tag{7.7}$$

Therefore the inviscid singularities (3.8) and (4.7) also hold near $t_*^!$ with the unchanged critical exponents but with a shifted divergence time $t_*^!$.

The relative diffusion is also modified and the results are summarized in Table 2 together with the diffusion laws for the steady case.

8. Concluding remarks

We have developed a general theory of fully developed turbulence which determines ψ and μ explicitly in agreement with experiments on the 3d energy cascade and the 2d enstrophy cascade, and also predicts μ and ψ of the 3d helicity cascade and the Bénard convection.

The basic assumptions used are

1) a reduction to the self-similar vortex stretching which is characterized by scaling exponents a and μ,
2) the variational principle of maximum intermittency entropy which enables us to determine the scaling exponents explicitly.

It is also an interesting problem to obtain a dynamical equation for determining the convergence ratio $\Delta^{-1}=b^{-a}$. The variational principle corresponds to the variational principle for the choice of a relevant invariant measure in dynamical-system theory. Further study of these assumptions would require an ergodic theory of turbulence based on the Navier-Stokes equations [16].

References

[1] D.J. Tritton, Physical Fluid Dynamics (Van Nostrand-Reinhold, New York, 1977).
[2] H.A. Rose and P.L. Sulem, J. Phys. 39 (1978) 441, and references cited therein.
[3] U. Frisch, P.L. Sulem and M. Nelkin, J. Fluid Mech. 87 (1978) 719.
[4] A.S. Monin and A.M. Yaglom, Statistical Fluid Mechanics, vol. 2 (MIT Press, Cambridge, 1975), Chap. 25.
[5] M. Nelkin, Phys. Fluids 24 (1981) 556.
[6] H. Fujisaka and H. Mori, Prog. Theor. Phys. 62 (1979) 54.
[7] K. Takayoshi and H. Mori, Prog. Theor. Phys. 68 (1982) 439.
[8] H.G.E. Hentschel and I. Procaccia, Phys. Rev. A27 (1983) 1266.
[9] H. Mori, Prog. Theor. Phys. 69 (1983) 756.
[10] H. Mori and K. Takayoshi, Prog. Theor. Phys. (to be published).
[11] R. Bolgiano, J. Geophys. Res. 64 (1950) 2226.
[12] B. Fornberg, J. Comp. Phys. 25 (1977) 1.
[13] R.H. Morf, S.A. Orszag and U. Frisch, Phys. Rev. Lett. 44 (1980) 572.
[14] P. Morel and M. Larcheveque, J. Atmos. Sci. 31 (1974) 2189.
[15] H. Mori and K. Takayoshi, Prog. Theor. Phys. 69 (1983) 725.
[16] D. Ruelle, Prog. Theor. Phys. Suppl. No.64 (1978) 339.

Computational Synergetics and Innovation in Wave and Vortex Dynamics

N.J. Zabusky

Department of Mathematics and Statistics, University of Pittsburgh
Pittsburgh, PA 15260, USA

ABSTRACT. It is demonstrated how the computer, used in a heuristic mode, has greatly augmented our understanding of the mathematics of nonlinear dynamical processes. Examples are given of recent work in soliton mathematics (waves) and contour dynamics - a boundary integral evolutionary method that is applicable to a wide class of 2D flows. The role of good graphics in enhancing the discovery, retention and communication of new mathematical properties of equations is illustrated.

I. INTRODUCTION

Man, by his make-up, seems limited in what he can do mathematically with non-linear problems in the natural sciences. Someday we may learn how the left and right side of the brain interact during the creative process and how they affect and are affected by the underlying electrochemical milieu of the nervous system. Then, perhaps, we will be able to quantify this limitation.

In recent years these limitations often led to choices of direction that were made for convenience, rather than for their probes into fundamental areas. This view was expressed with amazing prescience in 1946 by John von Neumann, when he asked [1]: "To what extent can human reasoning in the sciences be more efficiently replaced by mechanisms?" and: "What phases of pure and applied mathematics can be furthered by the use of large-scale, automatic computing instruments?" He noted that:

> Our present analytical methods seem unsuitable for the solution of the important problems arising in connection with non-linear partial differential equations and, in fact, with virtually all types of non-linear problems in pure mathematics. The truth of this statement is particularly striking in the field of fluid dynamics. Only the most elementary problems have been solved analytically in this field... .
>
> The advance of analysis is, at this moment, stagnant along the entire front of non-linear problems. That this phenomenon is not of a transient nature but that we are up against an important conceptual difficulty...yet no decisive progress has been made against them... which could be rated as important by the criteria that are applied in other, more successful (linear!) parts of mathematical physics... .
>
> ...we conclude by remarking that really efficient high-speed computing devices may, in the field of non-linear partial differential equations as well as in many other fields which are now difficult or entirely denied of access, provide us with those heuristic hints which are needed in all parts of mathematics for genuine progress... . This should ultimately lead to important analytical advances.

This view was discussed in an inspiring way by S. M. Ulam in a chapter of his 1960 monograph entitled, 'Computing Machines, a heuristic aid - "synergesis",' where the role of good graphics is discussed [2]. This medium, namely the visualization of mathematics will be the message of my talk. I will demonstrate to you how signatures that arise in well-chosen graphical displays can provide foci for ideation and concept formation in the well-prepared investigator. This synergy between computation and analysis will enhance the rate and depth of mathematical understanding of the sciences. At times, one picture which brings out evolving coherent patterns is worth more than a thousand equations. The appropriate graphical display (and I will discuss some below) when turned around in a rapid and interactive fashion will improve our ability to choose wisely among many directions and will provide a link to the more conventional approaches of theorem proving, analysis and asymptotic approximation.

My discussion will include two concrete problems: 1) Nonlinear lattics dynamics in one dimension - the Fermi-Pasta-Ulam (FPU) problem and its affect on the study of nonlinear waves; 2) Vortex dynamics in two dimensions (another one touched on by von Neumann) as described by the contour dynamical representation of the Euler equations.

II. WAVES

In the early fifties, Fermi, Pasta and Ulam [3] considered the equilibration of energy in nonlinear dynamical systems, and their natural intuition led them to the hypothesis:

H1. Given a one-dimensional lattice with fixed boundary conditions and with identical nearest-neighboring masses coupled by identical nonlinear springs. If the energy of the system is initially in a long-wavelength state, the energy will eventually be shared "equally" among all the degrees of freedom of the system.

In fact they observed the contrary, namely near-recurrence to the initial state and only the highest modes participating significantly in the dynamics. This led Kruskal and me to a second hypothesis:

H2. The near-recurrence phenomenon and the detailed modal history can be explained by a continuum model with periodic boundary conditions and progressive wave initial conditions.

This was verified computationally and for the "cubic" lattice Kruskal derived a Korteweg-de Vries (KdV) equation. We continued with the hypothesis:

H3. The Korteweg-de Vries equation

$$u_t + uu_x + \delta^2 u_{xxx} = 0 \qquad (1)$$

can describe propagation of small-but-finite amplitude long waves on a lossless "cubic" lattice excited by a progressive wave initial condition.

This was verified by solving (1) numerically, and Fig. 1 shows the trajectory of max $u(x,t)$ over an interval of time where "soliton" interactions are indicated.

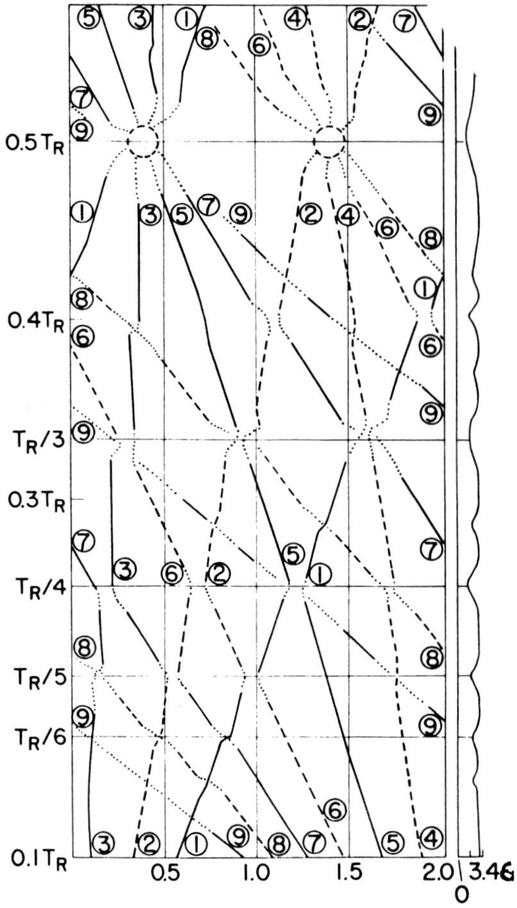

Fig. 1. Trajectories of maxima of the Korteweg-de Vries equation ($\delta = 0.022$) on a space-time diagram beginning at $t = 0.1t_R = 3.04t_B$

The modified KdV equation (u^2u_x replaces uu_x) had similar properties and both equations seemed to have a large number of corresponding conservation law forms. That is, both equations can be written in the form $T_t^{(n)} + X_x^{(n)} = 0$ where $T^{(n)}$ and $X^{(n)}$ were polynomial functions of u and its spatial derivatives, and $n = 1,2,\ldots$.

This milieu of computational and analytical results led R. Miura to prove the theorem.

THEOREM. If v is a solution of the modified KdV equation

$$Q(v) \equiv v_t - 6v^2v_x + v_{xxx} = 0 \tag{2}$$

then

$$u = v^2 + v_x \tag{3}$$

is a solution of the KdV equation

$$P(u) \equiv u_t - 6uu_x + u_{xxx} = 0. \tag{4}$$

Equation (3) is a Ricatti equation which can be transformed to a Sturm-Liouville problem $[\psi_{xx} + (\lambda - u(x,t))\psi = 0]$ and the problem is exactly linearized.

Much beautiful mathematics followed this pioneering step, including: identifying and solving the Toda lattice, the first of many exactly solvable discrete systems; an operator approach based on the invariance of λ and the conservation laws (P. D. Lax); the KdV equation can be written as a Hamiltonian system (C. Gardner, L. Fadeev, V. E. Zakharov); an ingenious direct method (R. Hirota); application to the nonlinear Schrödinger equation, a first step in generalization (V. E. Zakharov and A. B. Shabat); and an inverse (or isospectral) transform method for dealing with a wide class of nonlinear partial differential equations (M. Ablowitz, D. Kaup, A. Newell, H. Segur). In the last decade the exponential surge of beautiful mathematics and applications to natural phenomena has continued. For example, there have been several studies to investigate the validity of the KdV equation as a model for shallow-water waves. The solitary wave and soliton phenomena were first carefully observed by J. Scott Russel in 1834 for waves generated in canals. For a more complete account of these developments see Reference [4].

III. GRAPHICS

FPU plotted waveforms $y_n(t)$ and modal energies vs time. When it became clear that progressive waves with periodic boundary conditions contained the same effect, we began tracking trajectories of waveform extrema; this proved informative for it allowed us to see phase shifts arising from localized entity interactions. That is, as our analytical insight matured, the character of the graphical representation became focussed on particular phenomena. For example, a "Riemann-invariant" or progressive wave filtering was applied successfully to a cubic lattice with periodic boundary conditions that was strongly excited with a low-mode progressive wave [5].

The oblique projections of $\psi(x,t)$ developed by the Manchester group (Bullough, Eilbeck, Caudrey) have provided useful and artistically pleasing global summaries of nonlinear wave phenomena. The use of shading and color in still and cinematic displays can enhance the perception of small unexpected phenomena. Such displays also provide "parameters" to aid the retention and communication process. They enhance the mind's ability to recall and correlate old and new results. For problems of waves and fluctuating or turbulent fluids it would be desirable to develop algorithms to track and plot extremal lines, areas and volumes in one, two, and three dimensions. The trajectories of the centroids of these regions will probably be "chaotic" in places and small scales will be distorted by numerical errors. Thus, appropriate spatial and temporal correlations of the low-order moments of these regions will greatly enhance the signal-to-noise ratio and augment our ability to recognize patterns.

IV. VORTICES

Many computational studies of the evolution of the 2D Euler equations have been made in the last two decades [6,7,8,9]. In vorticity-stream function (ζ,ψ) form these are

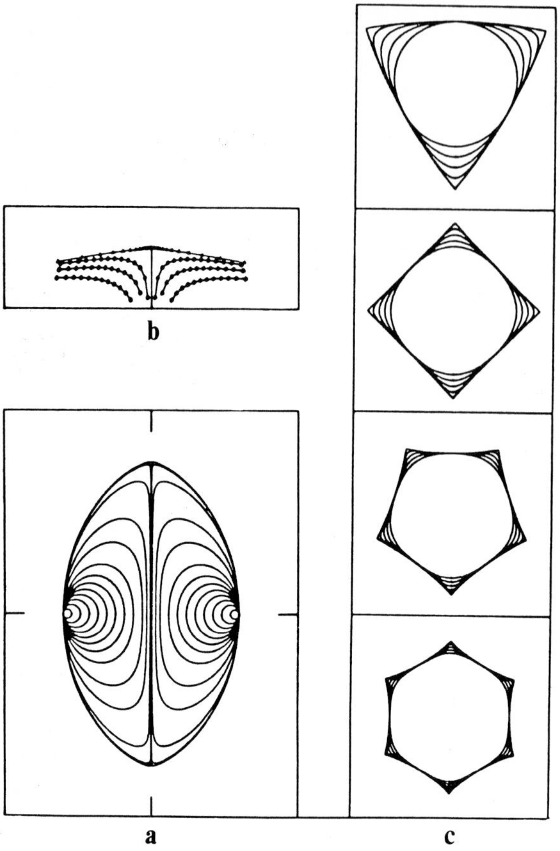

Fig. 2. Translating and Rotating V states. (a) Translating; (b) A magnified view of corner region for translating V states; (c) Rotating V states with limiting cases

$$\zeta_t + \psi_x\zeta_y - \psi_y\zeta_x = 0, \quad \Delta\psi = -\zeta. \tag{5}$$

Point-vortex interaction studies go back to Rosenhead and others in 1930's. In the last two decades finite-difference, spectral and vortex-in-cell studies have also elucidated dynamical phenomena. The contour dynamics (CD) approach [10], because of its clarity and precision, has recently opened the door to a deeper mathematical understanding of inviscid phenomena.

In CD we assume that the velocity field is created by piecewise-constant localized regions of vorticity or FAVR's. Thus, using Green's theorem, we may convert 2D domain integrals for the velocity (u,v) into 1D contour integrals.

$$(x_t,y_t)_{\partial D} \equiv (u,v)_{\partial D} = (2\pi)^{-1} \sum_{j=1}^{N_c} [\zeta_j] \int_{\partial D_j} \log \ell(d\xi,d\eta), \tag{6}$$

where $[\zeta_j]$ is the jump in vorticity (inside-outside) at ∂D_j and N_c is the total number of contours. A multitude of "free-boundary" steady states, that we call "V states", have been found [11]. Sets of singly connected rotating and

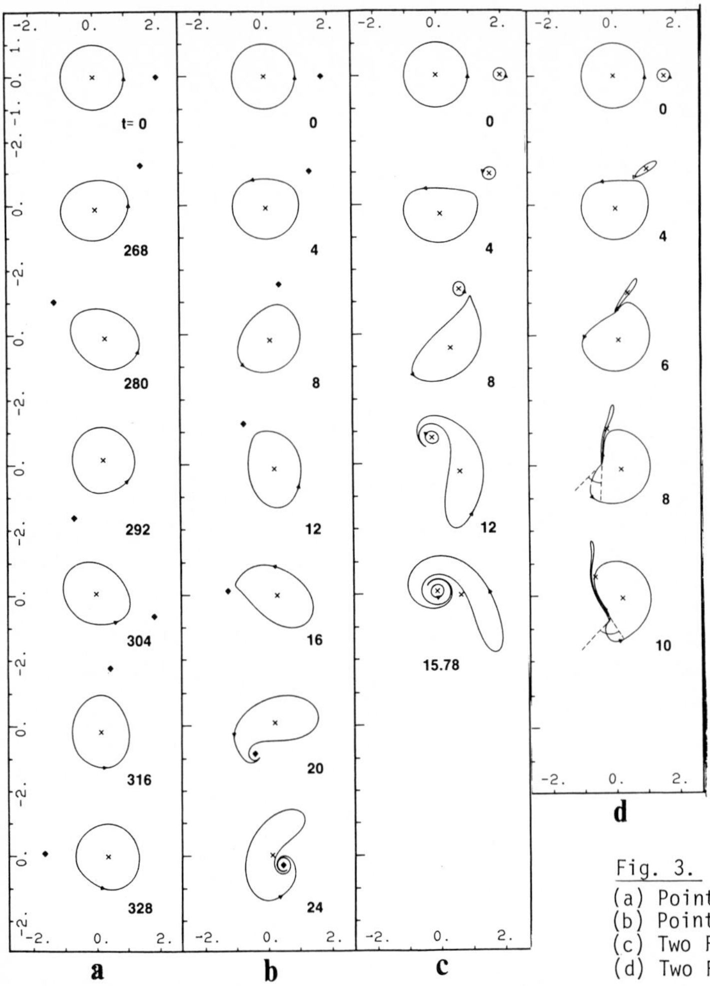

Fig. 3. Merger and nonmerger.
(a) Point vortex and FAVR;
(b) Point vortex and FAVR;
(c) Two FAVR's;
(d) Two FAVR's

symmetric translating V states including limiting states with 90° corners have been recently given by Wu, Overman, and Zabusky [12]. (See Fig. 2.) The linear stability of "wake-like" periodic configurations of vorticity have been investigated and the role of finite area on stabilization has been partially clarified [13,14]. Dynamical studies have been made, including merger (or "pairing") of isolated vortex regions [15] (see Fig. 3) and the coaxial scattering of translating V states [16]. In the latter paper we have also formed an asymmetric translating state through a dynamic process of "exchange" of FAVR's. That is

$$\overset{\rightarrow}{\begin{pmatrix} a \\ -a \end{pmatrix}} + \overset{\leftarrow}{\begin{pmatrix} -b \\ b \end{pmatrix}} \rightarrow \overset{\nearrow}{\begin{pmatrix} a^* \\ -b^* \end{pmatrix}} + \overset{\searrow}{\begin{pmatrix} -a^* \\ b^* \end{pmatrix}},$$

where the left side indicates the incoming head-on V states and the right side, the outgoing states. (See Fig. 4.) A close examination of diagnostics (e.g., perimeter and maximum curvature variation) shows that this state is "near" a V state. These processes occur in the laboratory and ocean when "coherent" vortex states interact. A computer-generated film illustrating this process will be shown.

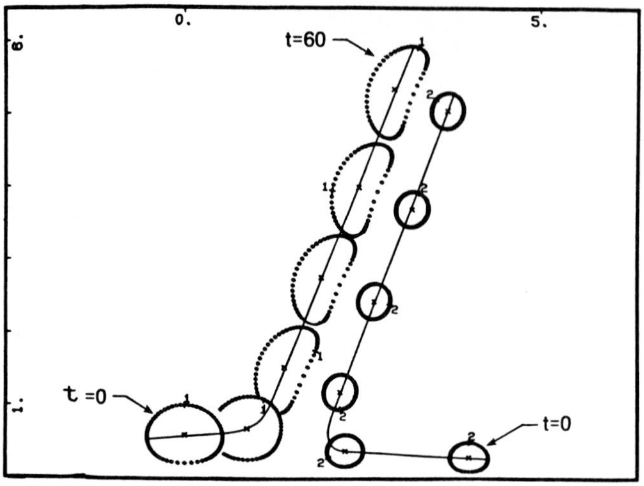

Fig. 4. Coaxial Scattering of Translating V states $\Gamma_2/\Gamma_1 = -1.0$

V. CONCLUSIONS

To me it is clear that the soliton paradigm validates von Neumann's foresight. Not only has it engendered much new activity in pure and applied mathematics, but it provides a new conceptual basis for applications in diverse areas of the natural sciences. For realistic systems that are near-integrable, the soliton concept provides an economy of thought in posing problems and obtaining solutions.

The present paper is an attempt to show concretely that the analytical-computational synergetic approach is a mode of working that is applicable generally in the natural sciences. It requires cogent analysis, new algorithms (like the contour dynamics representations), and good computation. But it also requires good graphics and modes of computer expression yet to be invented.

VII. ACKNOWLEDGMENTS

The work described in Sections IV and V was supported over several years by the Mathematics Division of the U.S. Army Research Office.

REFERENCES

1. H. H. Goldstine and J. von Neumann, On the principles of large scale computing machines, in "Collected Works of John von Neumann" (A. Taub, Ed.), Vol. 5, pp. 1-32. Macmillan, New York, 1963. The material in this paper was first given as a talk on May 15, 1946. In the opening sections the authors discuss the difficulty of nonlinear problems and describe how they propose to use the digital computer to break the deadlock. Also see Recent theories in turbulence, in "Collected Works of John von Neumann" (A. Taub, Ed.), Vol. 6, pp. 437-472. This paper was issued as a report in 1949. On p. 469, we find a lucid formulation of the synergetic approach.

2. S. M. Ulam, "A Collection of Mathematical Problems," Wiley-Interscience, New York, 1960. See Chap. VII, Sect. 8: "Physical Systems, Nonlinear Problems"; and Chap. VIII, Sect. 10: "Computing Machines as a Heuristic Aid-Synergesis."

3. S. M. Ulam, Introduction to "Studies of Nonlinear Problems" by E. Fermi, J. Pasta and S. M. Ulam in "Collected Papers of Enrico Fermi," Vol. II, Univ. of Chicago Press, Chicago, 1965.

4. N. J. Zabusky, "Computational synergetics and mathematical innovation," J. Comput. Phys. 43, (1981), 195-249.

5. N. J. Zabusky, Solitons and energy transport in nonlinear lattices, Comp. Phys. Comm. 5, 1 (1973).

6. N. J. Zabusky, Coherent structures in fluid mechanics. In The Significance of Nonlinearity in the Natural Sciences. Plenum Press, New York, 1977.

7. P. G. Saffman and G. R. Baker, "Vortex interactions," Ann. Rev. Fluid Mech. 11, 95-122 (1979).

8. A. Leonard, "Vortex methods for flow simulation," J. Comput. Phys. 37, 289-335 (1980).

9. H. Aref, "Integrable, chaotic, and turbulent vortex motion in two-dimensional flows," Ann. Rev. Fluid Mech. 15, 345-89 (1983).

10. N. J. Zabusky, M. H. Hughes, and K. V. Roberts, "Contour dynamics for the Euler equations in two dimensions," J. Comput. Phys. 30, 96-106 (1979).

11. G. S. Deem and N. J. Zabusky, "Vortex waves: Stationary 'V-states', interactions, recurrence, and breaking," Phys. Rev. Lett. 40, 859-62 (1978).

12. H. M. Wu, E. A. Overman, II, and N. J. Zabusky, "Steady-state solutions of the Euler equations in two dimensions: Rotating and translating V-states with limiting cases. I. Numerical algorithms and results," J. Comput. Phys. In press (1983).

13. P. G. Saffman and J. C. Schatzman, "Stability of a vortex street of finite vortices," J. Fluid Mech. 117, 171-85 (1982).

14. S. Kida, "Stabilizing effects of finite core on Karman vortex street," J. Fluid Mech. 122, 487-504 (1982).

15. E. A. Overman, II and N. J. Zabusky, "Evolution and merger of isolated vortex structures," Phys. Fluids 25, 1297-1305 (1982).

16. E. A. Overman, II and N. J. Zabusky, "Coaxial scattering of Euler-equation translating V-states via contour dynamics," J. Fluid Mech. 125, 187-202 (1982).

A Scalar Model of MHD Turbulence

A. Pouquet*

National Center for Atmospheric Research†, P.O. Box 3000, Boulder, CO 80307, USA

C. Gloaguen, J. Leorat, and R. Grappin

LAM, Observatoire de Meudon, F-92190 Meudon, France

A possible way to restrict the number of excited degrees of freedom in a turbulent flow is to constrain nonlinear interactions to be local in wave-number space. Such models were constructed by Denianskii-Novikov [1] and Bell-Nelkin [2] for Navier-Stokes. An extension of their work to the MHD case yields new results. Some of the properties of untruncated MHD flows which are recovered (dynamo effect above a critical magnetic Reynolds number, and growth of the relative velocity-magnetic field correlations) are reported elsewhere [3, 4, 5].

It is known from studies using statistical closures of three-dimensional turbulence [6] that, in the inertial range, the main contribution to the nonlinear transfer comes mostly from wave numbers differing at most by a factor of 2 or so. This provides the basic justification for reducing interactions to only nearest neighbors in Fourier space. The advantage of such a truncation is that very high Reynolds numbers can be modelled in this way, since it allows an exponential discretization in wave-number space. Taking the minimum wave number equal to unity ($k_1 \equiv 1$), k_n represents all wave numbers with modules in the range $[2^{n-1}/\sqrt{2}, \sqrt{2}\, 2^{n-1}]$. In the n shell, v_n and b_n are average amplitudes of the velocity and magnetic field. In [1] and [2], v_n is interpreted as a r.m.s. energy and must remain positive; however, such a restriction is not necessary since the v_n (and b_n) modes can be seen as modelling typical Fourier amplitudes of the corresponding fields. At zero viscosity ν and zero magnetic diffusivity λ the total energy $E_T=\sum (v^2+b^2)$ and total correlation $E_c=\sum \vec{v}\cdot\vec{b}$ are conserved.

We now require of our model that it fulfills the following structural properties of the primitive MHD equations: (i) quadratic nonlinearities with $\vec{z}^{\pm}=\vec{v}\pm\vec{b}$ symmetry preserved (see [3]); (ii) mode n interacts only with modes (n-1, n, n+1);

---

*Permanent address: Observatoire de Nice, BP52, Nice Cedex 06007, France.

†The National Center for Atmospheric Research is sponsored by the National Science Foundation.

iii) nonlinear time scale associated with mode n is $|k_n Z_n|^{-1}$; iv) conservation of total energy E_T and correlation E_c in the inviscid limit (with magnetic helicity identically zero).

The above conditions are sufficient to determine the model completely [3]; we find:

$$\frac{dZ_n^+}{dt} = \alpha[k_n Z_{n-1}^+ Z_{n-1}^- - k_{n+1} Z_{n+1}^+ Z_n^-] + \beta[k_n Z_{n-1}^+ Z_n^- - k_{n+1} Z_{n+1}^+ Z_{n+1}^-]$$

$$- \nu_\gamma k_n^2 Z_n^+ - \nu_m k_n^2 Z_n^- + C_n^u + C_n^b \; ; \qquad (1)$$

the Z_n^- equation is obtained by interchange of +/-. By setting $Z_n^+ = Z_n^-$ the non-magnetic case is recovered; α and β are two arbitrary constants; C^u and C^b are forcing terms included to study stationary solutions of the system. In all numerical calculations reported here, we shall take $\alpha=10^{-2}$, $\beta=1$, and all $C\equiv0$ but $C_1^u \equiv 1$. The rate at which phase volume shrinks is $(\nu+\lambda)\Sigma_n k_n^2$,where 2xN is the total number of modes. Thus, the more contraction takes place the more modes there are. Finally, note that the Kolmogorov (1941) spectrum which reads $x_n \sim k_n^{-1/3}$(x=v,b) is a stationary solution of the inviscid system (1).

Let us now focus on the chaotic properties of system (1).

With only 2x2 modes, it seems that a stationary fixed point will always be reached (although no global stability property has been found). The system of 2x3 modes has been studied extensively. The magnetic diffusivity is kept constant ($\lambda=0.1$) and the viscosity is varied. At high viscosity, the representative point of the system in phase space is attracted to a hydrodynamical ($b\equiv0$) stationary state (the coordinates of which depend on all the parameters). This region is followed by a domain where the solution appears chaotic but still with $b\equiv0$ (region labeled I). Two other regions of chaos, labeled II and III, have been found that have nonzero magnetic excitation. Between regions I and II ($0.140<\nu<0.175$) the solution is a hydrodynamical limit cycle that breaks down near $\nu=0.14$ with chaos and magnetic excitation appearing at the same time. Between regions II and III, there is a sequence of period-doubling bifurcations leading again to chaos (region III, $\nu<0.115$). The third region of chaos is observed down to $\nu=2\times10^{-2}$; for viscosities below this latter value, the Kolmogorov dissipation scale $k_D^{-1} \sim \nu^{+3/4}$ becomes significantly smaller than the smallest scale treated at that resolution and energy spectra do not have an exponentially decreasing tail. Higher resolutions are then required to model properly the turbulent regime, including the dissipation domain.

As the number of modes is increased, the chaotic aspect persists. Another feature of the flow becomes striking: its intermittency for high wave numbers. Temporal plots of the various velocity and magnetic modes show that they fluctuate with time. But for higher wave numbers, the fluctuations are more marked. For example, a numerical integration on 2x9 modes with $\nu=\lambda=10^{-3}$ shows bursts a hundred times above the average value for the last (n=9) mode, with quiescent periods between bursts lasting around fifty large-scale eddy turnover times. Such a spectral intermittency may result in correcting factors to the Kolmogorov scaling for the energy spectra and the higher order moments. However, the resolution is not yet sufficient to determine whether such corrections are present in the flow. The kurtosis (or normalized fourth-order moment $\langle x_n^4\rangle/\langle x_n^2\rangle^2$, where x=v,b) grows by a factor three between the lowest mode (n=1) and the n=7 mode. A calculation with 2x20 modes is presently under way to obtain clear-cut results on that issue.

The scalar model (1) presents many features of natural flows (chaos above a threshold, intermittency leading to possible corrections to the Kolmogorov scaling) and yet it is simple enough that this chaos may be characterized in a more quantitative way by computing the dimension of the underlying attractor as a function of Reynolds number. There are several definitions of dimensions concerning either metric or probabilistic properties of the attractor (see [7] for a review; also [8]). We shall be concerned with two of them: the pointwise dimension d_o and the dynamical dimension d defined as follows. Let $\vec{X}$ be a vector of phase space labeling a point on the attractor. Let $d_{ij} = |\vec{X}_i-\vec{X}_j|$ be the distance between the i and j points on the attractor (for example, the Euclidean distance). If now θ is the Heaviside function, we define, following [8]

$$C(\ell) = \frac{1}{N^2} \sum_{i,j} \theta(\ell-d_{ij}). \qquad (2)$$

Assuming $C(\ell)\sim\ell^d$, d is a dynamical dimension of the attractor. The pointwise dimensions d_o are obtained by keeping the point $i=i_o$ (say) fixed and averaging on N points. The pointwise dimension looks at how the volume of balls of increasing radii augments; it is a local definition but has the computational advantage that it requires only order N operations (as compared to N^2 for other definitions of dimension). The computation of the Lyapounov exponents, which will also yield a dimension of the attractor according to a conjecture found in [9] is under way and will be reported elsewhere. Numerically, one integrates forward in time the dynamical system and creates an array of N vectors on the attractor, where consecutive points in the array are separated by a fixed amount of time. This time interval is of order of the typical time (large-scale eddy turnover time) of the system. As the dimension of phase space augments, the recurrence time after which a representative point on the attractor will be approached again to within a small distance increases exponentially. This makes the computation of dimensions of underlying attractors of real flows a formidable task.

In the chaotic zone I, the attractor appears as a point on the large scales (of order 1), whereas, on scales 10^{-4} or smaller, it has a complicated three-dimensional structure. In that case, plotting ln C(1) as a function of ln 1, two distinct regions are found, which can be both approximated by a straight line of slope 0.10 in the large scales and 2.30 in the small scales with an intermediate region for distances in the 10^{-4} range.[1] This example shows that a counting algorithm for determining the structure of an attractor has enough information to unravel different types of structures at different scales.

For the 2x3 modes, with $\nu=0.11$ and $\lambda=0.10$, dimension of 2.25 is found, using 20×10^5 points for the pointwise dimension and 5×10^3 points for the definition of (2). Both calculations took 20 minutes of VAX. Varying slightly the magnetic Prandtl number ν/λ (from 1.1 to 1.0) yields a dimension of 2.30. It was checked that adding more modes to the system (from 2x3 to 2x4) while keeping ν and λ fixed did not change the dimension of the attractor.

Finally, a series of experiments at unit magnetic Prandtl number were performed, augmenting the number of modes and diminishing the viscosity while keeping the ratio of the Kolmogorov dissipative wavenumber k_D to the maximum wave number k_{max} fixed ($k_{max}=2^N$ for a 2xN modes system). The results are presented in Table 1.

Table 1. Variation of dimension of attractor as number of modes increases, keeping k_D/k_{max} fixed (see text).

2xn	6	8	10	12	14	16
$\nu=\lambda$	2.6×10^{-2}	10^{-2}	4×10^{-3}	1.6×10^{-3}	6.3×10^{-4}	2.5×10^{-4}
d	3.31	4.06	5.03	6.24	6.50	7.48

In the last three cases, 35×10^3 points were used. Calculations at higher resolutions are under way to see whether saturation of dimension occurs or not. If the pattern seen in Table 1 is preserved, this would mean that for a given number of modes, roughly half of them are relevant. For n=100, in this type of model, 50 modes would be sufficient to describe a flow of Reynolds number in the 10^6 range. The

[1]Note that this attractor is probably of a numerical nature since as the precision of the time integration of the ODE (1) is increased by a factor 10, the region of chaotic behavior shrinks by roughly a factor 10 also

question arises whether increasing the connectance [10] of the system (by including interactions with the next to nearest neighbors n±2) would leave this qualitative result unchanged.

The model presented in this paper presents many features of a conducting flow (dynamo effect, growth of velocity-magnetic field correlations) which is turbulent (chaotic behavior in time, spectral intermittency). Variation of dimension of the underlying attractor with Reynolds number is easily computed. Such a model may also be useful in a large-scale simulation of stellar atmosphere for parametrizing the small scale MHD turbulence not treated explicitly.

Acknowledgements: Computations of the dimension of attractors were performed at NCAR which is thanked for its hospitality.

References

1. V.N. Denianskii and Y.A. Novikov: J. Applied Math. Mech. 38 , 507 (1974)
2. T.D. Bell and M. Nelkin: Phys. Fluids 20 , 345 (1977)
3. C. Gloaguen-Lavolee: Un modele de turbulence MHD, Rapport de D.E.A., Observatoire de Meudon, France (1981)
4. C. Gloaguen: Doctorat de troisième cycle, Université Paris VII (1983)
5. C. Gloaguen, J. Leorat, A. Pouquet and R. Grappin: A Scalar Model of MHD Turbulence, preprint, Observatoire de Meudon (1983)
6. R.H. Kraichnan: J. Fluid Mech. 47 , 525 (1971)
7. J.D. Farmer, E. Ott and J.A. Yorke: The Dimension of Chaotic Attractors, to appear in Physica D, proceedings of Los Alamos conference of May 1982 (1983)
8. P. Grassberger and I. Procaccia: Phys. Rev. Lett. 50 , 345 (1983)
9. J. Kaplan and J. Yorke: Lecture Notes in Math. 730 , H.O. Peitgen and H.O. Walther Eds (Springer) p. 228 (1979)
10. C. Froeschle: Phys. Rev. A 18, 277 (1978)

The Analytic Structure of Turbulent Flows

U. Frisch

CNRS, Observatoire de Nice, B.P. 139
F-06003 Nice Cedex, France

ABSTRACT

The topics discussed include (i) the relation between the Painlevé property and integrability, (ii) high-frequency intermittency and complex-time singularities, (iii) possible real singularities of the 3-D Euler equations and how they can traced by going into the complex domain, (iv) short time behaviour of non-linear PDE's with entire initial data.

1. PAINLEVE AND INTEGRABILITY

Around the turn of the century it was realized, particularly by PAINLEVE /1/ and KOWALEVSKAYA /2/, that a complete understanding of the analytic structure in the complex time domain of a dynamical system governed by an ODE can be as helpful as having integrability in the sense of that time (i.e. reduction to "known" functions). Actually Painlevé proposed to extend the definition of integrability to include those ODE's whose only movable singularities (depending on initial data) are poles, so that the solution is uniform. This is the case of the celebrated Painlevé transcendants, the simplest of which is the solution of

$$\ddot{x} = 6x^2 + t \quad . \tag{1}$$

Let us briefly summarize the essential steps of what is now called testing the Painlevé property (see for details PAINLEVE /1/, HILLE /3/). Dominant balance requires that near a singularity t_* we have $x(t) \approx (t-t_*)^{-2}$. It is now assumed that $x(t)$ has an expansion near t_* of the form

$$x(t) = (t-t_*)^{-2} \sum_{j=0}^{\infty} u_j (t-t_*)^j \quad . \tag{2}$$

Substitution into (1) gives the relations ($j = 1, 2,\ldots$)

$$(j^2-5j-6)u_j = \sum_{\substack{0<r<j-1 \\ 0<s<j-1 \\ r+s=j}} u_r u_s + t_* \delta_{j,4} + \delta_{j,5} \quad . \tag{3}$$

Since the r.h.s of (3) involves only the u_n's ($n<j$), the relations can be solved recursively except at the "resonance" $j = 6$ where j^2-5j-6 vanishes. The interesting observation is that the r.h.s. vanishes too at $j = 6$; the fullfilment of this compatibility condition implies that an arbitrary constant can be introduced for u_6 ; together with the arbitrary choice of t_* this gives two arbitrary constants, as required for the general solution of a second-order equation.

Painlevé may have been right in stressing the connection between integrability and this sort of test. Indeed, recent work indicates that there may be a connection between the Painlevé property and integrability in one of the modern senses, e.g. existence of an inverse scattering solution for a PDE (NAKACH /4/, ABLOWITZ et al. /5,6/, WEISS et al. /7/, WEISS /8/, JIMBO et al. /9/, RAMANI et al. /10/, BOUNTIS /11/ and references therein). Conversely, it has been found that non-integrable dynamical systems (e.g. the Hénon-Heiles Hamiltonian system) tend to have "nasty" singularities, typically natural boundaries with a self-similar fractal structure (CHANG et al. /12/, WEISS /13/). Fractals in the complex domain do also arise in discrete dynamical systems, e.g. iteration of the logistic map (MANDELBROT /14/ ; see also these Proceedings).

I am not aware of the existence of any proof that the Painlevé property implies integrability but, at the moment, performing Painlevé tests (or generalized Painlevé tests as in ref. 10) seems to be the only non-numerical systematic method giving integrability hints. Painlevé tests on PDE's can be carried out either by reducing them to an ODE (through similarity assumptions) or, directly, by performing an expansion around the singular manifold in space-time (Weiss et al. /7/) : let $\mathcal{L}(\psi) = 0$ be a PDE with holomorphic coefficients in $\mathbb{C}^{d+1}$. One assumes a (movable) singular set $\mathcal{S}$ locally representable as $\phi = 0$. One then assumes an expression of the form

$$\psi = \phi^{\alpha} \sum_{j=0}^{\infty} u_j \phi^j \quad ; \tag{4}$$

here the u_j's are now functions defined on $\mathbb{C}^{d+1}$ and non-singular on $\mathcal{S}$. The algebra proceeds essentially in the same way as discussed above. This machinery has been applied by O. Thual and the author to the KURAMOTO /15/ equation

$$\partial_t v + v\partial_x v = -\partial_{xx} v - \nu\partial_{xxxx} v \quad . \tag{5}$$

We find that $\alpha = -3$ and a (compatible) resonance at $j = 6$. Two more compatible resonances would be needed for the Painlevé test to stand. Actually there are two complex resonances at $j = (1/2)(13 \pm i\sqrt{71})$. This suggests a fractal natural boundary /13/ ;

numerical integration of the steady state form of (5) gives evidence for such a natural boundary.

2. INTERMITTENCY AND COMPLEX TIME SINGULARITIES

There are reasons other than integrability searches to study dynamical systems in the complex domain. This is particularly true for turbulent flows. Indeed, the so-called "small-scale" properties of turbulent flows are connected with the analytic structure of the Navier-Stokes equations

$$\partial_t v + v.\nabla v = - \nabla p + \nu \nabla^2 v \qquad (6)$$

$$\nabla . v = 0 + \text{initial and boundary conditions.}$$

This is discussed in FRISCH and MORF /16/ and FRISCH /17/ and will only be summarized here.

The solutions of the Navier-Stokes equations (with viscosity) are likely to be (real) analytic in space and time ($t>0$). This can be proven for the Burgers equation and for the Navier-Stokes equations in two space dimensions (2-D). In 3-D, when the initial data are analytic in space, analyticity persists for at least a finite time but there is numerical evidence that it persists for ever (BRACHET et al. /18/, FRISCH /17/). Data about the small scales of turbulent flows are usually obtained by Fourier analyzing the signal $v(t)$ recorded by a probe. It is known that the high-frequency asymptotics of the Fourier transform $\hat{v}(\omega)$ of an analytic function $v(t)$ are governed by the complex singularities of the analytic continuation $v(t+i\tau)$. Roughly expressed, a singularity at $t_* + i\tau_*$ ($\tau_*>0$) with exponent ρ gives a contribution $\omega^{-\rho-1}e^{-\omega\tau_*}e^{i\omega t_*}$ to $\hat{v}(\omega)$ as $\omega\to\infty$. Clearly, only the singularities closest to the real domain matter for very high ω. These singularities are usually associated with dissipative structures in the real domain (which may be hard to visualize except in numerical simulations). When the signal $v(t)$ is high-pass filtered, each singularity gives a burst centred at t_* and with overall amplitude $e^{-\Omega\tau_*}$ (Ω is the filter frequency). Singularities with very small τ_*'s being rare, the strongest bursts are widely spaced ; this gives the high-pass filtered signal a strongly intermittent structure (see for example figure 1 of ref. /16/). The energy spectrum in the dissipation range at high frequencies is (up to algebraic prefactors) an average of $\exp(-2\omega\tau_*)$ weighted by the distribution (per unit time) $N(\tau_*)$ of imaginary parts τ_*. Experimental determination of $N(\tau_*)$ would be of interest. If $|\tau_*|$ has a positive lower bound the energy spectrum should be exponential.

Finally we note that complex singularities are important for the backscattering of high-frequency waves propagating through a random medium and can lead to localization (FRISCH and GAUTERO /19/).

3. REAL SINGULARITIES OF THE 3-D EULER EQUATIONS ?

As the viscosity is reduced (or equivalently, the Reynolds number is increased) some complex singularities may migrate to the real space-time domain. Real singularities, if they exist, provide a nice explanation for the scaling properties observed in fully developed turbulence. This scaling is very conspicuous on the longitudinal structure functions

$$<|\delta v(\ell)|^p> \equiv <|v(\vec{r}+\vec{\ell}) - v(\vec{r})|^p> \propto \ell^{\zeta_p} \qquad \ell = |\vec{\ell}| \ . \tag{7}$$

In (7) p is a positive real number ; the velocity components are measured parallel to $\vec{\ell}$; the average is over time ; the empirical scaling relation is obtained experimentally for separations ℓ which lie in the inertial range. The best available data for the ζ_p's may be found in ANSELMET et al. /20/ ; see also FRISCH /21/. The ζ_p's appear to vary non-linearly with p contrary to the prediction of simple black-and-white intermittency models such as the β model (FRISCH et al. /22/) but consistent with more elaborate models (MANDELBROT /23/). This is also consistent with a recent "multi-fractal" model due to G. PARISI (see Appendix of ref. 21). In this model one assumes the existence of a hierarchy of fractal sets $\mathcal{S}_h \subset \mathbb{R}^3$ ($\mathcal{S}_{h'} \subset \mathcal{S}_h$ for h'<h) of Hausdorff dimension d(h) such that the velocity field (in the zero viscosity limit) has one $\mathcal{S}_h$ singularity of exponent h. It is easily shown that this gives the required scaling with

$$\zeta_p = \min_h (ph + 3 - d(h)) \quad ; \tag{8}$$

i.e. ζ_p is the Legendre transform of the codimension 3-d(h)

All this of course is not a proof, not even strong evidence that real singularities exist : scaling behaviour can arise in many different ways. It is of independent interest to investigate the existence of singularities in the zero viscosity limit ; this can tell us for example if the Euler equations (Navier-Stokes with $\nu = 0$) lead to a well-posed problem for all times. Similar questions are easily resolved on Burgers'equation, a one-dimensional analog of the Navier-Stokes equation (SAFFMAN /24/, KIDA /25/, FOURNIER and FRISCH /26/) which reads

$$\partial_t v + v\partial_x v = \nu\partial_{xx} v \ . \tag{9}$$

One knows that for $\nu \downarrow 0$ the solutions develop real singularities (shocks) which do provide the explanation for the scaling properties ; Burgers equation is however pathological in at least two ways : compressibility and integrability. For the 2-D Euler equations it can be proven that real singularities will never occur unless present initially (see FRISCH /17/ and references therein) ; still there is good numerical evidence for scaling at high Reynolds numbers (BRACHET /27/). An extensive but still inconclusive numerical search for possible singularities has been made on the Taylor-Green 3-D vortical flow (ORSZAG /28/, MORF et al. /29/, BRACHET et al. /18/, FRISCH /17/). The initial conditions for the Taylor-Green flow are trigonometric polynomials (and therefore entire)

$$\begin{aligned} v_1 &= \cos x_1 \sin x_2 \cos x_3 \\ v_2 &= v_1(x_2, -x_1, x_3) \\ v_3 &= 0 \ (v_3 \text{ does not stay zero for } t > 0). \end{aligned} \tag{10}$$

Two methods have been developed to analyze the possible real singularities of the Taylor-Green flow. Both involve analytic continuation. First there is a Taylor expansion method. Since v(t,r) is an analytic function of t (at least for small t) it can be expanded :

$$\vec{v}(t,\vec{r}) = \sum_{n \geq 0} t^n \vec{v}_n(\vec{r}) \ . \tag{11}$$

The components of the v_n's are trigonometric polynomials of increasing degree which can be calculated recursively ; various norms of the solution (e.g. the enstrophy = mean square vorticity) can be expanded in powers of t (up to order t^{80} so far). Analysis of the data for the enstrophy indicates that the nearest singularities are on the imaginary time axis (approximately at $\pm$ 2.17 i). These singularities control the convergence radius of the Taylor series. A possible real singularity beyond the convergence disk must be obtained by analytic continuation techniques (e.g. by using Padé approximants). In spite of the high accuracy of the Taylor coefficients (up to 30 digits in Ref. /18/), accuracy in the analytic continuation deteriorates too quickly to be able to reach any definite conclusion. This deterioration may be due in part to the presence of natural boundaries.

The second approach, the analyticity strip method, is based on the tracing of complex space singularities as a function of real time (SULEM et al. /30/). The key observation is that for the Euler, Navier-Stokes and many other equations with analytic initial data the appearance at some real time t_* of a singularity in real space is necessarily preceded for $0<t<t_*$ by complex space singularities. The nearest complex singularity, at the edge of the analyticity strip, is within

a distance $\delta(t)$ of the real domain ; $\delta(t)$ vanishes at t_*. By analyzing the spatial Fourier transform $\hat{v}(t,\vec{k})$ of the velocity field one obtains $\delta(t)$. Indeed, for $k\delta>>1$ the Fourier amplitude decreases like $\exp(-\delta(t)k)$ (within algebraic prefactors and phase factors). Numerically, using high accuracy spectral methods it is possible to measure $\delta(t)$ from these exponential tails. Accurate measurements are possible as long as $\delta(t)$ remains larger than a few meshes ℓ_{min} (accuracy is controlled by $\exp(-\delta/\ell_{min})$. The method has been applied to a great variety of PDE's in 1+1 dimensions (SULEM et al. /30/, FUCITO et al. /31/, LIVI et al. /32/, BASSETTI et al. /33/), in 2+1 dimensions (MORF et al. /34/, SULEM et al. /30/, FRISCH et al. /35/, BRACHET /27/) and in 3+1 dimensions (BRACHET et al. /18/). When applied to the Taylor-Green problem, the method gives an exponential decrease of $\delta(t)$ (except for short times ; see Sec. 4) up to the time when accuracy is lost (about $t \approx 2.5$ when working with 256^3 Fourier modes). If such data are extrapolated, the prediction is that there is no real singularity. However, there is some evidence that the nature of the flow changes around $t = 4$ and that extrapolation may not be safe.

4. SHORT TIME ASYMPTOTICS

We wish to show that the short time behaviour of the width of the analycity strip $\delta(t)$ can be obtained by an asymptotic expansion when the initial conditions are entire so that $\delta(0) = \infty$. We shall restrict ourselves to trigonometric polynomials but the argument can be generalized.

Recent studies of the non-linear string (NLSt) equation

$$(\partial_{tt} - \partial_{xx} + m^2)\phi + g\phi^3 = 0 \tag{12}$$

with periodic boundary conditions and initial conditions of the form $\phi(0,x) = a \cos x$ and $\partial_t\phi\ (0,x) = 0$ give a logarithmic variation of $\delta(t)$ for short times

$$\delta(t) \propto -\operatorname{Log}(t/t_1) \qquad t \downarrow 0\ . \tag{13}$$

This was obtained in Ref. /31/ by numerical integration and given a theoretical interpretation which was later confirmed more systematically in Ref. /33/. A similar log law was obtained in Ref. /32/ for a continuous version of the Fermi-Pasta-Ulam (FPU) problem governed by

$$(\partial_{tt} - \partial_{xx})\phi = \beta\partial_x(\partial_x\phi)^3 \quad . \tag{14}$$

We shall now show that such log laws are very general (for trigonometric initial conditions) and are to be expected also for multi-dimensional problems such as the

Taylor-Green vortex. We first outline the argument on Burgers equation with zero viscosity and the simplest non-trivial trigonometric initial conditions

$$\partial_t v + v\partial_x v = 0 \quad ; \quad v(0,x) = v_0(x) = -\sin x \quad . \tag{15}$$

We now go into the complex x + iy domain. Because of the reflection symmetry $v_0(-x) = -v_0(x)$, it is easily checked that for any real t the analytic continuation of v on the imaginary space axis is pure imaginary. So we set

$$v(t, iy) = i\Psi(t,y) \quad . \tag{16}$$

From (15) we find that Ψ satisfies

$$\partial_t \Psi + \Psi\partial_y \Psi = 0 \qquad \Psi_0(y) = -\operatorname{sh} y. \tag{17}$$

At this point we could just solve (17) but we shall refrain from this because we are here using Burgers equation to simplify the algebra and not to take advantage of its integrability. Suppose, instead, that we expand $\Psi(t,y)$ in powers of t and we look at the dominant behaviour for large y>0, namely in the region where the singularities are expected for short times. To order t^0 we have $-\operatorname{sh} y \sim -(1/2)e^y$. Each additional power of t will have an additional factor of e^y. For small t, finite contributions to $t\Psi(t,y)$ are therefore expected to all orders if we choose te^y to be O(1). This suggests the asymptotic ansatz

$$\Psi(t,y) \sim (1/t)F(q) \qquad q = te^y \quad . \tag{18}$$

Substitution into (17) gives

$$-F + qF' + FF' = 0 \tag{19}$$

with the boundary condition $F(q) = -q/2 + O(q^2)$.

The ODE (19) can be solved exactly : the solution has a square-root branch point at $q_* = 2/e$. Even if the equation could be solved only numerically we could still expect the solution to have singularities for some q_*. For each such singularity there is a time-dependent singularity of the original equation at a $y_*(t)$ such that

$$q_* = t\, e^{y_*} \quad ; \tag{20}$$

hence the logarithmic law. This argument can be applied *mutatis mutandis* to the non-linear string equation (12) and to the FPU equation (14). The short time behaviour is then governed by the following ODE's

$$2F - 3F' + F'' + gF^3 = 0 \text{ (NLSt)} \quad , \tag{21}$$

$$2F - 3F' + F'' = \beta\partial_q(\partial_q F^3 \text{ (FPU)} \quad . \tag{22}$$

It is noteworthy that the non-time-differentiated linear terms in the original equations do not contribute to leading order. Hence, it would be equivalent to study the short time behaviour of

$$\partial_{tt}\phi + g\phi^3 = 0 \qquad \text{(NLSt)} \tag{23}$$

$$\partial_{tt}\phi = \beta\partial_x(\partial_x\phi)^3 \quad . \qquad \text{(FPU)} \tag{24}$$

Equation (23) was also obtained in Ref. /33/ through an asymptotic analysis in spatial Fourier space which is equivalent to our large y analysis (by the Paley-Wiener theorem an e^{ny} behaviour for large y corresponds to non-vanishing Fourier-space amplitude up to $k = \pm n$; see HÖRMANDER /36/ p. 21). The algebra in Fourier space is however more complicated.

Equation (21) has the Painlevé property and can be solved by elliptic functions (see Ref. /33/). Equation (22) is not Painlevé ; its singularities are likely to be branch points with a 3/2 exponent. The latter are obtained by first guessing that the 2G term in the l.h.s. of (22) is irrelevant to dominant order near the singularity (as can be checked a posteriori) ; the simplified equation is written in terms of $H = G'$; movable square-root singularities are found by expanding near $H_* = \pm(3\beta)^{1/2}$.

We stress that care must be taken before using the short time asymptotic ODE to study the Painlevé property in the original PDE. For example it is known that the viscous ($\nu > 0$) Burgers equation has the Painlevé property ; however, the asymptotic ODE is the same (19) as for $\nu = 0$ and this gives branch points. Actually the viscous term becames relevant very close to a branch point $iy_*(t)$ and an inner expansion is required ; this leads to an array of poles entending from $iy_*(t)$ to $i\infty$.

Finally the same type of short time asymptotic analysis can be applied in more than one space dimension. Let us briefly consider the 3-D Euler equation with the Taylor-Green initial conditions (10). We work in complex $\mathbb{C}^3$ space for large y_1, y_2 and y_3. The initial condition v_0 is proportional to $\exp(y_1+y_2+y_3)$. As a consequence of the incompressibility constraint it is found that the dominant contribution to tv_1 does not involve $\exp(2y_1+2y_2+2y_3)$ but only $\exp(2y_1+2y_2)$ + permutations. Similarly, the dominant contribution to $t^n v_n$ (in the sense of (11)) involves $\exp((n+1)y_1 + (n+1)y_2 + (n-1)y_3)$ + permutations. Thus, in generating v_{n+1} the interaction between v_0 and v_n is more efficient than any other interaction between v_r and v_s ($0<r<n$, $0<s<n$, $r+s = n$). This suggest that the short time small-scale dynamics of the flow can be linearized (as a consequence of incompressibility, not of short times!). A somewhat differently motivated linearization has been performed on the Taylor-Green flow (see Appendix D of Ref /18/) ; it gives complex singularities with a logarithmic law for $\delta(t)$ at short times.

REFERENCES

1 . P. Painlevé : Leçons sur la Théorie Analytique des Equations Différentielles (Hermann, Paris 1897)

2 . S. Kowalevskaya : Acta Math. 12, 177 (1889) and 14, 81 (1890)

3 . E. Hille : Ordinary Differential Equations in the Complex Domain (Wiley, New-York 1976)

4 . R. Nakach : "Self-Similar solutions of Nonlinear Evolution Equations of Physical Significance", in "Plasma Physics Nonlinear Theory and Experiments", ed. H. Wilhelmsson, pp. 456-474 (Plenum, New York 1977)

5 . M.J. Ablowitz, A. Ramani and H. Segur : Lett. al Nuovo Cimento 23, 333 (1978)

6 . M.J. Ablowitz, A. Ramani and H. Segur : J. Math. Phys. 21, 715 and 21, 1006 (1980)

7 . J. Weiss, M. Tabor and G. Carnevale : J. Math. Phys. 24, 522 (1983)

8 . J. Weiss : "The Painlevé Property for Partial Differential Equations. II. Bäcklund Transformations, Lax Pairs, and the Schwarzian Derivative", J. Math. Phys (in press) ; "The Caudrey-Dodd-Gibbon Equation and a Class of Integrable Systems", preprint, La Jolla Institute (1983)

9 . M. Jimbo, M.D. Kruskal and T. Miwa : Phys. Lett. 92A, 59 (1982)

10. A. Ramani, B. Dorizzi and B. Grammaticos : Phys. Rev. Lett. 49, 1539 (1982)

13. J. Weiss : "Analytic Structure of the Hénon-Heiles System" in "Mathematical Methods in Hydrodynamics", A.I.P. Conf. Proc. 88, A.I.P. (1982)

14. B.B. Mandelbrot : in "Nonlinear Dynamics" ed. R.H.G. Helleman pp. 249-259, Ann. N.Y. Acad. Sci. 357 (1980)

15. Y. Kuramoto : Suppl. Progr. Theor. Phys. 64, 346 (1978)

16. U. Frisch and R. Morf : Phys. Rev A23, 2673 (1981)

17. U. Frisch : "Fully Developed Turbulence, Singularities and Intermittency", Proc. Les Houches 1981 "Chaotic Behaviour in Deterministic Systems" (North Holland 1983)

18. M.E. Brachet, D.I. Meiron, S.A. Orszag, B.G. Nickel, R.H. Morf and U. Frisch : J. Fluid Mech. 130, 411 (1983)

19. U. Frisch and J.L. Gautero : "Backscattering and Localization of High-Frequency Waves in a one-Dimensional Random Medium", preprint, Observatoire de Nice (1983)

20. F. Anselmet, Y. Gagne, E.J. Hopfinger and R.A. Antonia : "High Order Velocity Structure Functions in Turbulent Shear Flows", preprint Institut de Mécanique de Grenoble (1983)

21. U. Frisch : "Fully Developed Turbulence and Intermittency", in Proc. Int. School of Phys. "Enrico Fermi", Course 88, ed. M. Ghil., to be published (North Holland 1984)

22. U. Frisch, P.L. Sulem and M. Nelkin : J. Fluid Mech 87, 719 (1978)

23. B.B. Mandelbrot : J. Fluid Mech 62, 331 (1974)

24. P.G. Saffman : "Lectures on Homogeneous Turbulence" in "Topics in Nonlinear Physics" ed. N. Zabusky, pp. 485-614 (Springer 1969)

25. S. Kida : J. Fluid Mech. 93, 337 (1979)

26. J.D. Fournier and U. Frisch : "L'équation de Burgers Déterministe et Statistique", J. Méc. Théor. Appl. Paris, in press (1983)

27. M. E. Brachet : "Simulation à Haute Résolution d'Ecoulements Bidimensionnels", preprint Observatoire de Nice (1983)

28. S.A. Orszag : "Statistical Theory of Turbulence" ; Proc. Les Houches 1973 "Fluid Dynamics", ed. R. Balian and J.L. Peube p. 235 (Gordon and Breach 1977)

29. R. H. Morf, S.A. Orszag and U. Frisch : Phys. Rev. Lett 44, 572 (1980)

30. C. Sulem, P.L. Sulem and H. Frisch : J. Comp. Phys. 50, 138 (1983)

31. F. Fucito, F. Marchesoni, E. Marinari, G. Parisi, L. Peliti, S. Ruffo, A. Vulpiani : J. Phys. (Paris) 43, 707 (1982)

32. R. Livi, M. Pettini, S. Ruffo, M. Sparpaglione, A. Vulpiani : "Relaxation to Different Stationary States in the Fermi-Pasta-Ulam Model" preprint Università di Firenze, Fisica (1983)

33. B. Bassetti, P. Butera and M. Raciti : "Complex Poles, Spatial Intermittencies and Energy Transfer in a Classical Nonlinear String", preprint Instit. Sci. Fis. Università degli studi di Milano (1983)

34. R.H. Morf, S.A. Orszag, D.I. Meiron, U. Frisch and M. Meneguzzi : Proc. 7th Intl. Conf. on numerical Methods in Fluid Dynamics (ed. R.W. MacCormack and W.C. Reynolds) ; Lecture Notes in Physics 141, 292 (Springer 1981)

35. U. Frisch, A. Pouquet, P.L. Sulem and M. Meneguzzi : "The Dynamics of two-Dimensional Ideal Magnetohydrodynamics", in "Two Dimensional Turbulence", J. Mec. Théor. Appl. (Paris) special issue p. 191 (1983)

36. L. Hörmander : Linear Partial Differential Operators (Springer 1963)

Additional references : H. Yoshida : "Necessary condition for the existence of algebraic first integrals, I and II", Celestial Mech. in press (1983) ; "Self-Similar Natural Boundaries of Non-Integrable Dynamical Systems in the complex t Plane, these Proceedings ;
"A type of second order linear ordinary differential equations with periodic coefficients for which the characteristic exponents have exact expressions", Celestial Mech. in press (1983).

Low Prandtl Number Fluids, a Paradigm for Dynamical System Studies

A. Libchaber

Groupe de Physique des Solides de l'Ecole Normale Supérieure, 24 rue Lhomond
F-75231 Paris Cedex 05, France

In the recent development of experiments related to the road to chaos [1], fluids have played an important role. In this paper, we want to stress the usefulness of low Prandtl number fluids in Rayleigh-Benard experiments. In such studies, the three basic equations are related to the mass, momentum and energy conservation [2], the two last ones being called the Navier-Stokes and the Fourier equations. The Prandtl number is a dimensionless number which characterizes the fluid and depends on the energy and the momentum transport. A small Prandtl number fluid has a small kinematic viscosity or a large heat diffusivity. The richness of the fluid behavior is associated with the nonlinear terms in the equations. For the low Prandtl number case, the nonlinear term in the Navier-Stokes equation is the dominant one. When this term, $(\vec{v}\vec{\nabla})\vec{v}$, is larger than the diffusion one, $\nu\Delta\vec{v}$, the Reynolds number for the velocity of convection is large. For a critical value of the Reynolds number, the fluid becomes unstable, through a Hopf bifurcation to a time-dependent state, called by Busse [2] the oscillatory instability (O.I. from now on). As the Prandtl number decreases, the relative importance of the nonlinear term increases, and the onset of the O.I. appears at lower Rayleigh numbers. One should notice that the O.I. is the only time-dependent transition occurring in the fluid, which can be derived from the three basic equations [2].

1 The Oscillatory Instability

We have tested the behavior of the oscillatory instability on three low Prandtl number fluids, helium ($P = 5\ 10^{-1}$), mercury ($P = 2.5\ 10^{-2}$) and potassium ($P = 5\ 10^{-3}$) and have qualitatively confirmed Busse and Clever's analysis. In particular, the onset value, in Rayleigh numbers, is reduced as one moves to lower Prandtl numbers. For a small aspect ratio cell $\Gamma = 2$ ($\Gamma = L/d$, L larger lateral size, d depth of the layer), we found a value of 70.000 for helium, 7.500 for mercury, and 4.500 for potassium.

In the case of electrically conducting fluids, a horizontal magnetic field adds a new term to the Navier-Stokes equation, related to the Lorentz force [4]. It leads to a change of the effective viscosity of the fluid, which increases with the magnetic field strength [5], thus increasing the fluid rigidity. One associates the magnetic field with another dimensionless number, the Chandrasekhar number Q, proportional to the square of the field amplitude. Recently, Busse and Clever [4] have studied the dependence of the O.I. on the magnetic field and have found a simple power law for the onset value

$$R_Q - R_0 \sim Q^{1,2}$$

where R_Q is the Rayleigh number for the onset of the O.I. for a given Q value, and R_0 its onset for Q = 0. We have confirmed this power law in mercury [6]. Thus for higher fields, the onset of the O.I. is displaced to higher Rayleigh numbers and one can then explore various fluid regimes [7].

The oscillatory instability defines in our experiment the first oscillating state for the fluid velocity and temperature fields. It is the first oscillator appearing in all our observed scenarios.

Let us now come to dynamical system study. This, in experiments, implies a small aspect ratio cell, which allows an ordered structure of rolls. We are then led to a physical system with a small number of degrees of freedom. Also the typical cell size is such that the length of the convective rolls is smaller than the O.I. wavelength. This avoids space resonances which would interfere in the evolution of the scenarios.

We will review now some of the observed phenomena occurring as one changes the aspect ratio, the Prandtl number or the Chandrasekhar number, thus the physical state of the fluid.

2 Frequency Lock-in

We have observed frequency locking in helium and mercury. It has always been present in cells of very small aspect ratio with two convective rolls. In this case, the rolls are identical and rotate in opposite directions. The evolution would go as follows. Increasing the Rayleigh number beyond the onset of the O.I., one reaches a transition to a quasiperiodic state, with frequencies f_1 and f_2. At onset, the amplitude of oscillator f_2 is very small, the two oscillators are quasi-independent, their nonlinear interaction being proportional to the product of the oscillators' amplitude. As one increases R_a, the amplitude of oscillator f_2 increases, thus the interaction. This leads to frequency lock-in between f_1 and f_2. When the frequency ratio is small ($f_1 = 2f_2$), one stays on the same Mathieu tongue as the constraint is increased. If the frequency ratio is large (typically 7 or 8), one crosses a number of locking tongues as R_a is increased. A type of Devil's staircase is then observed, the number of locking states observable depending on the stability of

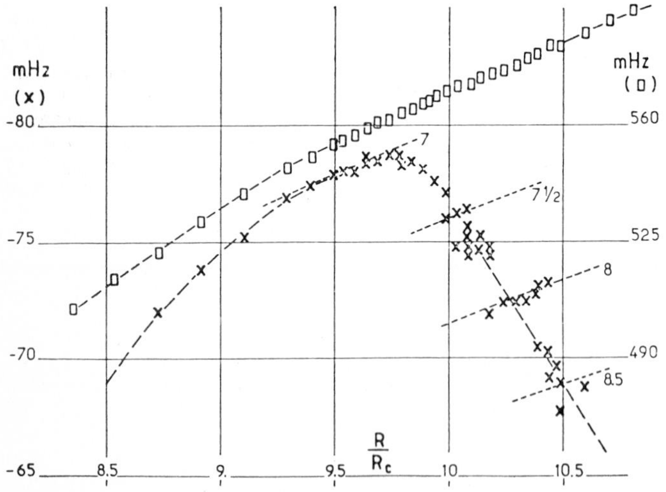

Fig.1 : evolution of frequencies f_1 and f_2 in a liquid helium cell ($d = 0.85$ mm, $L = 2.8$ mm, $\ell = 1.6$ mm). The two vertical axes indicate the two oscillators' frequencies

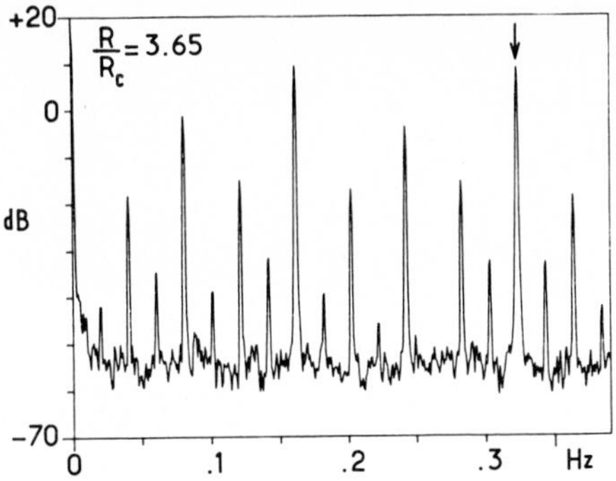

Fig.2 : Developed cascade in mercury (Γ = 4, Q = 22)

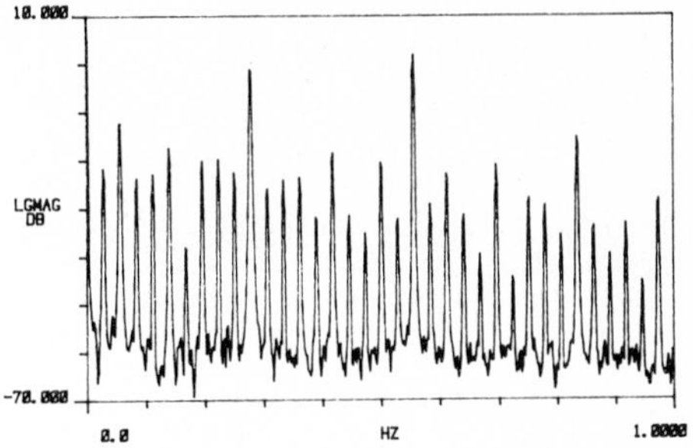

Fig.3 : Period 10T of the U sequence in mercury (Γ = 6, Q = 260)

the temperature regulation. We show on Fig.1 a typical example of a cascade of lock-in states in a liquid helium cell, where the following ratio can be seen : 7, 15/2, 23/3, 8, 17/2. Let us note that this set is consistant with the Farey theory. There if two lock-in states with ratio p/q and p'/q' appear, one should detect in between the state p+p'/q+q'.

Beyond the frequency locking state, a sequence of period doubling bifurcations generally occurs [8].

The behavior is different for larger aspect ratio cells (Γ = 4 or 6) where, in all our experiments, the sequence of period doubling started from one oscillator only [7]. The absence of the quasiperiodic state and of oscillator f_2 may be due to the break of translation symmetry for the rolls. Rolls adjacent to the later boundaries have a smaller circulation velocity than the other ones.

3 Period-doubling sequence

The complete scenario can be observed : the direct cascade, the reverse one in the chaotic region, and the universal sequence [9].

We observed the direct and reverse cascade in helium and mercury. Fig.2 is the Fourier spectrum of the developed cascade in mercury [7].

Three periods, in the universal sequence, could be seen in the chaotic region for a larger aspect ratio cell of mercury, under moderate magnetic fields (Γ = 6, Q = 260). One could clearly observe periods 10T, 3T, 9T, in this order of the U sequence, as the Rayleigh number was increased. Fig.3 shows the Fourier spectrum of period 10T.

4 Beyond period doubling, a quasiperiodic route

For low Prandtl number fluids the only instability competing with the O.I. is a stationary one, the skewed varicose instability. It corresponds to a periodic space modulation of the convective rolls. Increasing the Prandtl number up to the order of one [8], or increasing the magnetic field up to a large Q value for mercury [7], leads in both cases to an interesting region where the O.I. and the skewed varicose are in competition. What we found in experiments is that in both cases a quasiperiodic state would exist without frequency locking. From such a quasiperiodic state,

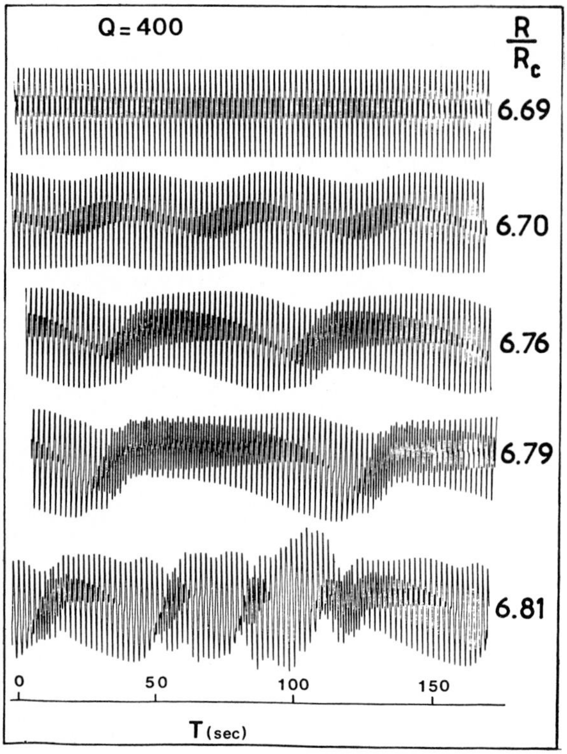

Fig.4 : Time recording of the temperature signal (Γ = 4, Q = 400). The quasiperiodic state starts for R/R_C = 6.7

various routes to chaos were found, including intermittency [8] and a three-frequency route [7] [8].

Let us describe here a simple route occurring in mercury just beyond the period-doubling one, as Q is increased. Fig.4 shows a time recording of the temperature signal evolution (cell Γ = 4, Q = 400). For R/R_C = 6.7, a Hopf bifurcation leads to a quasiperiodic state with a second oscillator f_2 of very long period. As the Rayleigh number is increased, the second oscillator period keeps increasing until a chaotic state is reached.

It is known [10] that the competition between an oscillatory instability and a stationary one may lead to a homoclinic orbit, thus an oscillation of long period. Our conjecture is that the stationary instability is the skewed varicose and that the observed "mode softening" is just an image of the homoclinic orbit. From a physical point of view, we can think in terms of an oscillatory exchange between the amplitude of the O.I. and the skewed varicose. The measurement being only sensitive to the time-dependent O.I., one sees its amplitude modulated by a low frequency. This frequency goes to zero as the Rayleigh number increases (homoclinic behavior).

References

1. J.P. Eckmann: Rev. Mod. Physics 53, 643 (1981).
2. F.H. Busse: Rep. Prog. Phys. 41, 1930 (1978).
3. F. Heslot: These Ecole Normale Supérieure, Paris 1983.

4. F.H. Busse, R.M. Clever: Stability of convection rolls in the presence of a horizontal magnetic field, preprint.
5. S. Chandrasekhar: "Hydrodynamic and Hydromagnetic Stability", Oxford, Clarendon Press, 1961.
6. S. Fauve, C. Laroche, A. Libchaber: Horizontal magnetic field and the onset of the oscillatory instability, preprint.
7. A. Libchaber, S. Fauve, C. Laroche: J. Physique Lettres 43, L211 (1982), Physica R-21D (1983).
8. A. Libchaber, J. Maurer in "Nonlinear phenomena at phase transitions and instabilities", T. Riste ed. Plenum Press (1982).
   J. Physique Lettres 41, L515 (1980).
9. P. Collet, J.P. Eckmann: "Iterated maps on the interval as dynamical systems" Birkhauser, Basel (1980).
10. J. Guckenheimer, P. Holmes: "Nonlinear oscillations, dynamical systems, and bifurcation of vector fields" Springer, New-York (1983).

Chaotic Attractors in Rayleigh-Benard Systems

M. Sano and Y. Sawada

Research Institute of Electrical Communication, Tohoku University
Sendai 980, Japan

1. Introduction

In these few years, a great step has been made in understanding fluid turbulence in terms of dynamical systems. Experimental evidence for deterministic origins of stochasticity and low-dimensional strange attractors has been confirmed by the efforts of many researchers [1,2,3]. However, it must be only a first step for understanding wide phenomena lying in developing turbulence.

In this paper we restrict ourselves to a particular problem in the onset of chaos by the transition from 2-torus to chaos. (See,e.g. [4]) We present here experimental observation of some routes to chaos utilizing the technique of Poincaré mapping. The observed phenomena are followings:

(a) Appearance of wrinkling on a torus.
(b) Transition from torus to chaos via intermittency.
(c) Doubling of torus.

2. Experimental Setup and Technique

In order to realize many types of low-dimensional chaos at least in the beginning of non-periodicity, we performed experiments for a small aspect ratio box. To assure a good horizontal temperature uniformity and stability at the boundary, the top and the bottom plates were made of two massive aluminium blocks. The side walls are made of 10-mm thick quartz of optical quality. The inner dimension of the convection cell are Lz=d=10-mm, Ly=15-mm and Lx=20$\sim$40-mm (variable). We are able to change aspect ratio (Γ=Lx/d) continuously as well as Prandtl number to realize many types of transition. The space around the cell was evacuated to diminish thermal disturbance at the side walls. The temperature of the top and the bottom plates was controlled within 2-3mK in long term operation by using a.c. balanced bridges, lock-in amplifiers and proportional, integral and derivative controllers.

By laser beam deflection technique the horizontal temperature gradient averaged along a horizontal line parallel to the short side of the cell was measured [3]. The displacement of the laser beam by deflection was measured by a 2-D position sensor. The output of the sensor was fed to a FFT analyser for spectral analysis with 4096 sampling points, alternatively, digitized and stored in a computer. The digital time series records X(ti) were utilized for constructing n-dimensional phase portraits of the orbit. The n-dimensional orbit was constructed be defining the following position vectors; y(ti)=(X(ti),X(ti+T),······,X(ti+(n-1)T)), where T is some fixed time interval [5,6,7].

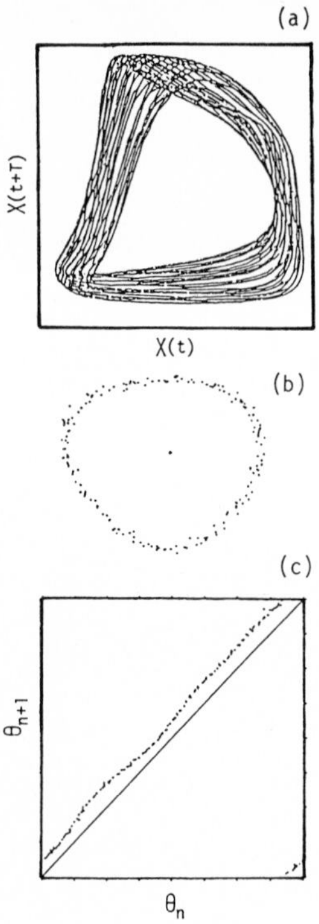

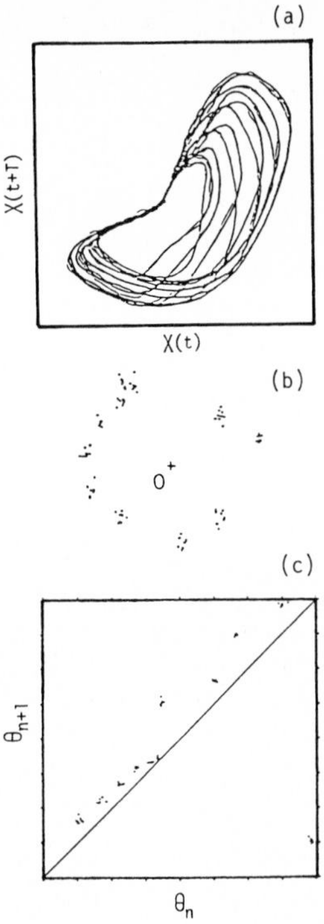

Fig.1 Quasi-periodic state Fig.2 Phase-locked state
(a) Two-dimensional projection of the orbit (b) Poincaré section
(c) One-dimensional return map (Γ =3.0)

3. Experimental Results

Two-dimensional projections of three-dimensional orbits are presented in Fig.1(a) and Fig.2(a). Figure 1 corresponds to a quasi-periodic state and Fig.2 corresponds to a phase-locked state. The rotation number, which is defined by $\rho = f_2/f_1$, is 1/11 for the latter case. Figures 1(b) and 2(b) show Poincaré cross sections of a three-dimensional orbit by a two-dimensional plane. One-dimensional return map, $\theta_{n+1}=f(\theta_n)$, is obtained by giving center 0 in a invariant closed curve and using a polar coördinate. (Fig.1(c) and Fig.2(c))

4. Wrinkling on a Torus

To study the mechanism which produces chaos from a torus, we studied the case of Γ=2.95 extensively. The surfaces of the torus and one-dimensional map are very smooth when the Rayleigh number is far below the onset point of chaos. However, further increase of Rayleigh number gives rise to much wrinkling on the torus as shown in Fig.3 (R=55.6Rc), which could be caused by increasing nonlinearity or coupling between two oscillators. For a much higher Rayleigh number (R=56.3Rc) the surface of the torus is broadened and a one-dimensional map is no longer adequate for describing this non-periodic behavior [8].

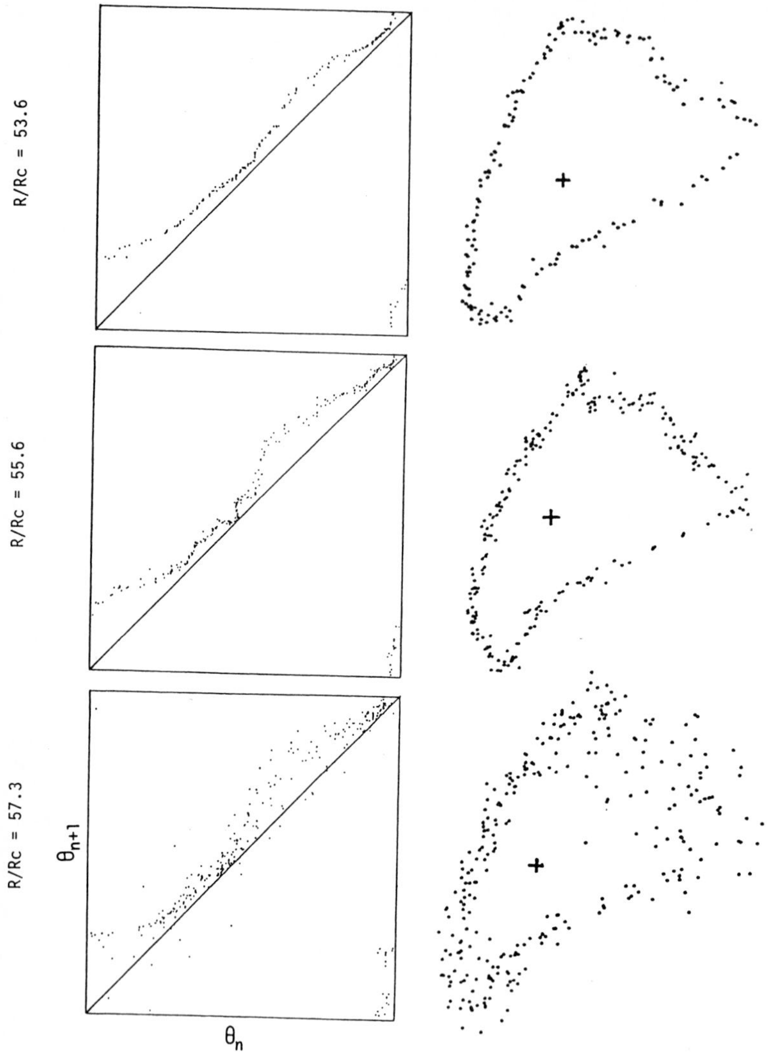

Fig.3 Poincaré section and one-dimensional return map of torus in the vicinity of transition point to chaos (Γ =2.95, P =5.5, water)

To determine the transition point to chaos and to obtain a quantitative measure for chaotic motion, we integrated noise power (N) from the power spectra and normalized it by total power (S). Figure 5 shows N/S vs. Rayleigh number. From Fig.5 we can estimate that the transition point is R $\sim$ 56Rc. It coincides with the Rayleigh number at which many wrinkles start appearing on the torus (Fig.3).

The broadening of the surface just above the transition point in this experiment could be due to the weakness of dissipation rate which attracts the orbits on a in-

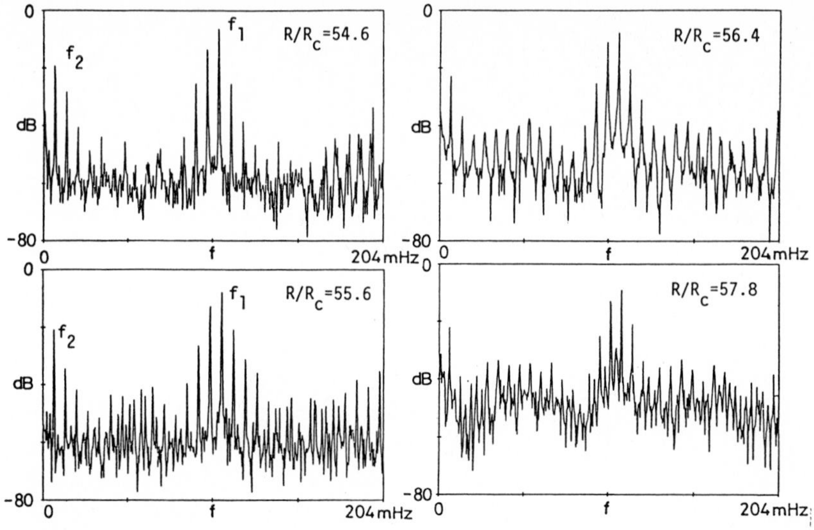

Fig.4 Power spectra correspond to transition from 2-torus to chaos

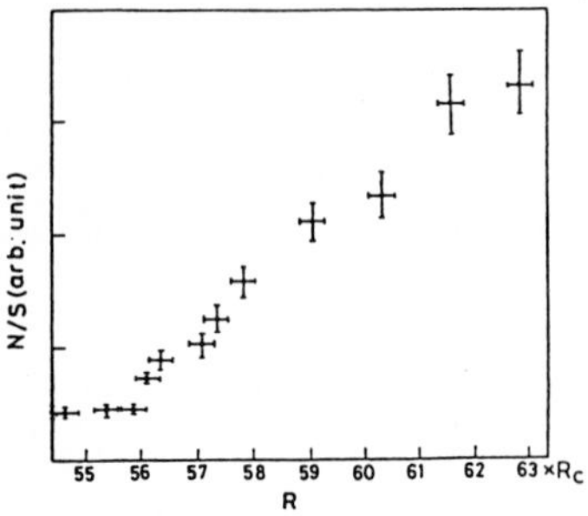

Fig.5 Integrated noise power vs. Rayleigh number

variant torus, then as soon as folding on the torus occurs the Poincaré section becomes fuzzy. It could be related to the appearance of a homoclinic orbit or heteroclinic orbit.

5. Torus Intermittency and Torus Doubling

Phase motion of a torus alone is enough to cause the transition to chaos from quasiperiodicity or from phase-locking. However, when we take account of the instability related to the radial motion of a map of an annulus, many different routes become possible.

By changing aspect ratio, we found two routes from torus to chaos relating to instability of radial motion. The first one is the route via intermittency and the second one is doubling of torus.

Torus intermittency was observed in the case of Γ =2.95 and P =6.0 by choosing different initial conditions from the preceding section. The distinctive features of this route we observed are very small rotation numbers (ρ = 1/70 ∿ 1/50) and an

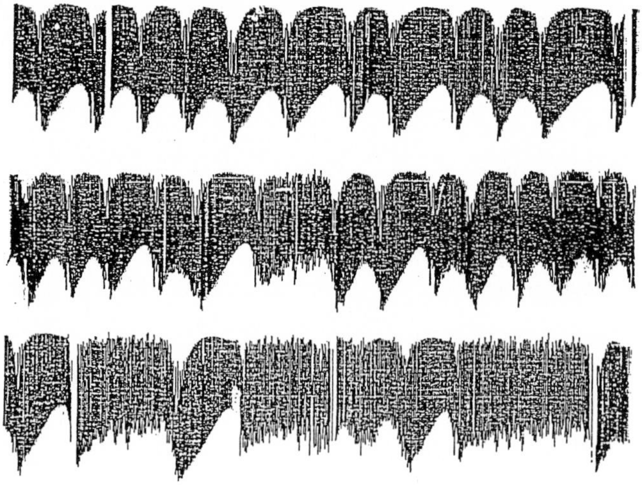

Fig.6 Time series for torus intermittency (Γ =2.95 ,P =5.5)

indefinite period of modulation as shown in Fig.6. With increasing Rayleigh number, chaotic bursts appear intermittently and mean duration of the laminar state becomes more and more shortened and finally it shows entirely chaotic oscillations. This type of transition may be caused by *saddle connection*, which was found by DAIDO [9] numerically. Laminar state with fluctuating f_2 and intermittent bursts may be considered as an indication of the existence of a saddle very close to invariant torus.

For Γ =3.5 and P =5.5, we observed doubling of the torus. In this case $\rho=f_2/f_1$ is close to 1/3 but not phase-locked. As shown in Poincaré section (Fig.7), we can see two disconnected invariant closed curves, which is the evidence of doubling of torus. The way to destroy torus in this route is very complicated [10,11]. To the authors' best knowledge, this is the first observation of torus doubling, although the collapse of torus is not identified with numerical calculations.

A quantitative analysis of the route to chaos presented in this paper is a future work.

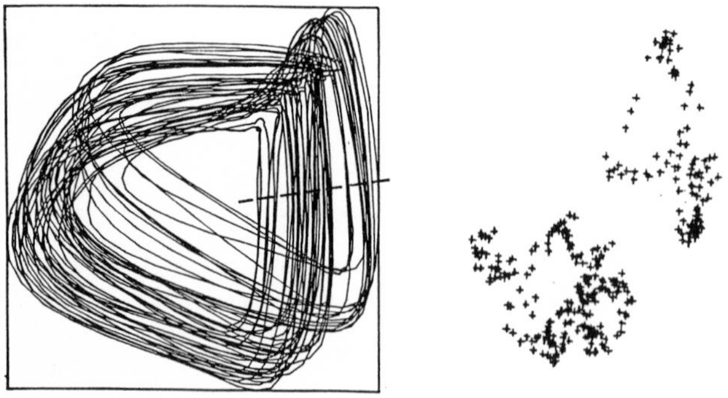

Fig.7 Phase portrait and Poincaré section for torus doubling (Γ =3.5, P =5.5). Complicated twisting of closed curves is caused by projection onto 2-D plane

We wish to thank Y.Kuramoto, H.Mori, I.Tsuda, H.Daido and K.Kaneko for illuminating discussions and suggestions. This work was partially financed by the Scientific Research Fund of the Ministry of Education, Science and Culture.

References

1. A.Libchaber, C.Laroche and S.Fauve: J. Physique Lett. 43 , 211 (1982)
2. J.P.Gollub and S.V.Benson: J. Fluid Mech. 100 , 449 (1980)
3. M.Giglio, S.Mussazzi and U.Perini: Phys. Rev. Lett. 47 , 243 (1981)
4. S.J.Shenker: Physica 5D , 132 (1982); D.Rand, S.Ostlund, J.Sethna and E.Siggia: Phys. Rev. Lett. 49 , 132 (1982); H.Daido: Prog. Theor. Phys. 68 , 1935 (1982); K.Kaneko: Prog. Theor. Phys. 69 , 1427 (1983); M.Sano and Y.Sawada: Phys. Lett. 97A, 73 (1983)
5. F.Takens: in "Dynamical Systems and Turbulence, Springer Lecture Notes in Mathematics", (ed. by D.Rand and L.S.Young, Springer, 1981)
6. J.C.Roux, R.H.Simoyi and H.L.Swinney: Physica 8D, 257 (1983)
7. M.Sano and Y.Sawada: to appear in "Turbulence and Chaotic Phenomena in Fluids", (ed. by T.Tatsumi, North-Holland Pub., 1983)
8. Clear wrinkling on a torus has been observed by P.Berge: Physica Scripta T1 , 71 (1982)
9. H.Daido: Prog. Theor. Phys. 70 , (1983) No.3 and RIFP Preprint
10. A.Arneodo, P.H.Coullet and E.A.Spiegel: Phys. Lett. 94A, 1 (1983)
11. K.Kaneko: Prog. Theor. Phys. 69 , 1806 (1983)

Onset of Chaos in Some Hydrodynamic Model Systems of Equations

H. Yahata

Faculty of Science, Hiroshima University, Higashisenda-machi, Naka-ku
Hiroshima 730, Japan

1. Introduction

We report on our model computational results on the transition to turbulence in the Taylor-Couette flow between the two concentric cylinders and the Rayleigh-Benard convection between the two horizontal parallel plates. These two fluid systems have several common features. In either case, the fluid is confined in a closed vessel; with gradual increase of an externally imposed mechanical or thermal force, a steady flow which appears accompanied by a regular spatial structure becomes turbulent, or more precisely, temporally nonperiodic (chaotic) after only a few successive transitions each giving rise to a periodic or quasi-periodic regular fluid motion; the regular spatial flow pattern persists to a considerable degree even after the motion becomes chaotic. Recent precise experimental observations performed on these systems by quite a few research groups provided us with rich information on both temporal and spatial behaviors exhibited by a fluid during the course of transitions from a steady to a turbulent flow.

The present work is an attempt to elucidate mechanisms operative during the transition process of the above-mentioned type using several model systems of truncated ordinary differential equations for the flow mode variables. We make these model systems of evolution equations from the Navier-Stokes equations with the aid of the Galerkin method, taking advantage of the periodic spatial structure persistent in the flow. The underlying reasons for this procedure are the Hopf hypothesis [1] and the strange attractor hypothesis [2] in the sense that fluid turbulence is in some cases described in terms of a finite dimensional attractor in phase space of a dynamical system at least in so far as the Reynolds number remains small. Although the dimension of the strange attractor is usually considered to be relatively small, it is pointed out that the number of the modes for obtaining the correct approximate solutions to the Navier-Stokes equations is, if not infinite, enormously large [3] . In the present work we use for each case a small model system consisting of no more than several dozens of mode variables to describe some characteristic aspects of the fluid motion near the onset of turbulence under the condition that the overall spatial structure of the flow is coarsely determined by the Galerkin basis functions. In the circumstances the qualitative behavior of the system evolution generally depends on the number and the kind of the retained mode variables and furthermore the choice of the Galerkin basis functions, since the spatial structure of the flow including the spatial initial condition is given by that of some superposition of the expansion basis functions.

2. Taylor-Couette Flow

We consider the Couette flow of an incompressible viscous fluid confined between the two concentric cylinders with only the inner cylinder rotating. Experimental results on this system are summarized as follows [4]. As the rotating velocity of the inner cylinder is gradually increased, a circular steady flow in the azimuthal direction becomes unstable and steady toroidal cells called the Taylor vortices appear stacked periodically in the axial direction; with futher increase of the rotating velocity, they lose their axial symmetry and turn into wavy vortices with the propagating waves in the azimuthal direction; the number of their waves along the circumference of the cylinders takes some definite value depending on an initial condition in a complicated way; when the rotating velocity is increased still further, the regular flow pattern of the wavy vortices disappears and the fluid motion becomes highly irregular, although the periodic spatial structure of the Taylor vortices is still persistent.

We deal with the Taylor vortex flow between the two cylinders of infinite length whose inner and outer radii are R_1 and R_2 respectively. The Reynolds number is defined by $Rn=\Omega R_1 d/\nu$, where $d=R_2-R_1$, Ω is the angular velocity of the inner cylinder and ν is the kinematic viscosity. We express the nondimensional variables in terms of the length scale d and the time scale d^2/ν. Let r,θ and z denote the radial, azimuthal and axial coordinates respectively. On account of the periodic spatial structure of the Taylor vortices, the disturbance velocity u_r and u_θ superimposed on the steady circular flow are assumed to be expanded as

$$u_r(r,\theta,z,t)=\sum_{\varepsilon=\pm}\sum_{\ell=0}^{\infty}\sum_{m=-\infty}^{\infty}\sum_{i=1}^{\infty} C(1,(\varepsilon,\ell,m)i;t)\ \phi_{1,i}(r)\ \exp(im\theta)\psi_{\varepsilon,\ell}(z)$$

$$u_\theta(r,\theta,z,t)=\sum_{\varepsilon=\pm}\sum_{\ell=0}^{\infty}\sum_{m=-\infty}^{\infty}\sum_{i=1}^{\infty} C(2,(\varepsilon,\ell,m)i;t)\ \phi_{2,i}(r)\ \exp(im\theta)\psi_{\varepsilon,\ell}(z) \qquad (1)$$

where $\psi_{+,\ell}(z)=-\sqrt{2}\cos(\ell az)$ and $\psi_{-,\ell}(z)=\sqrt{2}\sin(\ell az)$. The periodicity of the Taylor vortices is specified by the axial wave number a and the azimuthal waviness m. We hereafter use the symbol $(\varepsilon, \ell, m)i$ to denote the mode specified by ε, ℓ, m and i. Experimental results suggest us to assume the velocity u_z in the form: $u_z \sim \sin a(z-b\sin(m\theta-0.5\cos(az)-\omega t))$ where ω is the angular velocity of the wavy vortices and b is some constant. Hence, if this holds, its Fourier decomposition shows that u_z consists of the modes sin az, $\exp(i2km\theta)\sin \ell az$, $\exp(i(2k-1)m\theta)\cos \ell az$; k, ℓ = 1, 2,.... It should be noted here that the modes sin az and $\exp(im\theta)\cos az$ were shown by Davey, DiPrima and Stuart [5] to represent the fundamental modes for the wavy vortex flow near its onset value of the Reynolds number. In the present work we deal with a four-fundamental-wave (m=4) model which consists of the mode variables: (+,1,0)i, (+,2,0)i, (-,1,4)i, (+,2,8)i, (i=1, ..., 6); and (+,0,0)i, (i=1, ..., 8). This truncation contains 82 variables in all. The radial basis functions $\phi_{1,i}$ and $\phi_{2,i}$ we employ are the beam (or Chandrasekhar-Reid) and the trigonometric functions respectively. Following the procedures given in [6], we obtain the evolution equations for the mode amplitudes $C(p;\lambda i;t)$ with $\lambda=(\varepsilon,\ell,m)$:

$$\partial_t C(p,\lambda i;t)+\Sigma_{q,j}\ L(p,\lambda i:q,\lambda j)C(q,\lambda j;t)$$

$$=\Sigma_{p'p''}\ \Sigma_{\lambda'\lambda''}\ \Sigma_{j'j''}\ U(p\lambda i;\ p'\lambda'j',\ p''\lambda''j'')\ C(p',\lambda'j';t)\ C(p'',\lambda''j'';t). \qquad (2)$$

To take into account a slight violation of the axial spatial periodicity of the vortices near the upper and lower boundaries of the actual cylinders, we add the small terms $0.05\ \Sigma_{q,j}\ (L(p,(1,0)i:\ q,(1,0)j)+L(p,(2,0)i:\ q,(2,0)j))C(q,(3-\ell,0)j)$ to the lhs of (2) for each mode $(+,\ell,0)i$; ℓ=1,2 and i=1, ..., 6.

We consider the system for the case $R_1/R_2=0.875$ and a=2.67 [4]. To study the transition to turbulence, we integrated (2) on computer for several values of $s=Rn/(Rn)_c$, where $(Rn)_c=118.16$ is the onset value of the steady Taylor vortex flow. Using the time series data thus obtained, we computed the power spectral densities (PSD) for

$$U(t)=\Sigma_{i=1}^{6}(\mathrm{Re}(C(1,(1,4)i)+C(1,(2,8)i))+C(1,(1,0)i)+C(1,(2,0)i))\phi_{1,i}(0)$$

which represents the local radial velocity at $r=(R_1+R_2)/2$. We give the results in Figs. 1 and 2(a). In addition to these, we give in Fig. 2(b) the PSD for the axisymmetric part X(t) of the velocity U(t).

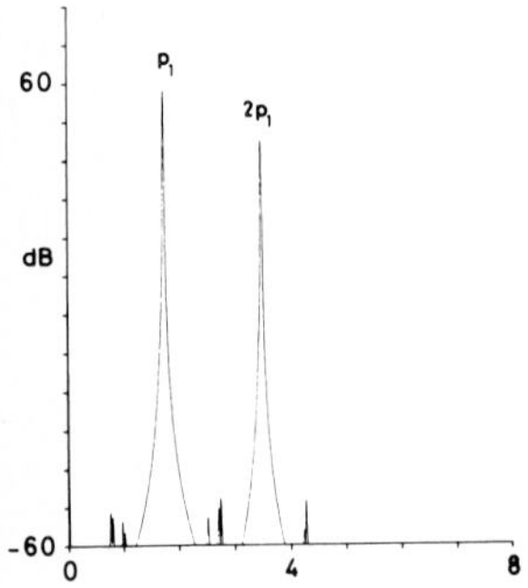

Fig. 1 The PSD for the local radial velocity U(t) at s=6.376. The abscissa measures the angular frequency in the unit Ω

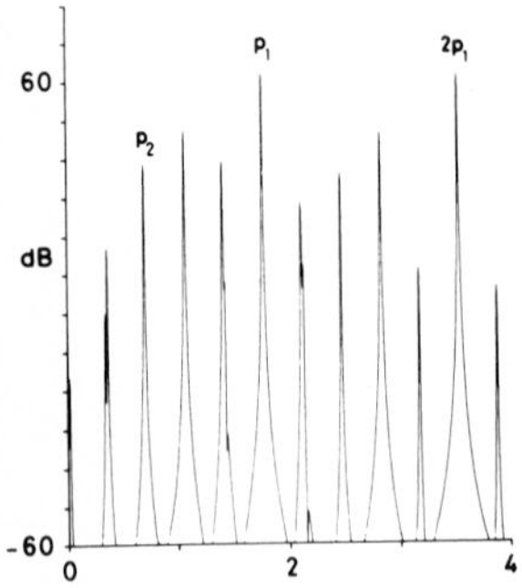

(a1)

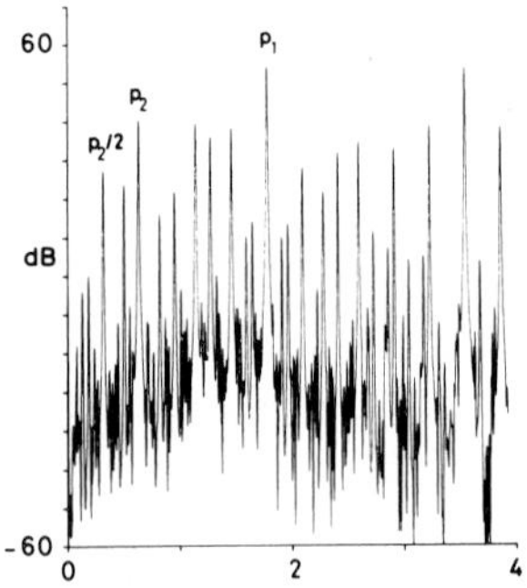

(a2)

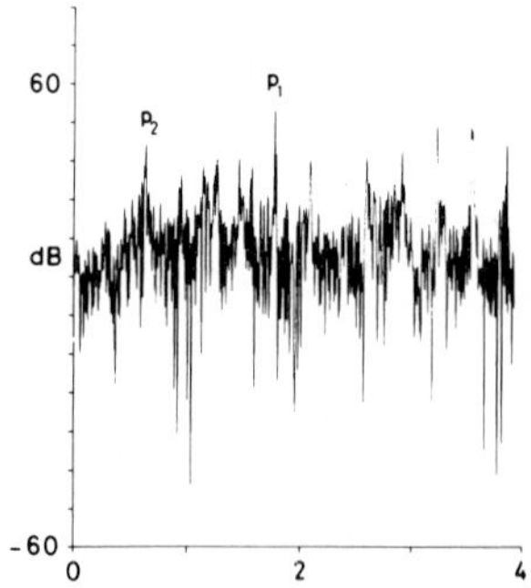

(a3)

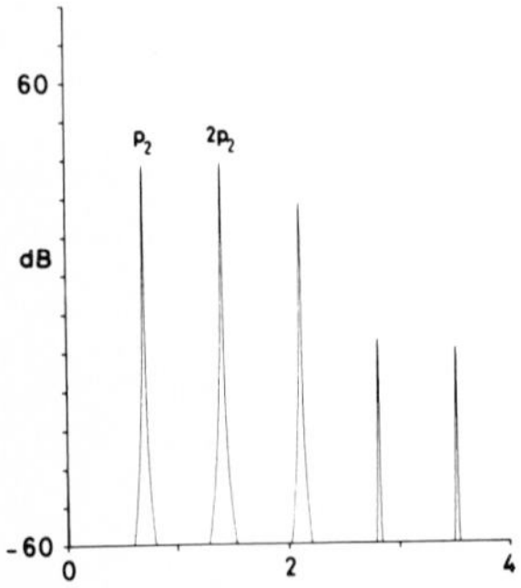

(b1)

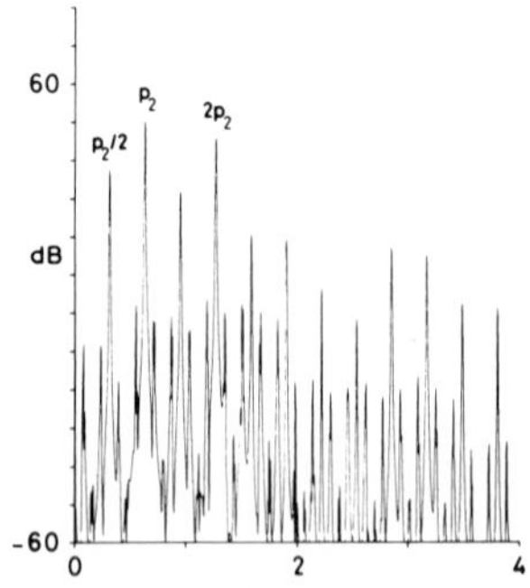

(b2)

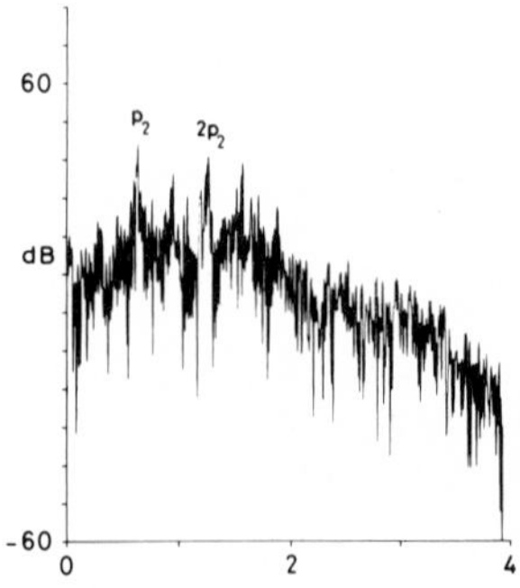

(b3)

Fig. 2 (continued)

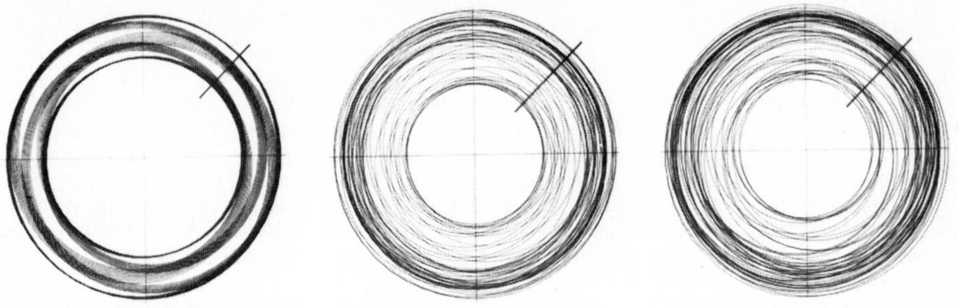

(c1) (c2) (c3)

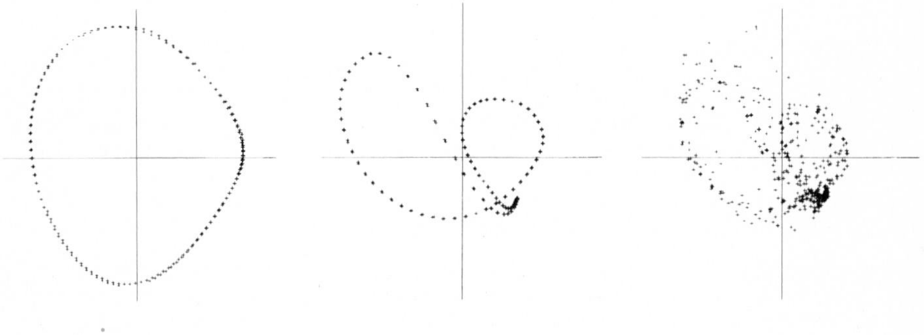

(d1) (d2) (d3)

Fig. 2 (ai):i=1,2,3 The PSD for U(t); (bi):i=1,2,3 the PSD for X(t); (ci):i=1,2,3 the phase prtraits in the subspace (Re(C(1,(1,4)1;t)), Im(C(1,(1,4)1;t))); (di):i=1,2,3 the surfaces of section whose abscissa and ordinate are the cut marked in (c) and the variable Im(C(1,(2,8);t)) respectively. The values of s are denoted by the suffix i: s=9.563, 11.157 and 11.257 for i=1, 2 and 3 respectively. The abscissa of the PSD measures the angular frequency in the unit Ω .

Figure 1 shows the result at s=6.376 where the motion is singly periodic with the fundamental frequency p_1=1.75. According to experimental results [4], however, the fundamental frequency takes a value near 1.33 ; and the spectral components with many other higher multiple values (i.e. three times or more) of the fundamental make considerable contribution to the PSD. When s is increased to s=9.563, Fig. 2(a1) shows that the motion is quasi-periodic with the two fundamental frequencies p_1=1.77 and p_2=0.705 . The second frequency p_2 in our model characterizes the axisymmetric mode oscillations. Figure 2(c) shows orbital profiles mapped to the two-dimensional subspace (Re(C(1,(1,4)1;t)), Im(C(1,(1,4)1;t))); and Fig. 2(d) shows two-dimensional surfaces of section given by the intersection of orbit with a plane spanned by the diagonal in the space of Fig. 2(c) and the coordinate Im(C(1,(2,8)1;t)). The orbit at s=9.563 takes a toroidal form in Fig. 2(c1) and the associated section in Fig. 2(d1) represents a well-defined closed orbit whose rotation number is found to be 0.398 (=p_2/p_1). When s is further increased to s=11.157, subharmonic bifurcation gives rise to a component $p_2/2$ and involves doubling of torus as shown in Figs. 2(b2) and 2(d2). With a slight increase of s to s=11.257, although sharp peaks still remain, the noise level has increased conspicuously in the PSD as shown by Figs. 2(a3) and (b3), which, together with the surface section given by Fig. 2(d3) consiting of randomly scattered points, leads us to conclude that the motion has become chaotic. In contrast to this, experimental results indicate that the two-torus characterizing the wavy vortex flow with increase of s gradually loses its well-defined profile and turns chaotic without undergoing a subharmonic bifurcation [7] . In this sense, a more realistic model system is needed to describe the process of transition to turbulence in the Taylor vortex flow. On the other hand, several authors have recently reported that the transition to chaos via a route involving doubling of torus occurs in two-, or three-dimensional equations [8] . Although the role of a periodic force or noise is there emphasized for the case where chaos arises without the accumulation of successive doubling of torus, distortion and consequent disruption of the torus in our present model appears to be due to the multidimensional character of the surface section.

3. Rayleigh-Benard Convection

We next consider convection in a small-aspect-ratio rectangular box heated from below. Experimental results show that a periodically oscillating convection with increase of vertical thermal gradient turns into turbulence in a quite different manner depending on the box aspect ratio, the fluid Prandtl number and the experimental initial conditions. Let x, y and z denote the Cartesian coordinates with z directed upward. We hereafter use the following notations: d is the distance between the two parallel horizontal boundary plates, κ is the thermal diffusivity, α is the thermal expansion coefficient, g is the gravitational constant, T_d is the difference between the temperatures of the two horizontal plates; ρ_0 and T_0 are the density and the temperature at the lower boundary (z=-d/2) respectively. Within the framework of the Boussinesq approximation, we have the equations of motion obeyed by the disturbances of the velocity u (u_x, u_y, u_z), the temperature $\theta=T-T_0+(T_d/d)(z+d/2)$, the density $\delta\hat{\rho}$ and the pressure δp. In the present work we deal with a model characterized by the parameters : $\rho/\rho_0=1-\alpha_1(T-T_0)-\alpha_2(T-T_0)^2$, $\nu=\nu_0(1-b(T-T_0))$, $\kappa=\kappa_0$; where the suffix 0 denotes the value at z=-d/2. The nonvanishing b and/or α_2, if relatively small, will be shown to give rise to a qualitatively different behavior of the system evolution. Using the length scale d, the time scale d^2/κ_0, and the temperature scale $\kappa_0\nu_0/g\alpha_1 d^3$, we obtain the nondimensional equations of motion for this model

$$\partial_t u_i-\sigma(\partial_j^{\,2}-h_1(R/R_c)\partial_j z\partial_j)u_j$$
$$-\sigma\lambda_i(1-2h_2(R/R_c)z)\theta+\partial_i(\delta p/\rho_0)=-u_j\partial_j u_i, \qquad i=x,\ y \text{ and } z;\ \lambda=(0,0,1);$$
$$\partial_t\theta-\partial_j^{\,2}\theta-R\lambda_j u_j=-u_j\partial_j\theta,$$
$$\partial_j u_j=0, \tag{3}$$

where $h_1=b(T_d)_c$, $h_2=(T_d)_c\alpha_2/\alpha_1$; the suffix c denotes the value at the onset of steady convection. The system is characterized by the nondimensional parameters: the Rayleigh number $R=g\alpha_1 d^3T_d/\kappa_0\nu_0$, the Prandtl number $\sigma=\nu_0/\kappa_0$, and the aspect ratios $\Gamma_x=L_x/d$ and $\Gamma_y=L_y/d$, where L_x and L_y are the horizontal dimensions of the box. We assume that the velocity fields u satisfy the rigid boundary conditions (b.c.); that the temperature field satisfies the conducting and insulating b.c. at the horizontal and lateral walls respectively.

We consider the case where σ is small; the aspect ratios Γ_x and Γ_y are both small and unequal to each other: $\Gamma_x<\Gamma_y$. This leads us to assume that the basic convection pattern is the rolls whose axes are along the x direction; that the oscillatory motion appears associated with the roll modes with vertically directed axes [9]. Hence, we expand the velocity fields in superposition of these two kinds of the mode variables whose axes are perpendicular to each other:

$$u_x=\partial_y\zeta_2,\ u_y=-\partial_z\zeta_1-\partial_x\zeta_2,\ u_z=\partial_y\zeta_1, \tag{4}$$

$$\zeta_1(x,y,z,t)=\Sigma_{i,k,m}\ C(1;i,k,m;t)\ \chi_i(x/\Gamma_x)\ \phi_k(y/\Gamma_y)\ \Phi_{1;m(i,k)}(z), \tag{5}$$

$$\zeta_2(x,y,z,t)=\Sigma_{i,k,m}\ C(2;i,k,m;t)\ \Phi_{2;i(k,m)}(x/\Gamma_x)\ \phi_k(y/\Gamma_y)\ \chi_m(z). \tag{6}$$

We further expand the temperature field:

$$\theta(x,y,z,t)=\Sigma_{i,k,m}\ H(i,k,m,t)\ \psi_i(x/\Gamma_x)\ \psi_k(y/\Gamma_y)\ \chi_m(z). \tag{7}$$

The above basis functions over the domain (-1/2, 1/2) satisfy at $x=\pm1/2$ the b.c.:

$$\phi_m=\partial_x\phi_m=0,\ \Phi_{p;i(j,k)}=\partial_x\Phi_{p;i(j,k)}=0,\ \chi_m=\partial_x\psi_m=0. \tag{8}$$

Following the procedures given in [10], we obtain a system of time evolution equations for the mode variables (4)-(7). For a truncated system consisting of several dozens of variables, we find various kinds of models which yield different dynamical behavior depending on the choice of the basis functions. We here study the transition from periodic to chaotic convection for the two cases of model systems. In both cases, ϕ_m, ψ_m and χ_m are assumed to be respectively determined by the eigenvalue equations:

$$\partial_x^4\phi_m=-\alpha_m^2\partial_x^2\phi_m,\ \partial_x^2\psi_m=-\beta_m^2\psi_m,\ \text{and}\ \partial_x^2\chi_m=-\gamma_m^2\chi_m, \qquad m=1,\ 2,\dots. \tag{9}$$

The definitions of $\Phi_{p;i(j,k)}$ will be given later for each case. We use a truncated system of equations consisting of the mode variables {A}={C(1;ie,lo,le), C(2;io,lo,lo), H(ie,2e,le), H(ie,le,le); i=1, ..., 6} and {B}={C(1;ie,lo,lo), C(2;io,lo,le), H(ie,2e,lo), H(ie,le,lo); i=1, ..., 6} where the suffices e and o indicate the evenness and oddness of the corresponding basis functions. The truncation contains 48 variables; and the basic flow pattern is the two rolls with the x-directed axes. The evolution equations are symbolically written in the form

$$\begin{aligned}&\partial_t\{A\}=\{A\}+(h_1,h_2)\{B\}+\{A\}\{B\}\ ,\\ &\partial_t\{B\}=(h_1,h_2)\{A\}+\{B\}+\{A\}\{A\}+\{B\}\{B\}\ .\end{aligned} \tag{10}$$

If $h_1=h_2=0$, the mode variables {A} and {B} will generally oscillate with the different frequency components. This division of the variables into {A} and {B} is due to the spatial symmetry of the system in the vertical direction. Therefore, the terms containing h_1 and/or h_2 (as well as R/R_c), if its magnitide exceeds some threshold value, break this symmetry and set off synchronization of oscillations between these two groups of the variables {A} and {B}. If this occurs, the whole mode variables will oscillate with the same frequency components. Since these asymmetry effects arise in actual experimental situations for many reasons, we should regard h_1 and h_2 at best as the model parameters rather than as the material constants. It should be noted here that these asymmetry effects play an important role in determining the convection pattern even over the steady-state range of R [11].

In the present work we deal with the case where $\sigma=2.5$, $\Gamma_x=2.0$ and $\Gamma_y=3.5$. In fact, under these external conditions Gollub and Benson have observed experimentally that convection whose basic flow is the two rolls with the x-directed axes persists over quite a wide range of R even up to the chaotic state [12]. The onset value of R for steady convection in an infinite layer is $R_{c\infty}=1707.8$. We integrated (10) on computer for several values of $s=R/R_{c\infty}$.

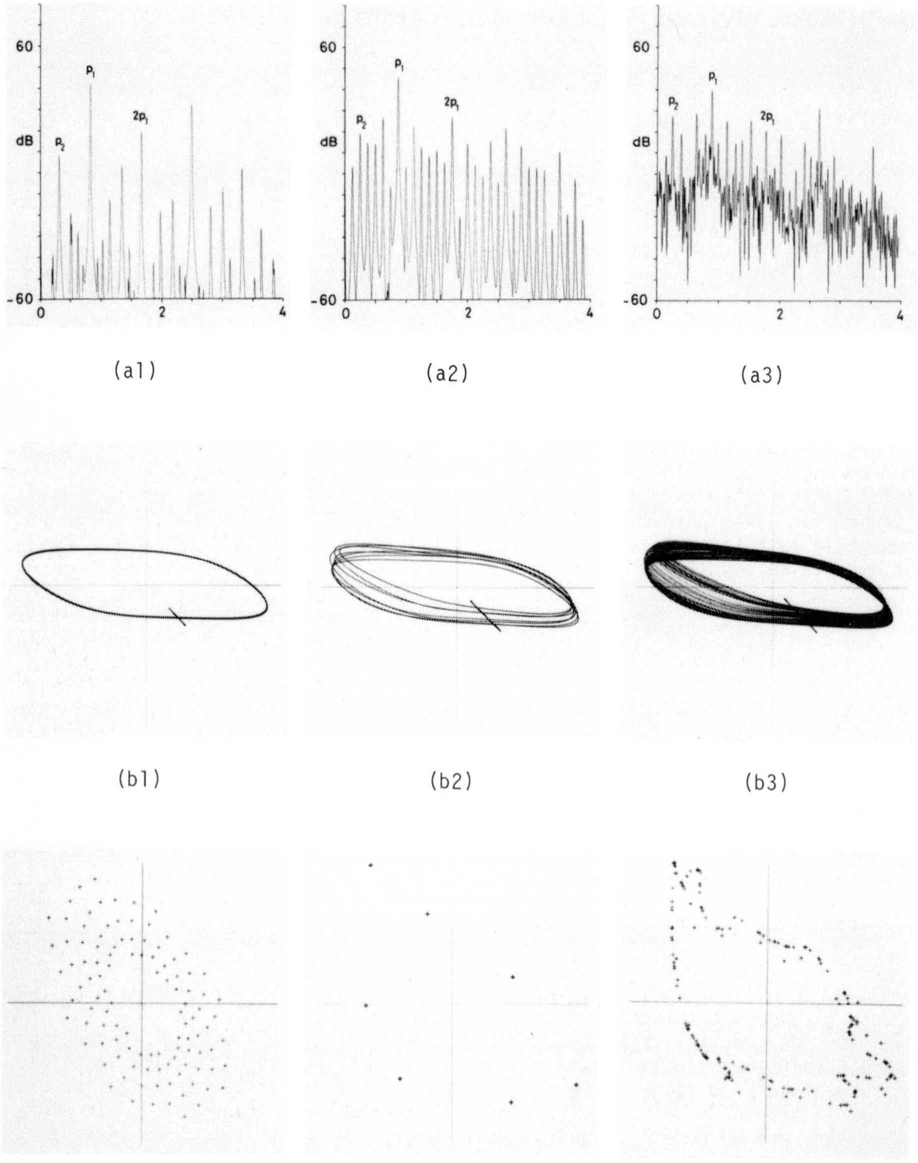

Fig. 3 (continued)

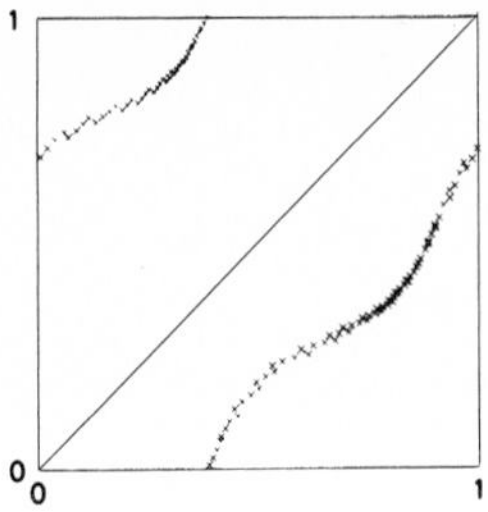

(d1)

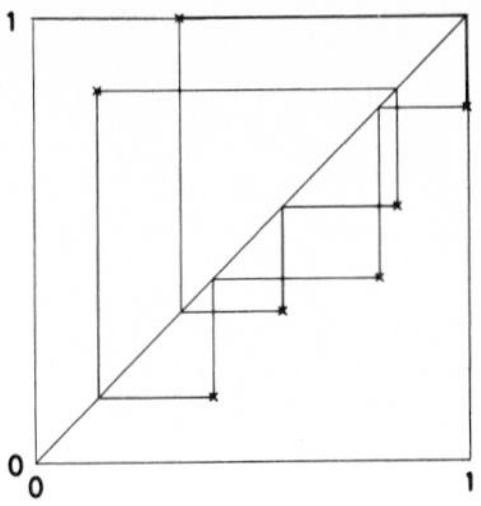

(d2)

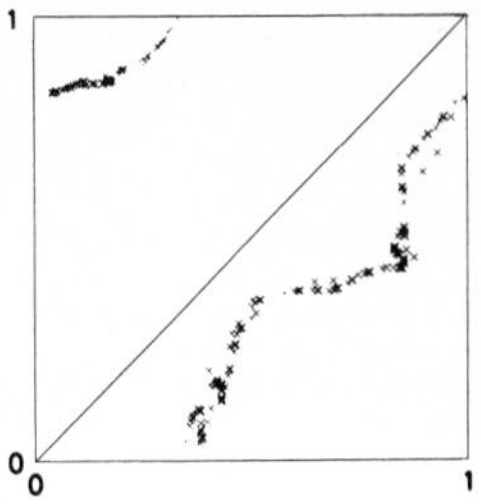

(d3)

Fig. 3 (ai):i=1,2,3 The PSD for the basic roll mode C(1;le,lo,le) where the abscissa measures the frequency in the unit $13.1\kappa_0/d^2$; (bi):i=1,2,3 the phase portraits in the subspace (C(1;le,lo,le),C(2;lo,lo,lo)); (ci):i=1,2,3 the surfaces of section whose abscissa and ordinate are the cut marked in (b) and the variable H(le,le,le); (di):i=1,2,3 the return maps for the argument values between the two temporally neighboring points in the planes of (c). The values of s: s=39.77, 40.41 and 40.51 for i=1, 2 and 3 respectively

For the first case we consider a model system whose basis functions $\Phi_{p;i(j,k)}$ are determined in the same form as ϕ_i; i.e., $\Phi_{p;i(j,k)}=\phi_i$ irrespective of p,j,k for i=1, 2,... For the case where $h_1=h_2=0$, we have found the motion turns into chaos when the three fundamental frequencies become rationally dependent[10]. We here consider a model characterized by the parameter values: $h_1=0.01$ and $h_2=0$. We give the results by the PSD for C(1;le,lo,le) in Fig. 3(a); and the phase portrait mapped to the subspace (C(1;le,lo,le), C(2;lo,lo,lo)) in Fig. 3(b). The motion at s=39.77 is quasi-periodic with the two fundamental frequencies p_1 and p_2. When s is increased to s=40.41, a phase-locked state with $p_1/7=p_2/2=p_L$ appears and the associated profile is a closed orbit on the torus. With further increase of s to s=40.51 the PSD Fig. 3(b3) exhibits marked increase of the noise level. To clarify the transition in a little more detail, we give in Fig. 3(c) surfaces of section given by the intersection of orbit with a plane spanned by the cut in the space of Fig. 3(b) and the coordinate H(le,le,le). In addition, computing the time series $\{x_n;\ n=1,\ 2,...\}$ whose sequential points are the argument values (divided by 2π) between the temporally neighboring sectional points of Fig. 3(c), we plot each pair of the neighboring values $(x_n,\ x_{n+1})$ on the two-dimensional plane $[0,1)\times[0,1)$. We give the return maps thus obtained in Fig. 3(d). Although the map is everywhere monotonic at s=39.77; it comes to exhibit noninvertible character at s = 40.51 as a result of highly distorted surface of section given in Fig. 3(c3). These results suggest us to conclude that the motion at s=40.51, which arises after breakdown of the phase-locked state, is chaotic. This route to turbulence was observed experimentally in the Rayleigh-Benard system [12] and has been studied theoretically for low-dimensional map systems [13].

For the second case, we consider a model whose basis functions $\Phi_{p;i(j,k)}$ are determined by the eigenvalue equation for p=1, 2:

$$(D_{p;j,k})^4\Phi_{p;i(j,k)}=-(\lambda_{p;i(j,k)})^2(D_{p;i(j,k)})^2\Phi_{p;i(j,k)}$$

where $(D_{p;j,k})^2=(\partial_x/G_{3-p})^2-(\gamma_j/G_p)^2-(\alpha_k/\Gamma_y)^2$, $G_1=\Gamma_x$ and $G_2=1$. In a way similar to

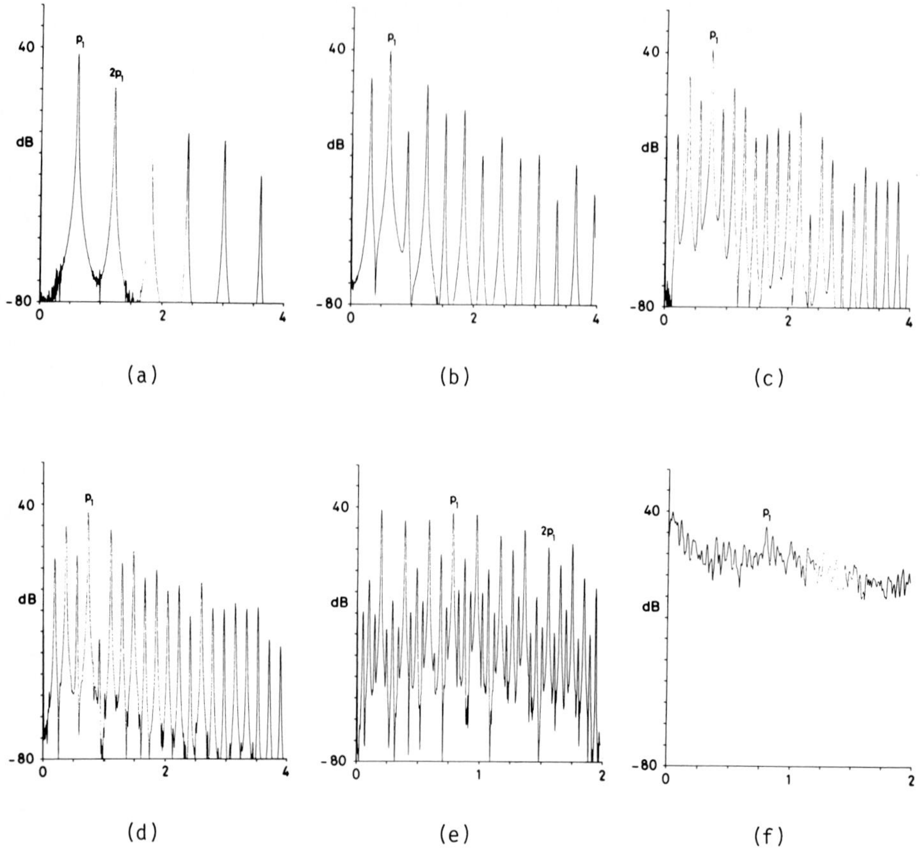

Fig. 4 The PSD for the vertical roll mode $C(2;1o,1o,1e)$ where the abscissa measures the frequency in the unit $13.1\kappa_0/d^2$. The values of s: (a) 22.83; (b) 26.09; (c) 27.72; (d) 29.35; (e) 32.61; (f) 34.24

the first case, we make a 48-mode system of the form (9). We give the results for the case $h_1=0$ and $h_2=0.01$ by the PSD for the vertical roll mode $C(2;1o,1o,1e)$ in Fig. 4. The PSD at s=22.83 Fig. 4(a) shows the motion is singly periodic with the fundamental frequncy p_1 and its multiples. When s is gradually increased, the motion undergoes period-doubling bifurcations one after another and comes to develop at s=32.61 the subharmonic components at least up to $p_1/16$. However, our PSD still lacks in the resolution sufficient to determine precisely the ratio between each pair of successive subharmonic peak heights which is predicted to be one of the universal constants for the period-doubling cascade to chaos [14] . With a slight increase of s to s=34.24, the PSD becomes dominated by continuum with some remaining broad peaks. This period-doubling route to chaos was actually observed by many research groups for the Rayleigh-Benard convection under various experimental conditions [15] ; however, further work will be needed to clarify to what extent the transition occurs in the one-dimensional way.

References

1. E. Hopf: Comm. Pure and Appl. Math. 1, 303(1948).
2. D. Ruelle and F. Takens: Comm. Math. Phys. 20, 167(1971).
3. C. Foias, O.P. Manley, R. Temam and Y.M. Treve: Phys. Rev. Lett. 50,1031(1983)
4. P.R. Fenstermacher, H.L. Swinney and J.P. Gollub: J. Fluid Mech. 94, 103(1979)
5. A. Davey, R.C. DiPrima and J.T. Stuart: J. Fluid Mech. 31, 17(1968).
6. H. Yahata: Prog. Theor. Phys. 66, 879(1981).
7. A. Brandstater, J. Swift, H.L. Swinney, A. Wolf, J.D. Farmer, E, Jen and J.P. Crutchfield: Preprint.
8. A. Arneodo, P.H. Coullet and E.A. Spiegel: Phys. Letters 94A, 1(1983); K. Kaneko, Prog. Theor. Phys. 69, 1806(1983).
9. R.M. Clever and F.H. Busse: J. Fluid Mech. 65, 625(1974).
10. H. Yahata: Prog. Theor. Phys. 68, 1070(1982).
11. F.H. Busse: Rep. Prog. Phys. 41, 1929(1978).
12. J.P. Gollub and S.V. Benson: J. Fluid Mech. 100, 449(1980).
13. D.G. Aronson, M.A. Chory, G.R. Hall and R.P. McGehee: Comm. Math. Phys. 83,303(1982).
14. M.J. Feigenbaum: Phys. Letters 74A, 375(1979); M. Nauenberg and J. Rudnick: Phys. Rev. B24, 493(1981).
15. J.P. Gollub, S.V. Benson and J. Steinman: Ann. N.Y. Acad. Sci. 357,22(1980); A. Libchaber and J. Maurer: J. de Phys. 41,C3-51(1980); M. Giglio, S. Musazzi and U. Perini: Phys. Rev. Lett. 47,243(1981); A. Libchaber, C. Laroche and S. Fauve: J. de Phys. 43,L-211(1982).

Part VIII

Chemical and Optical Systems

Instabilities and Chaos in a Chemical Reaction

H.L. Swinney, R.H. Simoyi[1], and J.C. Roux[2]

Department of Physics, University of Texas
Austin, TX 78712, USA

Phase space portraits, Poincaré sections, and maps have been obtained from experiments on the Belousov-Zhabotinskii chemical reaction. For some parameter values the dynamical behavior is described by a strange attractor, that is, the largest Lyapunov exponent is positive. A period-doubling sequence, the U sequence, and a tangent bifurcation have been observed as the flow rate is varied.

1. Introduction

The first evidence for chaos in a chemical reaction was obtained in 1977 in experiments on the Belousov-Zhabotinskii [BZ] reaction by SCHMITZ et al.[1] and on the horseradish-peroxidase reaction by DEGN and OLSEN[2]. Experiments in the past few years by HUDSON and co-workers [3-5] in Virginia, by ROUX, VIDAL, BACHELART, and ROSSI[6-9] in Bordeaux, and by our group [9-13] in Texas have provided definitive evidence for the existence of deterministic nonperiodic behavior (chaos) in a nonequilibrium chemical reaction. Although chaotic behavior has been studied in a detailed quantitative way only for the BZ reaction, complex dynamical behavior has been observed in many reactions (see, e.g., [14]), and it is likely that chaos is fairly common in chemical systems. A variety of models (differential equations and maps) that were developed to describe chemical reactions have been found to exhibit chaos (e.g., see [10, 15-18]). In this paper we will summarize some of the results of our experiments; details are reported elsewhere [9-13].

2. The System

The BZ reaction is by far the most extensively investigated and best understood oscillating chemical reaction. Oscillations in this reaction, which involves the cerium-catalyzed bromination and oxidation of malonic acid by a sulfuric acid solution of bromate, were discovered in 1959 in the Soviet Union by BELOUSOV[19] and were subsequently studied by ZHABOTINSKII[20], another Russian chemist. Until the past few years all experiments were conducted in closed ("batch") reactors; the quantitative information obtainable from such

---

[1] Permanent address: University of Zimbabwe, Department of Chemistry, P.O. Box MP167, Mount Pleasant, Harare, Zimbabwe

[2] Permanent address: Centre de Recherche Paul Pascal, Université de Bordeaux I, Domaine Universitaire, 33405 Talence Cedex, France

experiments is limited since the state of the system is continuously changing as the system evolves toward thermodynamic equilibrium. Our experiments and most other recent experiments have been conducted with stirred flow reactors; the system can be maintained in a steady state away from equilibrium by continuously pumping the chemicals through the reactor. The reactor is vigorously stirred so diffusion can be neglected; hence the reaction can be modeled by a set of coupled nonlinear ordinary differential equations. The control parameter for our experiments was the flow rate; the flow rates of the input chemicals were changed while the temperature, stirring rate, and relative concentrations were held fixed[10]. All of the data to be discussed in this paper were obtained in only a 10% range of flow rate[11-12]; other kinds of dynamical behavior in addition to those discussed here have been observed for other flow rates and chemical concentrations[1-10, 13, 14].

3. Strange Attractors

Phase portraits have been constructed from measurements of the time dependence of the concentration of one of the chemical species, the bromide ion. The bromide ion potential $B(t_k)$, proportional to the log of the concentration, was measured in successive time intervals, $t_k = k\Delta t[k=1,2,\ldots,32768]$. Phase portraits of dimension m were then constructed from the vectors $\{B(t_k), B(t_k+\tau),\ldots,B(t_k+(m-1)\tau\}$ (see [12-13]). A phase portrait obtained for a nonperiodic state of the BZ reaction is shown in Fig. 1(a). The post-transient limit set shown in Fig. 1(a) is an attractor; that is, orbits are found to return rapidly to this set following a perturbation[12].

The Poincaré section in Fig. 1(b) indicates that the attractor in Fig. 1(a) is essentially two-dimensional, although the dimension must be slightly greater than two since the attractor is strange (chaotic), as we will show. The smooth curve in Fig. 1(b) can be parameterized, and the coordinates X_n of successive points along this curve then provide a sequence which defines a one-dimensional map, $X_{n+1} = f(X_n)$, as shown in Fig. 1(c). The points fall on a single-valued curve, indicating that the system is deterministic. This map and its evolution with changing flow rate are well described by $X_{n+1}=f(X_n)=AX_n\exp[-(A/e)X_n]$, as will be discussed elsewhere; for this function the maximum value of $f(X_n)$ is always unity, as in the maps constructed from the data. Similar one-dimensional maps with a single extremum have been obtained for the BZ reaction in other experiments covering a wide variety of concentrations and flow rates [4,5,7,16,21].

Is Fig. 1(a) a strange attractor? A strange attractor can be defined operationally to be an attractor whose largest Lyapunov exponent is positive. (This exponent describes the average rate of separation of nearby points in phase space.) The Lyapunov exponent for the attractor in Fig. 1, determined from the map in Fig. 1(c), has a value 0.3±0.1[12], definitely positive; hence the attractor is strange. Further evidence that this attractor characterizes deterministic chaos is provided by the folding of the attracting sheets that has been observed in the Poincaré sections[12]. Low-dimensional strange attractors have also been observed very recently in several experiments on weakly turbulent fluid flows[22-24].

4. Period Doubling and the U Sequence

It is now well-known that well-behaved one-dimensional maps of the interval with a single extremum exhibit universal dynamics[25]. As the bifurcation parameter λ for such a map is increased, the fixed point corresponding to a

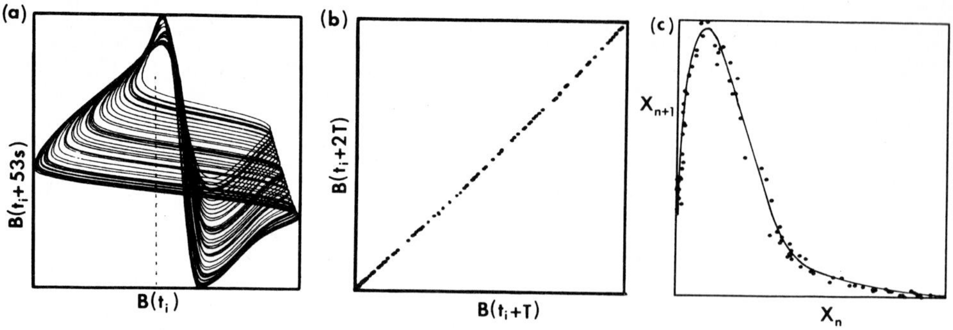

Fig. 1. Data obtained for a nonperiodic state observed in the BZ reaction [10-12]. The power spectrum of the measured observable, the bromide ion potential $B(t_k)$, contains broadband noise that is several orders of magnitude above the instrumental noise level. (a) A two-dimensional phase portrait, $B(t_k+\tau)$ vs. $B(t_k)$, where τ = 53 sec (the average period of an orbit is about 130 sec). (b) A Poincaré section given by the intersection of orbits in a three-dimensional phase portrait with a plane normal to the paper passing through the dashed line in (a); the third axis, which is normal to the paper, is given by the chemical concentration at twice the basic delay time, i.e., by $B(t_k+2\tau)$. (c) A one-dimensional map constructed by plotting as ordered pairs (X_n,X_{n+1}) the coordinates X_n of successive intersections of orbits with the Poincaré section in (b) (from [10-12])

periodic state (a "1-cycle") eventually loses its stability to a 2-cycle in a pitchfork bifurcation. There exists an infinite sequence of such period-doubling transitions that converges to a 2^∞-cycle at finite $\lambda=\lambda_c$; FEIGENBAUM[26] showed that for maps with a quadratic extremum the convergence rate of the sequence (the ratio of successive intervals in λ between transitions) is a map-independent universal number, δ=4.6692016...

Period doubling has been observed in the experiment which yielded that one-dimensional map shown in Fig. 1(c). Oscillations of the chemical concentrations with periods 1, 2, and 4 are illustrated at the top of Fig. 2; oscillations with periods 8 and 16 were also observed. In the past two years period doubling has been observed in more than a dozen other experiments on quite diverse physical systems (see refs. in [11] and [13]). However, since the sequence converges so fast, it will always be difficult to see more than a

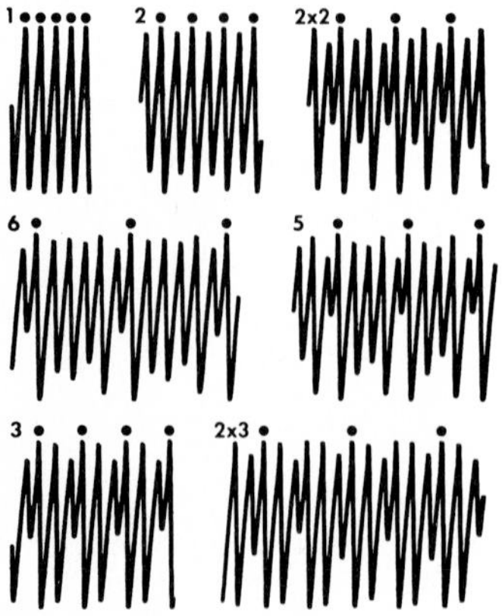

Fig. 2. Some periodic states of the U sequence observed in measurements on the BZ reaction for different flow rates of the chemicals through the reactor. The bromide ion potential is plotted as a function of time for states with periods T (where T ≈ 115 sec), 2T, 2x2T, 6T, 5T, 3T, and 2x3T. The dots above the time series are separated by one period.

few period doublings; thus, e.g., even if a very clever experimenter could increase his control parameter resolution by $(4.669)^2=21.8$, which would be a remarkable feat, he would see only two more doublings. Therefore, it will be very difficult to obtain accurate values of δ from experiments.

However, there is far more than period doubling of the 1-cycle in the universal dynamics of one-dimensional maps. A method called symbolic dynamics uses the single extremum property of the map (the extremum need not be quadratic) to predict the existence of a U(universal) sequence of periodic intervals beyond the accumulation point for period doubling of the 1-cycle[25,27]. The periodic states may be uniquely classified by the order in which points on the X_n axis are visited. Only certain periodic states are allowed by the theory; each allowed state occurs only once (i.e., for one λ interval) as λ is varied. The periodic intervals as a function of λ are dense. There are no chaotic intervals, but the points (λ values) for which the behavior is chaotic have positive measure. At any given λ not more than one periodic state is stable if the map has a negative Schwarzian derivative[25], as is the case for our maps.

The theory[25,27] predicts that beyond λ_c fundamentals of all integer periods appear and undergo their own period-doubling sequences. The larger the integer, the larger the number of allowed states; for example, there are three distinct 5-cycles and 27 distinct 9-cycles. The theory also predicts (at least locally in λ) the ordering of the periodic intervals.

We have observed nine of the fundamental U-sequence states, including the only 3-cycle, the only 4-cycle, and all three 5-cycle states. Period doubling has been observed for the wider of these intervals; for the widest interval, period 3, periods 6, 12, and 24 were observed. Figure 2 shows a fundamental 6-cycle, a 5-cycle, and the 3-cycle, and also the first harmonic of the 3-cycle.

Each periodic interval is predicted to appear at a tangent bifurcation. A third-iterate map, X_{n+3} vs. X_n, constructed for chaotic data near the onset of period 3,is in fact nearly tangent at three points along the $X_{n+3}=X_n$ diagonal, as predicted.

In summary, the observations of the map iteration patterns, the order of occurrence of the periodic states, the period doubling of the fundamental periodic intervals, and the tangent bifurcation at the onset of period 3 are all in accord with the theory for one-dimensional maps. Thus, even though the reaction is quite complex, involving more than 25 chemical species, the dynamical behavior for some parameter range is modeled in detail by a one-dimensional map.

Acknowledgments

This research was conducted in collaboration with A. Wolf, W.D. McCormick, and J. Turner. Support is provided by Robert A. Welch Foundation Grant F-805.

References

1. R. A. Schmitz, K. R. Graziani, and J. L. Hudson: J. Chem. Phys. 67, 3040 (1977)
2. L. F. Olsen and H. Degn: Nature 267, 177 (1977)
3. J. L. Hudson, M. Hart, and D. Marinko: J. Chem. Phys. 71, 1601 (1979)
4. J. L. Hudson and J. C. Mankin: J. Chem. Phys. 74, 6171 (1981)
5. J. L. Hudson and O. E. Rossler: in Chaos in Chemical Systems, ed. by V. Hlavacek (1983)
6. C. Vidal, J. C. Roux, S. Bachelart: Ann. N.Y. Acad. Sci. 357, 377 (1980)
7. J. C. Roux, A. Rossi, S. Bachelart, and C. Vidal: Physica 2D, 395 (1981)
8. Y. Pomeau, J. C. Roux, A. Rossi, S. Bachelart, and C. Vidal: J. Phys. Lett. (Paris) 42, L271 (1981)

9. J. C. Roux: Physica 7D, 57 (1983)
10. J. S. Turner, J. C. Roux, W. D. McCormick, and H. L. Swinney: Phys. Lett. 85A, 9 (1981). See also J. C. Roux, J. S. Turner, W. D. McCormick, and H. L. Swinney: in Nonlinear Problems: Present and Future, ed. by A. R. Bishop, D. K. Campbell, and B. Nicolaenko (North-Holland, Amsterdam, 1982), p. 409
11. R. H. Simoyi, A. Wolf, and H. L. Swinney: Phys. Rev. Lett. 49, 245 (1982)
12. J. C. Roux, R. H. Simoyi, and H. L. Swinney: Physica 8D, 257 (1983)
13. H. L. Swinney: Physica 7D, 3 (1983)
14. I. Epstein: Physica 7D, 47 (1983)
15. O. E. Rossler: Z. Naturforsch. 31a, 259 (1976)
16. K. Tomita and I. Tsuda: Prog. Theor. Phys. 64, 1138 (1980)
17. I. Tsuda: Phys. Lett. 85A, 4 (1981)
18. A. S. Pikovsky: Phys. Lett. 85A, 13 (1981)
19. B. P. Belousov: Sb. Ref. Radiats. Med. 1958 (Medgiz, Moscow, 1959), p. 145
20. A. M. Zhabotinskii: Dokl. Akad. Nauk. SSSR 157, 392 (1964); Biofizika 9, 306 (1964)
21. H. Nagashima: J. Phys. Soc. Japan 49, 2427 (1980)
22. A. Brandstater, J. Swift, H. L. Swinney, A. Wolf, J. D. Farmer, E. Jen, and J. P. Crutchfield: Phys. Rev. Lett. 51, 1442 (1983)
23. J. Guckenheimer and G. Buzyna: Phys. Rev. Lett. 51, 1438 (1983)
24. B. Malraison, P. Attens, P. Berge, and M. Dubois: J. Phys. Lett. (Paris) 44 (1983)
25. P. Collet and J. P. Eckmann: Iterated Maps of the Interval as Dynamical Systems (Birkhauser, Boston, 1980)
26. M. J. Feigenbaum: J. Stat. Phys. 19, 25 (1978); Physica 7D, 16 (1983)
27. N. Metropolis, M. L. Stein, and P. R. Stein: J. Combinatorial Theory A 15, 25 (1973)

Optical Turbulence

K. Ikeda

Department of Physics, Kyoto University, Kyoto 606, Japan

O. Akimoto

The School of Allied Health Sciences, Yamaguchi University, Kogushi
Ube 755, Japan

1. Introduction

Laminar flow in fluid systems becomes unstable and turns to turbulent flow when the velocity gradient exceeds the damping of fluctuations due to viscosity. The origin of turbulence lies in nonlinear coupling of fluid elements which is involved in the inertia term in the Navier-Stokes equation. Light propagating in matter also has a similar character. When the light remains weak, the Maxwell equations are nearly linear, so that no peculiar behavior can be seen. As the light intensity is increased, however, nonlinear optical processes via interactions between light and matter generate photons of new frequencies successively, and a transition is finally induced to *optical turbulence*, a very complicated space-time variation of the electromagnetic field. The present paper is a brief review of recent studies on this phenomenon and its underlying mathematics.

2. Dynamics of Nonlinear Optical Resonators

When a dielectric medium is set in a Fabry-Perot resonator, as shown in Fig.1(a), or in a ring resonator, as shown in Fig.1(b), and is irradiated with a laser light, it exhibits a bistable response. For the same incident light, if it is sufficiently intense, the intensity of the transmitted light from the medium has two alternative values for which the system is stable, being accompanied by a hysteresis between them [1]. This phenomenon, called optical bistability, has its origin in the nonlinearity of light-matter interactions coupled with a partial feedback of the transmitted light by mirrors. The feedback is attended by a time delay due to propagation of light. And it is this *delayed feedback* coupled with nonlinearity that makes the transmitted light chaotic or turbulent under suitable conditions [2].

2.1. Formulation

Consider a Fabry-Perot resonator shown in Fig.1(a). Let $\hat{E}^{\pm}(t,z)$ be the complex envelope of the electric fields propagating in two opposite directions. They are connected with the refractive index of the medium, $n(t,z)$, through the Maxwell equation

$$(\pm\partial/\partial z + c^{-1}\partial/\partial t)E^{\pm} = (-\alpha/2 + in)E^{\pm}. \tag{2.1}$$

Here the electric field has been scaled in a dimensionless form by $E^{\pm}=\{k|n_2|(1-e^{-\alpha\ell})/\alpha\}^{\frac{1}{2}}\hat{E}^{\pm}$, where α is the absorption coefficient of the medium and n_2 the quadratic coefficient of the nonlinear refractive index for constant field.

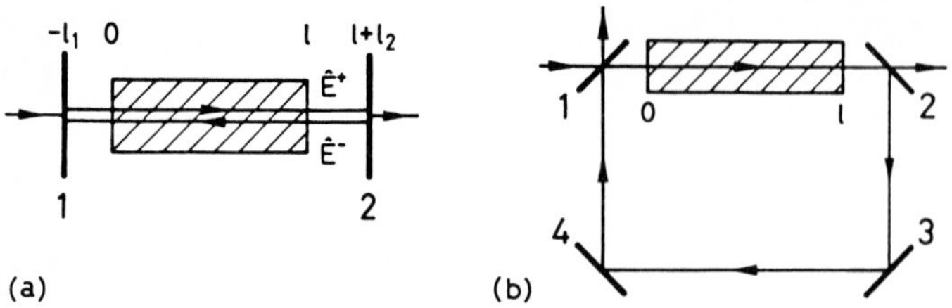

Fig.1. Optical resonators containing a nonlinear dielectric medium: (a) Fabry-Perot resonator, and (b) ring resonator. Mirrors 1 and 2 are semitransparent with reflectivity R

The refractive index n(t,z), on the other hand, obeys the Debye relaxation equation

$$\gamma^{-1}\partial n/\partial t = -(n-n_0) + q(|E^+|^2 + |E^-|^2), \tag{2.2}$$

where γ is the relaxation rate, n_2 the linear part of the refractive index, and $q \equiv \mathrm{sgn}(n_2)\alpha(1-e^{-\alpha\ell})^{-1}$.

Equations (2.1) and (2.2) are subject to the boundary conditions

$$E^+(t,-\ell_1) = A + \sqrt{R}E^-(t,-\ell_1), \qquad E^-(t,\ell+\ell_2) = \sqrt{R}E^+(t,\ell+\ell_2) \tag{2.3}$$

at the position of mirrors 1 and 2, respectively, where A is the amplitude of the laser light incident through mirror 1. Under these conditions, eqs.(2.1) and (2.2) can be integrated with respect to z and are reduced to the set of coupled equations:

$$E(t) = A + BE(t-t_R)\exp\{i[\phi(t)-\phi_0]\}, \tag{2.4a}$$

$$\gamma^{-1}\dot{\phi}(t) = -\phi(t) + \psi(t) + (1+B)|E(t-t_R)|^2, \tag{2.4b}$$

$$\delta^{-1}\dot{\psi}(t) = -\psi(t) + \sum_{i=1}^{4} a_i|E(t-t_R^{(i)})|^2, \tag{2.4c}$$

where $\delta = \alpha c/2$. The three independent variables in these equations are: the electric field at z=0

$$E(t) \equiv E^+(t,0), \tag{2.5a}$$

the phase lag of the electric field across the medium

$$\phi(t) \equiv q\int_0^{\ell} dz'[n(t+(z'-\ell_1)/c,z') + n(t-t_R+(\ell_1-z')/c,z')], \tag{2.5b}$$

and the additional photon density at z=0 induced by the interaction of E^+ and E^-

$$\psi(t) \equiv q\int_0^{\ell} dz'[Re^{\alpha(z'-2\ell)}|E(t-t_R+2z'/c)|^2 + e^{-\alpha z'}|E(t-t_R-2z'/c)|^2]. \tag{2.5c}$$

In eqs.(2.4a-c), parameter B defined by $B \equiv Re^{-\alpha\ell}$ (<1) characterizes the dissipation of the field, and $t_R \equiv 2(\ell_1+\ell_2+\ell)/c$ is the time required for light to make a round trip in the resonator. The other parameters are

$$a_1=(1-e^{-\alpha\ell})^{-1}, \quad a_2=-e^{-\alpha\ell}a_1, \quad a_3=Ba_1, \quad a_4=-Be^{-\alpha\ell}a_1,$$
$$t_R^{(1)}=2\ell_1/c, \quad t_R^{(2)}=t_R-2\ell_2/c, \quad t_R^{(3)}=t_R+2\ell_2/c, \quad t_R^{(4)}=t_R-2\ell_1/c. \tag{2.6}$$

Note that eqs.(2.4a-c) involve multiple time delay, t_R and $t_R^{(i)}$.

In a ring resonator, on the other hand, there is no backward propagating field E^-. Owing to this, $\psi(t)$ vanishes and eqs.(2.4a-c) are largely simplified to the equations with a single time delay [2]:

$$E(t)=A+BE(t-t_R)\exp\{i(\phi(t)-\phi_0)\}, \tag{2.7a}$$

$$\gamma^{-1}\dot{\phi}(t)=-\phi(t)+|E(t-t_R)|^2. \tag{2.7b}$$

Parameter B should here read $Re^{-\alpha\ell/2}$.

To see as clearly as possible how a nonlinear resonator displays chaotic behavior, let us further simplify eqs.(2.7a,b) by taking the limit of large dissipation, $B\ll1$. For a full nonlinearity to be taken into account, however, A^2B must be kept constant. Then, eq.(2.7a) is approximated by

$$|E(t)|^2=A^2[1+2B\cos(\phi(t)-\phi_0)]. \tag{2.8}$$

[The derivation is straightforward by iterative use of eq.(2.7a).] By eliminating E(t) from eqs.(2.7b) and (2.8), one finally obtains the one-variable delay-differential equation for ϕ [2]:

$$\gamma^{-1}\dot{\phi}(t)=-\phi(t)+A^2[1+2B\cos(\phi(t-t_R)-\phi_0)]. \tag{2.9}$$

As this equation is probably the simplest one that describes the essence of the dynamics of nonlinear resonators, the discussion below will be made on the basis of this equation.

2.2. Instabilities and the Transition to Chaos

In the limit of $t_R\gamma\gg1$, eq.(2.9) is formally reduced to the difference equation

$$\phi(t)=\mu f(\phi(t-t_R)), \tag{2.10}$$

where

$$f(\phi)=\pi[1+2B\cos(\phi-\phi_0)], \tag{2.11}$$

and $\mu=A^2/\pi$. If a solution ϕ that is constant for $nt_R\leq t<(n+1)t_R$ is considered, eq.(2.10) is equivalent to the map of a discrete variable

$$\phi_{n+1}=\mu f(\phi_n) \tag{2.12}$$

with $\phi_n\equiv\phi(nt_R)$, a map of the class that has extensively been studied by Feigenbaum [3]. When μ is increased beyond the first bifurcation point, the stationary solution of this map becomes unstable and is replaced by a periodic solution of period $2t_R$. This period successively doubles itself, until the transition to chaos takes place.

By using eq.(2.12), the Ljapunov exponent of this map can be calculated as

$$\lambda = \langle d\phi_{n+1}/d\phi_n \rangle$$

$$= \ln(2\pi\mu B) + \langle \ln|\sin(\phi_n - \phi_0)| \rangle, \qquad (2.13)$$

where $\langle \cdots \rangle$ is a long-time average. As the second term on the right-hand side can be regarded as a quantity of the order of unity, λ becomes positive if $2\pi\mu B \gg 1$ is satisfied. Thus the onset of chaos is conditioned by

$$t_R\gamma \gg 1 \quad \text{and} \quad A^2B \gg 1. \qquad (2.14)$$

The latter inequality represents the situation that the input power A^2 exceeds the dissipation B^{-1} so that the system can no longer maintain the stationary state. Namely, A^2B plays a role similar to the Reynolds number in fluid systems.

The inequalities (2.14) are by no means the necessary condition for the occurrence of chaos. In fact, it has been proved that the transition to chaos can be seen also under another limiting condition $t_R\gamma \ll 1$ and $A^2B \ll 1$ [4,5]. Especially, the existence of differently polarized, but degenerate lights brings about chaos of the electric polarization under these conditions [5].

The above discussions for a ring resonator are well applied also to a gapless Fabry-Perot resonator ($\ell_1 = \ell_2 = 0$) [6]. In a Fabry-Perot resonator with gaps, on the other hand, transition of the torus-to-chaos type is expected to appear because of multiple time delay. It has been pointed out that this type of transition in fact occurs when the incident laser beam has a non-uniform intensity distribution in the transversal direction [7,8].

Under the condition that optical grating, i.e. spatial oscillation of the nonlinear refractive index, is induced in the medium, the Bragg reflection of light from it forms the origin of feedback. In such a case, the occurrence of chaos is possible without mirrors [9,10]. As to the theoretical studies of various instabilities and chaos in a ring resonator, see Ref.11.

The first experimental observation of optical turbulence was made by a group in Tucson. By using a hybrid optical device in which the feedback is given by an electric circuit instead of mirrors, they succeeded in observing period-doubling bifurcations and the transition to chaos [12]. Subsequently, similar observations were reported also in acoustic systems [13] and in acousto-optical systems [14]. The first experiment in an all-optical system was achieved by a group in Kyoto, who observed the transition from period-two to chaos in a ring resonator consisting of single-mode optical fiber [15]. Quite recently, a group in Edinburgh reported an observation of period-two and period-four in a ring resonator containing NH_3 which is regarded as an ensemble of two-level atoms [16].

3. Bifurcations and Chaos in Systems with Delayed Feedback

Since the system described by eq.(2.9) is regarded as a system of infinite degrees of freedom, it is impossible to describe all of its dynamical properties by a simple one-dimensional map like (2.12), though approximately. In fact, it is being revealed both theoretically and experimentally that the delay-differential equation of the class

$$\dot{x}(t) = -x(t) + \mu f(x(t-t_R)) \qquad (3.1)$$

exhibits such peculiar bifurcation phenomena that can never be seen in low-dimensional maps [17,18,19,20,21]. This section is devoted to giving the outline of them.

3.1. Hierarchy of Multistable Bifurcated States

As the bifurcation parameter μ is increased, the solution of eq.(3.1) undergoes a sequence of successive bifurcations and finally a transition to chaos. This phenomenon is well understood by referring to the bifurcation scheme of the corresponding difference equation

$$x(t) = \mu f(x(t-t_R)), \tag{3.2}$$

which in turn is directly connected with that of the map $x_{i+1}=\mu f(x_i)$.

The first bifurcation [17,18,19,20] At the first bifurcation point, the stationary solution of eq.(3.1) becomes unstable and a square-wave-like solution of period $\sim 2t_R$ ($\equiv T_0$) appears. This solution, called the fundamental solution, corresponds to the following solution of eq.(3.2): $x(t)=x_1$ for $t \in I^{(2p)}$ and $x(t)=x_2$ for $t \in I^{(2p+1)}$, where x_1 and x_2 are two values in a period-two solution of the map $x_{i+1}=\mu f(x_i)$, satisfying $x_2=\mu f(x_1)$ and $x_1=\mu f(x_2)$, and $I^{(q)}=(qt_R, (q+1)t_R]$ (p,q: integer). It can be proved that eq.(3.1) admits, in addition to the fundamental solution, its harmonics of odd degree as solutions. The domains in the parameter space (t_R,μ) that enable the harmonics to exist stably are shown in Fig.2. As can be seen in this figure, a number of domains overlap each other for $\mu > \mu_1$, if t_R is sufficiently large, indicating that the harmonics coexist there. The degree of the coexisting harmonics has an upper limit proportional to t_R. The counterparts in eq.(3.2) of the harmonic solutions are constructed as follows: $x(t)=x_1$ for $t \in I_{2j-1}^{(2p)}$ or $I_{2j}^{(2p+1)}$ and $x(t)=x_2$ for $t \in I_{2j}^{(2p)}$ or $I_{2j+1}^{(2p+1)}$, where $I_j^{(q)} = (qt_R+(j-1)t_R/k, qt_R+jt_R/k]$ $(j=1,2,\dots,k)$. As the period of this solution is $2t_R/k$, it corresponds to the k-th harmonic.

Higher-order bifurcations [18,20] With further increase of μ, each harmonic forms a route to chaos. On the fundamental branch, the transition goes through usual period-doubling bifurcations. On the higher-harmonic branch, on the other hand, a cascade-like generation of multistable periodic solutions takes place at each bi-

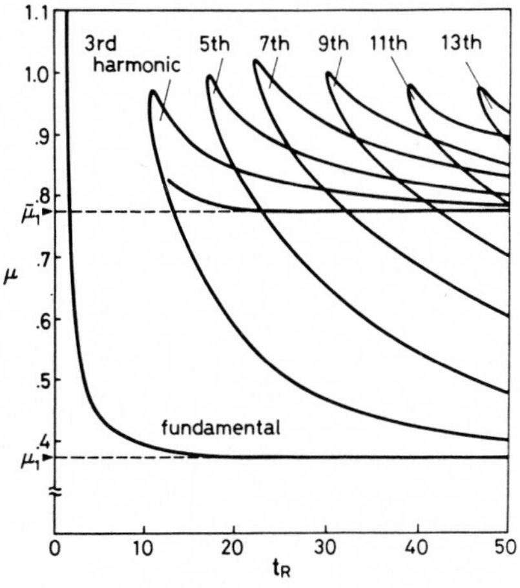

Fig.2. Existence domains of harmonic solutions in the parameter space (t_R,μ). The feedback function has been chosen as $f(x)=\pi(1-\sin x)$

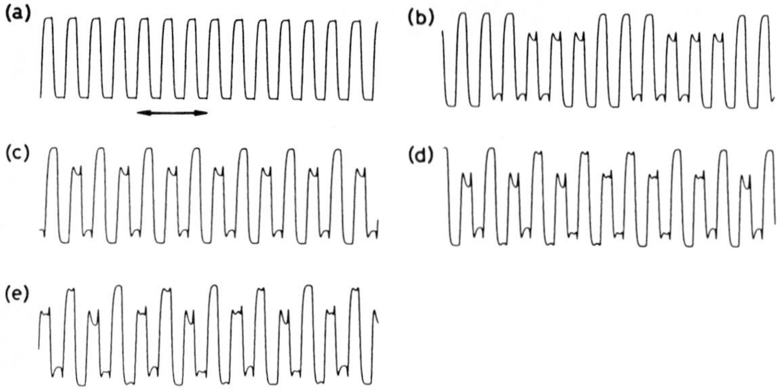

Fig.3. Various isomers for the third harmonic: (a) period-two, (b) period-four [111], (c) period-four [131], (d) period-eight [131], and (e) period-eight [175]. The horizontal arrow in (a) indicates the fundamental period, which is common for all cases

furcation point. For example, the third harmonic shown in Fig.3(a) can bifurcate into either Fig.3(b) or (c) at the second bifurcation point. This phenomenon can be explained by referring to the bifurcation scheme of eq.(3.2) as follows. Consider the third harmonic as an example. When μ is increased beyond the second bifurcation point, x_1 and x_2, the period-two solution of the map $x_{i+1}=\mu f(x_i)$ bifurcates into sets (x_1,x_3) and (x_2,x_4), respectively, making a period-four solution. Correspondingly, bifurcated solutions of eq.(3.2) can be constructed by assigning x_1 or x_3 to three subsections $I_1{}^{(2p)}$, $I_3{}^{(2p)}$ and $I_2{}^{(2p+1)}$. Each solution thus obtained is represented by the set of the values assigned to the above three subsections; for example, the solution $x(t)=x_1$ for $t\in I_1{}^{(2p)}$ or $I_3{}^{(2p)}$ and $x(t)=x_2$ for $t\in I_2{}^{(2p+1)}$ is represented by [113]. Some of possible 2^3 solutions then turn out identical with each other due to translation, so that only two solutions [111] and [131] are independent. These two solutions of eq.(3.2) correspond to the two bifurcated solutions shown in Fig.3(d) and (c), respectively. Bifurcated solutions of geometrically different forms are called *isomers*.

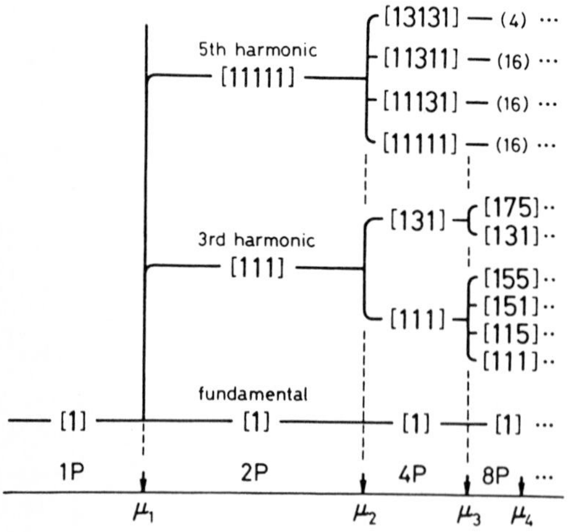

Fig.4. Hierarchy of multistable bifurcated states. The numbers in the round brackets are for possible independent period-eight isomers

A similar phenomenon appears at each higher-order bifurcation point. Beyond the third bifurcation point, for example, the above [111] and [131] solutions bifurcate into four and two isomers of period eight, respectively. Some examples are shown in Fig.3(d) and (e). Unlike the usual period doubling, the bifurcation of isomers does not proceed without limit. At a certain step of period doubling it terminates and every isomer breaks into a chaos. Figure 4 is a diagram illustrating how the bifurcations of isomers successively proceed. It shows that *the route to chaos itself exhibits a series of successive bifurcations.*

3.2. Merger of Multistable Chaotic States [17,19,20]

For the difference equation (3.2), the phenomenon of inverse bifurcations is known: an infinite number of chaotic bands generated at the Feigenbaum point merge, with further increase of μ, into fewer bands successively. Correspondingly, the multistable chaotic isomers in eq.(3.1), which appear at some terminal bifurcation point and still maintain traces of periodic structure, merge into fewer chaotic isomers with smaller periodicity. For example, suppose that each of period-eight isomers breaks into a chaotic solution at some termination point. The chaotic solution just appeared still maintains traces of the isomer out of which it has been generated. This skeleton structure of chaotic solutions inversely bifurcates like $8P \to 4P \to 2P$, reducing the number of coexisting isomers. Finally, chaotic solutions with period-two skeleton remain for *each* harmonic. Then comes the final stage at which these solutions merge into a single *developed chaos* which has no traces of periodic structure any more. This final merging point is represented by $\bar{\mu}_1$ in Fig. 2. It corresponds to the final inverted bifurcation point of the map $x_{i+1}=\mu f(x_i)$ at which the last two chaotic bands merge into a single one.

Finely seen, merger of chaotic harmonics into a developed chaos does not take place at a single point of μ. It is accomplished through a series of fine transitions called higher-harmonic bifurcations. Let us start from the chaos of the k-th harmonic (k: odd) and increase μ slowly in the vicinity of $\bar{\mu}_1$. Then, the period of the skeleton, T_0/k, changes discontinuously like $T_0/k \to T_0/(k+2) \to T_0/(k+4) \to \ldots$, ascending a staircase of harmonic chaos. A developed chaos does not appear until

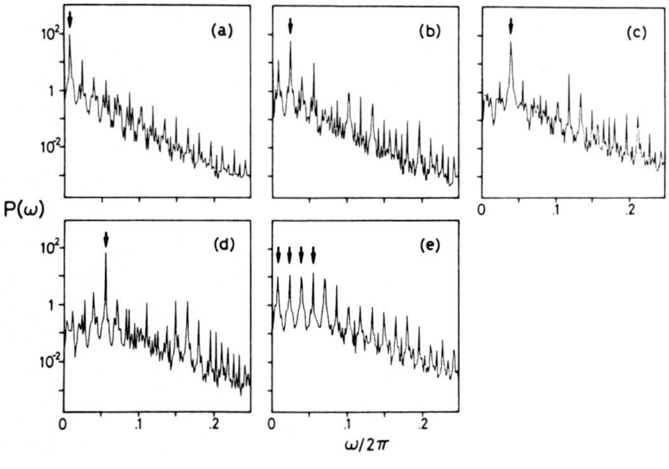

Fig.5. Power spectrum of harmonic chaos: (a) fundamental, (b) third harmonic, (c) fifth harmonic, (d) seventh harmonic, and (e) developed chaos

the harmonic chaos of the maximum degree has become unstable. Fig.5(a)-(d) show the power spectrum of each harmonic chaos which appears in the course of higher-harmonic bifurcations starting from the fundamental chaos. The position of the highest peak, represented by the arrow in the figure, shifts like ω_0 ($=2\pi/T_0 \cong \pi/t_R$) $\rightarrow 3\omega_0 \rightarrow 5\omega_0 \rightarrow 7\omega_0$. The chaos of the seventh harmonic breaks into a developed chaos, the power spectrum of which is shown in Fig.5(e). This spectrum has peaks of nearly equal height at the frequencies of all the harmonics that have appeared before. This implies that the orbit in a developed chaos wanders among the ruins of harmonic attractors with equal probability.

3.3. Fully Developed Chaos — Statistical Description [21]

When the feedback function f is a periodic function as that in eq.(2.11), another transition takes place for sufficiently large μ, converting the developed chaos into a *fully developed chaos* showing totally complicated time variation. The chaotic state at this final stage is so complicated that its statistical properties prove to obey rather simple rules. The dynamics in this state is quite different on two time scales: the longer one $t \gg t_c$ and the shorter one $t \lesssim t_c$, where t_c is a characteristic time of the order of $(\mu B)^{-1}$.

The longer time scale On the longer time scale, the second-order correlation function of x is approximated by

$$R(t) \equiv \langle (x(t)-\langle x\rangle)(x(0)-\langle x\rangle)\rangle$$
$$= R(0)\exp(-t), \tag{3.3}$$

so that the variation of x obeys a colored Gaussian process. This fact means that the delay force $f(x(t-t_R))$ in eq.(3.1) acts as a random force whose correlation time is smaller than t_c. The variance R(0), which is of the order of μB, is determined by the variation of x on the shorter time scale, as will be shown below.

The shorter time scale The variation of x is no longer Gaussian on the scale of $t \lesssim t_c$, but it largely depends on the shape of f. For f in eq.(2.11), it turns out possible to calculate the generating function

$$\Phi(\xi,t) = \langle \exp[i\xi(x(t)-x(0))]\rangle \tag{3.4}$$

by using the following stochastic model. Consider a semi-infinite series of coupled rotors each of which rotates on a unit circle. Let them be coupled to each other in the way that the angular frequency of the n-th rotor is equal to a projection of the vector of the (n+1)th rotor onto a certain direction, i.e.,

$$\dot{\psi}_n = \sin\psi_{n+1} \qquad (n=0,1,2,\ldots). \tag{3.5}$$

The initial phases $\psi_n(0)$ are taken at random. Then, the generating function on the short time scale is given by

$$\Phi(\xi,t/2A^2B) = \langle\langle \cos[\xi(\psi_0(t)-\psi_0(0))]\rangle\rangle, \tag{3.6}$$

an ensemble average with respect to the initial phases. The variance R(0) is related to Φ as

$$R(0) = (2A^2B)^2 \int_0^\infty \Phi(1,t)dt. \tag{3.7}$$

From the above coupled rotor model it is also clear that the variation of x on the shorter time scale is non-Gaussian. In fact, the flatness factor

$$F^{(n)} \equiv \langle (x(t)-x(0))^{2n} \rangle / \langle (x(t)-x(0))^2 \rangle^n \tag{3.8}$$

calculated by using the above Φ approaches $(2n)!/2^n(n!)^2$ in the limit of $t \to 0$. This value is much smaller than $(2n-1)!!$, the value for a Gaussian model. It should be noted that the above rotor model is integrable; that is, the orbit is originally not chaotic, but quasi-periodic. The chaotic behavior is involved only through the initial phases taken at random.

4. Conclusion

Recent studies on optical turbulence which was found in nonlinear optical resonators have been reviewed. It has been shown that the underlying dynamics of this phenomenon is well described by a rather simple delay-differential equation and that it exhibits various phases that can never be seen in systems of a few degrees of freedom, such as successive bifurcations of new types, chaotic states of different complexities, and so on. These studies may also contribute to understanding the turbulent state in more complex infinite-dimensional systems such as fluids.

Physically, the occurrence of turbulence in optical systems can be attributed to a cascade-like generation of photons of new frequencies through nonlinear optical processes. The traditional nonlinear optics has been constructed on the assumption that the amplitude of those successively generated photons rapidly converges for higher-order processes. However, the discovery of optical turbulence demonstrates that this assumption breaks down under some conditions.

References

1. *Optical Bistability*, ed. by C.M. Bowden, M. Ciftan, and H.R. Robl (Plenum, New York, 1981); E. Abraham and S.D. Smith, Rep. Prog. Phys. 45, 815 (1982).
2. K. Ikeda, Opt. Commun. 30, 257 (1979); K. Ikeda, H. Daido, and O. Akimoto, Phys. Rev. Lett. 45, 709 (1980).
3. M.J. Feigenbaum, J. Stat. Phys. 19, 25 (1978); and ibid. 21, 669 (1979).
4. K. Ikeda and O. Akimoto, Phys. Rev. Lett. 48, 617 (1982); L.A. Lugiato, L.M. Narducci, D.K. Bandy, and C.A. Pennise, Opt. Commun. 43, 281 (1982).
5. H.J. Carmichael, G.M. Savage, and D.F. Walls, Phys. Rev. Lett. 50, 163 (1983).
6. W.J. Firth, Opt. Commun. 39, 343 (1981); E. Abraham and W.J. Firth, to be published.
7. J.V. Moloney, D.L. Kaplan, H.M. Gibbs, and R.L. Shoemaker, Phys. Rev. A 25, 3442 (1982).
8. P. Davis and K. Ikeda, in preparation.
9. H.G. Winful and G.D. Cooperman, Appl. Phys. Lett. 40, 298 (1982).
10. Y. Silberberg and I. Bar-Joseph, Phys. Rev. Lett. 22, 1541 (1982).
11. J.C. Englund, R.R. Snapp, and W.C. Schieve, to appear in Prog. in Optics.
12. H.M. Gibbs, F.A. Hopf, D.L. Kaplan, and R.L. Shoemaker, Phys. Rev. Lett. 46, 474 (1981).
13. M. Kitano, T. Yabuzaki, and T. Ogawa, Phys. Rev. Lett. 50, 713 (1983).
14. J. Chrostowski, Phys. Rev. A 26, 2023 (1982).
15. H. Nakatsuka, S. Asaka, H. Itoh, K. Ikeda, and M. Matsuoka, Phys. Rev. Lett. 50, 109 (1983).
16. R.G. Harrison, W.J. Firth, C.A. Emshary, and I. Al-Saidi, Phys. Rev. Lett. 51, 562 (1983).
17. K. Ikeda, K. Kondo, and O. Akimoto, Phys. Rev. Lett. 49, 1467 (1982).
18. K. Kondo, K. Ikeda, and O. Akimoto, to be published.
19. F.A. Hopf, D.L. Kaplan, H.M. Gibbs, and R.L. Shoemaker, Phys. Rev. A 25, 2172 (1982).
20. M.W. Derstine, H.M. Gibbs, F.A. Hopf, and D.L. Kaplan, to be published.
21. K. Ikeda and O. Akimoto, paper presented at the 5th Rochester Conference on Coference and Quantum Optics, 1983.

Part IX

Anomalous Fluctuations

Scaling Theory of Relative Diffusion in Chaos and Turbulence

M. Suzuki

Department of Physics, Faculty of Science, University of Tokyo, Hongo
Bunkyo-ku, Tokyo 113, Japan

1. Introduction

The characteristic feature of chaos and turbulence is studied in the present paper from a general point of view of the scaling theory of transient phenomena proposed by the present author [1∿5]. For this purpose, we investigate here the relative diffusion in chaos and turbulence, namely

$$y(t) = \langle (x_1(t)-x_2(t))^2 \rangle \quad , \tag{1.1}$$

where $x_j(t)$ denotes the trajectory of the j-th particle. The essence of chaos and turbulence is characterized by the instability of trajectories and consequently by the exponential growing of the relative diffusion y(t) in the initial time region, for a small initial deviation y(o); namely

$$y(t) \simeq y(0)e^{\gamma t}; \ \gamma > 0 \quad . \tag{1.2}$$

The positivity of the growing rate (or Lyapunov exponent) γ expresses the instability of the system. In ordinary unstable situations [1∿4], such instability occurs at a special point. In chaos and turbulence, however, such unstable situations appear in a wide region (almost everywhere).

Now, not only the linear region corresponding to (1.2), but also the nonlinear region is very interesting from our view-point of the scaling theory [1∿5]. Furthermore, there is a possibility to classify chaotic or turbulent phenomena with use of the scaling function of the relative diffusion y(t) in the nonlinear scaling regime [1∿5].

2. Scaling Property of the Relative Diffusion in Turbulent Electric Fields

In this section we study how two charged particles separate in time in a fluctuating electric field E(x,t), when x and t denote the position and time, respectively. This problem is closely related to clumps [6,7] in turbulent plasma.

Our starting equation of motion is given by

$$\dot{x}(t) = v(t) \text{ and } \dot{v}(t) = \frac{q}{m}E(x(t),t) \quad , \tag{2.1}$$

where q and m denote the charge and mass, respectively. In our closed Gaussian-White approximation [5], we obtain the following nonlinear differential equation for the relative diffusion y(t)

$$\kappa^2\tau_0^3 \frac{d^3}{dt^3} y(t) = 1-\exp(-\kappa^2 y(t)) \quad , \tag{2.2}$$

where κ^{-1} denotes the correlation length ξ of electric field spatial fluctuations and τ_0 is the diffusion characteristic time [7] defined by

$$\tau_0 = (4\tau_c(\kappa qE/m)^2)^{-1/3} \quad . \tag{2.3}$$

Here τ_c denotes the effective correlation time of electric field temporal fluctuations, and E is the average magnitude of electric fields. The initial condition for (2.2) is given by

$$y(t) = y_0(t)+O(t^3);\ y_0(t) = (r(0)+g(0)t)^2 \quad , \tag{2.4}$$

where $r(0) = x_1(0)-x_2(0)$ and $g(0) = v_1(0)-v_2(0)$.

From our fundamental equation (2.2), we find that the solution y(t) of (2.2) shows exponential growing of the form

$$y(t) \simeq \kappa^{-2}\delta e^{t/\tau_0} \ , \ \delta = \frac{1}{3}(a^2+2ab+2b^2) \tag{2.5}$$

with $a \equiv \kappa r(0)$ and $b \equiv \kappa g(0)\tau_0$, in the initial time region. In the second non-linear region, we can show [5] that y(t) takes the following scaling form

$$y(t) \simeq \kappa^{-2} f^{(sc)}(\tau) \tag{2.6}$$

where τ denotes the scaling variable defined by

$$\tau = \delta e^{t/\tau_0} \quad . \tag{2.7}$$

For details of the scaling function, see Refs. 5 & 6. In the final time region, we have [6]

$$y(t) \simeq \frac{1}{6}\kappa^{-2} t^3 \tau_0^{-3} \quad . \tag{2.8}$$

Thus, the relative diffusion is characterized by the time τ_0 given by (2.3).

3. Relative diffusion in Turbulence

The purpose of the present section is to discuss phenomenologically the relation between the relative diffusion of two test particles and the energy spectrum E(k) in the Kolmogorov inertia range [8] of the form

$$E(k) \sim k^{-n} \quad ;\ n = \frac{1}{3}(5+\mu) \ , \tag{3.1}$$

where μ denotes the intermittency exponent [9].

Here we introduce a Gaussian stochastic model of turbulent velocity fields with the following correlation

$$\langle v(x,t)v(y,s)\rangle = v^2 S(\kappa^2(x-y)^2,\ \omega|t-s|), \tag{3.2}$$

where κ and ω denote the inverse spatial and temporal correlation lengths, respectively. Then, our stochastic model is defined by the following multiplicative process $dx_i(t)/dt = v(x_i(t),t)$. The relative diffusion y(t) defined by $y(t) = \langle(x_1(t)-x_2(t))^2\rangle$ satisfies the following equation

$$\frac{d}{dt}y(t) = 4\int_0^t ds\ \{C_{11}(t,s)-C_{12}(t,s)\} \ , \tag{3.3}$$

where $C_{ij}(t,s) = \langle v(x_i(t),t)v(x_j(s),s)\rangle$ and

$$C_{ij}(t,s) = \int_{-\infty}^{\infty} dx \int_{-\infty}^{\infty} dy \langle v(x,t)v(y,s)\delta(x-x_i(t))\delta(y-x_j(s))\rangle$$

$$\simeq \int_{-\infty}^{\infty} dx \int_{-\infty}^{\infty} dy \langle v(x,t)v(y,s)\rangle\langle\delta(x-x_i(t))\delta(y-x_j(s))\rangle$$

$$= v^2\langle S(\kappa^2(x_i(t)-x_j(s))^2 \; , \; \omega|t-s|)\rangle$$

$$\simeq v^2 S(\kappa^2\langle(x_i(t)-x_j(s))^2\rangle, \; \omega|t-s|) \quad . \tag{3.4}$$

Thus, we obtain

$$\frac{dy(t)}{dt} = 4v^2\int_0^t ds \; \{S(\kappa^2 y_{11}(t,s), \; \omega|t-s|)-S(\kappa y_{12}(t,s), \; \omega|t-s|)\} \tag{3.5}$$

with $y_{ij}(t,s) \equiv \langle(x_i(t)-x_j(s))^2\rangle$. If the correlation time $t_c \equiv 1/\omega$ is large, i.e., $\omega t \ll 1$, then we arrive at

$$\frac{d^2}{dt^2}\, y(t) \simeq 4v^2\{S(0)-S(\kappa y(t))\} \quad , \tag{3.6}$$

with $S(y) \equiv S(y,0)$. As the velocity correlation $S(y)$ is related to the energy spectrum (3.1), we obtain

$$S(y) \sim y^{\varphi} ; \quad \varphi = \tfrac{1}{2}(n-1) = \tfrac{1}{6}(2+\mu) \tag{3.7}$$

for large y. Therefore, we can derive the following asymptotic relative diffusion

$$y(t) \sim t^{2/(1-\varphi)} \sim t^{4/(3-n)} \sim t^{12/(4-\mu)} \tag{3.8}$$

for large t in the inertia range.

These results agree with those obtained more intuitively by HENTSCHEL and PROCACCIA [10]. For the Kolmogorov spectrum $n = 5/3$ (i.e., $\mu = 0$), we obtain $y(t) \sim t^3$, namely the Richardson t^3 law [8].

When $\omega t \gg 1$, we have to take into account the temporal correlation effect, by integrating (3.5) with respect to time explicitly. For this purpose, we assume that

$$S(y,t) = S(y)g(t/t_c(y)) \tag{3.9}$$

for a typical case. Then, we obtain approximately

$$dy(t)/dt \simeq v^2\{S(0)-S(\kappa^2 y(t))\}t_c(y(t)) \quad . \tag{3.10}$$

Here we assume [10] that $t_c(y) \sim y^{\varphi}$ for large y. Then we arrive at the conclusion [10]

$$y(t) \sim t^{1/(1-2\varphi)} \sim t^{4/(3-n-\mu)} \sim t^{3/(1-\mu)} \tag{3.11}$$

for $t \gg t_c$. This agrees with that obtained by TAKAYOSHI and MORI [11]. However, the result (3.8) may be compatible with Richardson's data [10].

4. Scaling Property of the Relative Diffusion in Chaos

The relative diffusions $y(t)$ for continuous time t and $y(n)$ for a discrete step n are also very useful to study the essential aspect of chaos. For simplicity, we discuss here the chaos of the following general difference equation

$$x(n+1) = T(x(n)) \tag{4.1}$$

for a mapping T. Then the relative diffusion of the two trajectories $x_1(n)$ and $x_2(n)$ is defined by

$$y(n,\delta) \equiv \langle (x_1(n)-x_2(n))^2 \rangle \tag{4.2}$$

with the initial difference

$$\delta^2 \equiv \langle (x_1(0)-x_2(0))^2 \rangle \quad . \tag{4.3}$$

In the initial time regime, the relative diffusion shows the following exponential growing

$$y(n,\delta) \simeq \delta^2 e^{2n\lambda} \quad , \tag{4.4}$$

for a chaotic regime, where λ denotes the Lyapunov exponent [12]. In this sense, the Lyapunov theory is a linear theory and our purpose is to extend his theory to the nonlinear region and to find the scaling property of the relative diffusion.

From our general arguments [1∿5], it is possible, in principle, to derive the scaling behavior

$$y(n,\delta) \simeq f^{(sc)}(\tau) \tag{4.5}$$

in the limit (so-called scaling limit) that

$$\tau \equiv \delta e^{n\lambda} = \text{fixed} \quad , \tag{4.6}$$

and

$$\delta \to 0 \text{ and } n \to \infty \quad . \tag{4.7}$$

If the domain of $x(n)$ is a finite region, the scaling function $f^{(sc)}(\tau)$ is shown to be periodic, which is a big contrast to other situations. In ordinary unstable phenomena [1∿4] such as laser, superradiance and spinodal decomposition, the scaling function $f^{(sc)}(\tau)$ is finite and monotonic, while it diverges monotonously in turbulence, as was shown before.

Thus, the type of scaling functions can clarify and classify the characteristic feature of unstable, chaotic and turbulent phenomena.

First, we discuss a simple exactly soluble map (ULAM and VON NEUMANN [12])

$$x(n+1) = 4x(n)[1-x(n)] \quad . \tag{4.8}$$

The solution of (4.8) is given by

$$x(n) = \sin^2(2^n \sin^{-1}\sqrt{x(0)}) \quad . \tag{4.9}$$

From this expression, we obtain the following scaling result

$$f^{(sc)}(\tau) = 1-\cos\tau \quad , \tag{4.10}$$

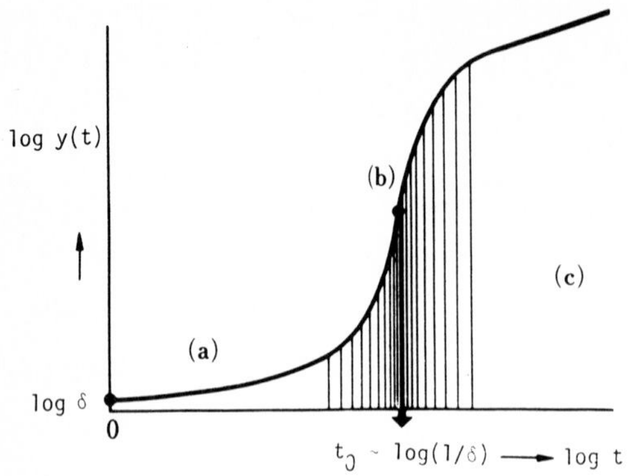

Fig. 1 Typical change of the relative diffusion y(t) with respect to time t: (a) initial regime, (b) nonlinear scaling regime and (c) final regime, where δ denotes the initial deviation of two trajectories

where

$$\tau = 2^n\delta; \ \delta^2 = \langle|\sin\sqrt{x_1(0)} - \sin\sqrt{x_2(0)}|^2\rangle \quad . \tag{4.11}$$

That is, the scaling function $f^{(sc)}(\tau)$ is periodic in the present case.

In order to discuss general situations, it will be useful to make use of perturbational expansions from an exactly soluble limit. It should be noted, however, that such a perturbational expansion is not always analytic. In fact, for the following map

$$x(n+1) = (4-\varepsilon)x(n)[1-x(n)] \quad , \tag{4.12}$$

the perturbational expansions of the solution and Lyapunov exponent λ are not analytic.

For example, the Lyapunov exponent shows the following singular behavior

$$\lambda(\varepsilon) = \log 2 + \lambda_1\sqrt{\varepsilon} \tag{4.13}$$

near $\varepsilon = 0$. This was also briefly mentioned in Ref. 12. Hopefully, a new expansion parameter $\sqrt{\varepsilon}$ will be useful and $\lambda(\varepsilon)$ might be expanded as

$$\lambda(\varepsilon) = \log 2 + \lambda_1\sqrt{\varepsilon} + \cdots \lambda_n(\sqrt{\varepsilon})^n + \cdots \quad , \tag{4.14}$$

and the scaling function might be expanded in a similar way. These will be reported in detail elsewhere.

5. Conclusion

The scaling property of relative diffusion is very useful to understand the essence of unstable, chaotic and turbulent behaviors.

Acknowledgements

The author would like to thank Professor Balescu, Professor H. Mori and Professor J. Krommes for useful discussions.

The present work is partially financed by the Scientific Research Fund of the Ministry of Education.

References

1. M. Suzuki, Prog. Theor. Phys. 56, 77, 477 (1976) and 57, 380 (1976)
2. M. Suzuki, J. Stat. Phys. 16, 11 and 380 (1977)
3. M. Suzuki, Adv. Chem. Phys. 46, 195 (1981)
4. M. Suzuki, Physica 117A, 103 (1983)
5. M. Suzuki, in "Global Analysis of Nonlinear Transient Phenomena near the Instability Point" in Statistical Mechanics and Chaos in Fusion Plasma,edited by C.W. Horton, Jr. and L.E. Reichl, John Wiley & Sons
6. H. Dupree, Physics Fluid 15, 334 (1972)
7. J.H. Misguich and R. Balescu, Plasma Physics 24, 289 (1982)
8. A.S. Monin and A.M. Yaglom, Statistical Fluid Mechanics. MIT Press, Cambridge, Vol. 1 (1971)
9. M. Nelkin, Phys. Rev. A9, 388 (1974)
10. H.G.E. Hentschel and I. Procaccia, Phys. Rev. 27A, 1266 (1983)
11. F. Takayoshi and H. Mori, Prog. Theor. Phys. 68, 439 (1982)
12. P. Collet and J.-P. Eckmann, Iterated Maps on the Interval as Dynamical Systems Birkhäuser, Boston, 1980

1/f Resistance Fluctuations

M. Nelkin

School of Applied and Engineering Physics and the Materials Science Center
Cornell University
Ithaca, NY 14853, USA

Abstract

The essential experimental properties of 1/f noise in solids are reviewed. The lack of universality and deviation from power law behavior suggested by recent experiments are emphasized. Theoretical questions relating to the equilibrium origin of 1/f resistance fluctuations, and to random walk models for excess low frequency noise are also discussed.

1. Introduction

When a small dc voltage V_0 is put across a resistor R in a constant current circuit, the voltage across the resistor fluctuates. The spectrum of these voltage fluctuations is observed to have the form

$$S_V(f) = 4kTR + V_0^2 S_1(f) , \tag{1}$$

where $S_1(f)$ is independent of the dc voltage V_0. If a small dc current I_0 is passed through the same resistor in a constant voltage circuit, the spectrum of current fluctuations is observed to have the form

$$S_I(f) = 4kT R^{-1} + I_0^2 S_1(f) , \tag{2}$$

with the same function $S_1(f)$ appearing in both (1) and (2). The first term in (1) and (2) is the equilibrium Johnson noise which is well understood. A naive use of Ohm's law suggests that

$$S_1(f) = S_R(f)/R^2 ,$$

describes resistance fluctuations. I will refer to the observed $S_1(f)$ as the spectrum of resistance fluctuations in the remainder of this paper. In a wide variety of materials [1][2], $S_1(f)$ is observed to vary approximately as 1/f over several decades of frequency, f. The phenomenon described by (1) and (2) in this case is known as 1/f resistance fluctuations. This is a special case of observed stochastic processes in nature whose spectral density varies approximately as 1/f. Most other examples are, however, much less well documented, and are probably hopeless to explain physically. I will concentrate on resistance fluctuations, and argue that we have enough experimental information to treat 1/f resistance fluctuations as a soluble problem in condensed matter physics.

I will first describe the principal experimental properties of 1/f resistance fluctuations with an emphasis on recent data. Of primary importance is the dependence of noise magnitude on material properties such as temperature and surface treatment. Also important are the indications that the dependence on frequency is not accurately of power law form. I will then discuss several theoretical questions of interest. I will consider whether resistance fluctuations can be observed at equilbrium without the application of a dc voltage, and discuss the connection to a proper statistical mechanical definition of equilibrium resistance fluctuations. I will then consider random walk models for hopping conduction, and discuss how excess low-frequency current fluctuations can arise in the presence of a dc current. Finally I will comment briefly on some recent theoretical explanations of 1/f noise.

2. Experimental Properties

It is convenient to introduce a dimensionless noise amplitude

$$A = f\,N\,S_1(f)\,, \tag{3}$$

where N is the total number of charge carriers in the sample. Experimentally A is independent of the dc voltage at low voltage, is weakly frequency dependent, is approximately independent of sample size, but varies strongly with temperature in some cases. A typical value of A is about 10^{-3}. Experimental information is available for carbon resistors, semiconductors, metal films, metal-insulator composites, and a variety of more complicated devices. That A is independent of the dc voltage suggests that the resistance is the property which is fluctuating. That it is independent of sample size suggests a volume effect. I return later to the temperature dependence. Some authors [2] consider the value of 10^{-3} to represent a universal property of bulk conducting material. I argue later that recent experiments strongly disagree with this interpretation.

Consider first the frequency dependence. Typically 1/f noise is observed over two or three decades of frequency above about 1 Hz. More recently a new ac technique [3] allows measurements over more than six decades above about 10^{-3} Hz. Most data are consistent with a spectrum proportional to $f^{-\alpha}$, with the exponent α varying in the range from 0.9 to 1.2. The exponent α depends on temperature for metal films [1], and varies from one material to another. Some more recent data [4] on silicon suggest a deviation from power law behavior. None of the data, however, show a flattening off at low frequencies. Any theoretical explanation must account for the long time scales observed.

The essential question in terms of material properties is universality. Is 1/f noise at typically observed levels a universal property of bulk conduction? Two recent experiments have given a negative answer. The Cornell group [3] has studied niobium films with a 1/f noise level below the level of detectability with its new sensitive ac technique. The Illinois group [4] has found that silicon wafers with a carefully cleaned surface also have 1/f noise below the level of detectability. This still leaves open the ubiquity of the phenomenon. Why is it almost always present, and why does the amplitude A usually have a "normal" value of about 10^{-3}? Before trying to answer this, I will discuss some more of the observed properties.

Silicon-on-sapphire wafers have a "normal" 1/f noise level [4]. Temperature-dependent features in the spectrum suggest a thermal activation mechanism. The observed correlation between the Hall fluctuations and resistance fluctuations suggest carrier number fluctuations. These are probably due to some kind of surface traps. In the semiconductor HgCdTe, recent experiments [5] show that the low frequency noise can be substantially reduced by sputtering, and then returns to its original value after exposure of the sample to air for about 30 minutes. Earlier suggestions [2] that the noise in semiconductors had a universal component due to bulk mobility fluctuations now seem doubtful.

Most metal films [1] have a "normal" 1/f noise level, and there is some experimental indication [6] that the effect in metals is a volume rather than a surface effect. In metals, where the number of carriers is comparable to the number of atoms, it is not plausible that number fluctuations due to trapping should be an important mechanism. Metal films typically show [1] a strong temperature dependence of the noise amplitude, and a systematic variation of the exponent α with temperature. An interpretation in terms of thermal activation is suggested, and is discussed further in the following section. A further clue is that the observed conductivity fluctuations are definitely not scalar [7]. There have been some proposals [8][9] of a particular defect motion which can lead to tensor mobility fluctuations, but this requires more detailed investigation.

Only in the special case of a stationary Gaussian process is a stochastic process fully characterized by the knowledge of its spectrum. To establish that a process is Gaussian, however, requires the measurement of an infinite number of higher order correlations. It is profitable to ask whether 1/f resistance fluctuations are consistent with a stationary Gaussian process. Earlier evidence was ambiguous [10], but pointed towards an affirmative answer. Recent experiments [11] confirm this. This suggests that 1/f noise is due to a linear superposition of contributions due to independent charge carriers, and that the Gaussian behavior is a consequence of the central limit theorem. It tells us nothing about the random process associated with the motion of an individual carrier. Unfortunately there does not seem to be any feasible experiment which can give us more information on this subject.

3. Qualitative Explanations

A random process with a single characteristic time τ would be expected to give a Lorentzian spectrum. A very old idea concerning 1/f noise is that an appropriately chosen distribution of characteristic times could give a linear superposition of Lorentzian spectra which is nearly 1/f. This idea has been recently extended in an interesting way [1]. Suppose that the characteristic time τ is associated with an activation energy E via

$$\tau = \tau_0 \exp\,(E/kT)\ , \tag{4}$$

where τ_0 is a microscopic time scale, and $E \gg kT$. Suppose further that the distribution D(E) of activation energies is broad with a width substantially greater than kT. It is then easily shown that the dimensionless noise amplitude A of (3) can be written as

$$A = C\ kT\ D(\tilde{E}), \tag{5}$$

where

$$\tilde{E} = -kT\ \ln(2\pi f \tau_0). \tag{6}$$

Equations (5) and (6) are a scaling law relating the frequency dependence and the temperature dependence of the noise amplitude. The measured temperature dependence of the noise amplitude at fixed frequency can be used to infer the distribution D(E). From this the temperature dependence of the effective exponent α can be calculated. This works in reasonable agreement with experiment for metal films [1], silicon wafers [4], and Josephson junctions [12].

Because of the logarithm in (6), temperature variation can scan the function D(E) much more efficiently than can frequency variation. In the limit that D(E) is constant, the distribution of activation energies is uniform, and the noise is 1/f with a weak temperature dependence. For a silver film [1] the observed D(E) is peaked at an energy of about 1 ev with a width of about 0.15 ev. Such characteristic activation energies suggest defect motion. The spread in activation energies could come from a locally noncrystalline environment in the neighborhood of the defects. A detailed proposal of a possible mode of defect motion has been made [8]. This is a promising direction for future investigation.

In general D(E) is expected to be a function with a broad maximum. Over a limited frequency range this will give a noise amplitude A, which is approximately of power law form as a function of frequency. Over a wide enough frequency range, however, the frequency dependence of A should reflect the actual shape of D(E). Since A is proportional to f S(f), the deviations of the spectrum from 1/f will not be accurately of power law form. With new techniques [3] that allow measurements over six decades of frequency these deviations from power law behavior may be directly observable. In any case they are important in principle.

An earlier explanation for 1/f noise in metal films was equilibrium temperature fluctuations [13]. The spectrum in this case is determined from a boundary value problem in heat diffusion. This gives a spectrum which flattens out at low frequency, has a weak temperature dependence, and has characteristic spatial correlations. Experimentally there is no flattening at low frequency, the temperature dependence is strong, and the noise [14] shows no spatial correlations. Thus equilibrium temperature fluctuations, though surely present, well understood, and dominant in certain special cases [15], are irrelevant as a general explanation of 1/f noise, even in the limited domain of metal films.

4. Equilibrium Resistance Fluctuations

Although equilibrium temperature fluctuations have not proven relevant to 1/f noise, a different proposal relating to equilibrium fluctuations [16] by the same authors is interesting and probably correct. Equations (1) and (2) suggest that the resistance of the sample is fluctuating. The applied dc voltage doesn't cause these resistance fluctuations, but only probes them. If this is so, how could these resistance fluctuations be observed at equilibrium with no applied votage? In an RC circuit the total Johnson noise power integrated over all frequencies is independent of the resistance, but the noise power integrated over a finite frequency band below the RC cutoff is proportional to the resistance. CLARKE and VOSS [16] proposed that this band-limited noise power should exhibit the same slow fluctuations at equilibrium that are probed by a dc voltage. More precisely let V(t) be the voltage filtered through a band-pass filter, and let $P(t) = [V(t)]^2$. Then the spectrum of fluctuations of P(t) divided by the square of the average value of P(t) should give the same function $S_1(f)$ measured by the usual nonequilibrium experiment

This is a difficult experiment since it requires measuring fluctuations of fluctuations, but it has been done in a few cases [16] in agreement with the Clarke and Voss idea. From a theoretical point of view the proposal is intuitively reasonable, but has no general statistical mechanical basis. Two special cases have been examined. The first is a nonlinear Langevin model such as would apply for equilibrium temperature fluctuations [17]. The second [18] is a random walk model [19] developed for the calculation of 1/f noise in the presence of traps. In both of these models the slow resistance fluctuations which are probed by a dc voltage also appear in the fluctuations of the band-limited Johnson noise power. This suggests that there should be a more general statistical mechanical basis for the Clarke-Voss idea, but we have not yet found it. In the second model [18] we can give a simple physical interpretation to the equilibrium resistance fluctuations. They correspond directly to the equilbrium fluctuations in the number of untrapped carriers.

It should be emphasized that the quantity measured in the equilibrium noise experiment is a rather subtle four time correlation function

$$\emptyset(t) = \langle V(0)\ V(t_1)\ V(t)\ V(t + t_2) \rangle \ . \qquad (7)$$

In (7) t_1 and t_2 are of the order of $1/\Delta\omega$, where $\Delta\omega$ is the bandwidth of the band-pass filter, and t is of the order of 1/f, where f is the frequency at which the resistance fluctuation is measured. In order for (7) to describe the fluctuations of interest we require the inequalities

$$0 < t_1 \ll t \quad \text{and} \quad 0 < t_2 \ll t \ .$$

These results put an important constraint on any mechanism for 1/f noise. Any theory which obtains 1/f noise from an intrinsically nonequilibrium mechanism associated with current flow is likely to be wrong. On the other hand it can be seen from (7) that 1/f noise at equilibrium can not be observed in the simplest statistical properties.

5. Random Walk Models

I turn now to explicit random walk models for hopping conduction. For independent carriers, the current fluctuations can be calculated in terms of the velocity correlation function, which can in turn be expressed in terms of the mean square displacement. The resulting expression is

$$S_I(f) = -2Ne^2 \omega^2 L^{-2} \int_0^\infty \cos \omega t \left\langle [x(t)-x(0)]^2 \right\rangle dt \ , \tag{8}$$

where N is the number of carriers, L is the length of the sample, $\omega = 2\pi f$, and x(t) is the x component of the position of a charge carrier at time t. Exploiting the equivalence between multi-state trapping models and continuous time random walk (CTRW) developed by KEHR and HAUS [20], we [19] adapted TUNALEY's [21] CTRW model for 1/f noise to calculate the current noise in terms of the distribution of trap release times and the density of traps. I will not give details here. The treatment is phenomenological, and puts in the distribution of trap release times by hand so as to give the observed nearly 1/f spectrum. The new feature is that the current noise is given as an explicit nonlinear functional of the trap density. The formalism automatically builds in the observed (I^2/N) dependence so that we can consider explicitly the magnitude of the current noise as a function of trap density. For low trap density the model reduces to the familiar linear superposition of Lorentzian spectra [1]. As the trap density is increased, the spectrum changes only slightly, and the noise magnitude at a fixed frequency saturates. This suggests a possible origin of the apparent universality of observed noise amplitudes. For sufficiently dirty samples the amplitude will be about the same, but for clean enough samples it can be much lower. There are some signs that the experimental picture is following this pattern.

From the theoretical point of view the preceding model does a very crude averaging of the properties of a disordered medium. The natural next step is to take the disorder seriously, and to calculate the mean square displacement for a biased random walk on a disordered lattice. The current fluctuations we want to study are due to the combined effects of disorder and the bias. Alternatively we could study an unbiased walk, but would have to calculate the four time current correlation analagous to (7). So far we have considered only a random walk on a one-dimensional chain with random jump rates [22]. There are no internal states corresponding to traps in this model system, and we do not expect to obtain a 1/f spectrum for the current fluctuations. It does allow us to calculate explicitly, however, the combined effects of disorder and a dc current on current fluctuations in a simple model system.

There is an extensive recent literature on random walks on disordered one-dimensional chains. Most of it deals with unbiased walks. The most important results for our purposes are summarized by MACHTA [23]. It is useful to distinguish between bond disorder and site disorder. For bond disorder there is a long time tail proportional to $t^{-3/2}$ in the velocity correlation function. For site disorder this long time tail is absent [24]. More recently there have been some papers [25][26] on the biased disordered random walk. These have considered only bond disorder, and have not calculated the mean square displacement. We [22] have calculated both bond and site disorder, but the site disorder problem is simpler, and we have written up the details of this case only. The primary new result concerns the combined effect of site disorder and a weak bias on the mean square displacement. This can be combined with (8) to calculate the spectrum of current fluctuations at low frequencies, which is given by

$$S_I(f) = 4kT\,R^{-1} + B\,I_0^{\,2}\,N^{-1}\,\Delta_c\,f^{-1/2}, \tag{9}$$

where Δ_c is the variance in the jump rates, and B is a constant. The phenomenological $(I_0^{\,2}/N)$ dependence is again automatically recovered, but the frequency dependence is $1/\sqrt{f}$, which does not agree with the observed 1/f dependence. Physically this is a long time tail effect of the familar sort, but for the site disorder problem it requires a bias in order to exhibit it. Related studies are continuing, and will be extended to more realistic configurations including internal states which act as traps.

6. Discussion

Finally I would like to discuss some other recent theoretical attempts to understand 1/f noise. Two recent papers by MARINARI et al. [27][28] have obtained a nearly 1/f spectrum from a one-dimensional random walk with a random bias. At each site there is a probability π to jump to the right, and a probability $1-\pi$ to jump to the left. This probability is selected from a uniform distribution in the range from 0 to 1. The mean square displacement of the random walker is found numerically to go as $\log^4 t$, in accord with theoretical expectations for this case. The spectrum of fluctuations of the displacement x(t) is found numerically to go approximately as $\log^4 f\ f^{-1}$, as suggested by a scaling argument. This is a very interesting result, but care must be taken in comparing it with the random walk models discussed in the preceding section of this paper. In the model of Marinari et al., x(t) can not be taken as the displacement of a charge carrier if one wants to recover agreement with the phenomenology of 1/f noise. It must be taken as some more abstract quantity such as the resistance of the sample. It is intriguing to search for physical models in which the resistance of a material undergoes this kind of random walk.

Since our subject this week is Chaos and Statistical Mechanics, it is fitting to close with a mention of a possible connection between 1/f noise and deterministic chaos. MANNEVILLE [29] has presented a very interesting discontinuous map which generates a strongly intermittent signal with a spectrum which is nearly 1/f. More recently PROCACCIA and SCHUSTER [30] have generalized this model and given a functional renormalization group treatment of a class of such maps. That the signals are far from Gaussian is not a serious problem. The observed nearly Gaussian 1/f voltage or current noise could be synthesized from a linear superposition of such independent events, and the Gaussian behavior could arise from the central limit theorem. Unfortunately this prevents one of the most interesting features of the deterministic models from being observable. A more serious problem is that there is no known physical mechanism for such deterministic models in electrical conduction. For the present I see no reason why these interesting papers should be relevant to any observed 1/f spectra.

To conclude, 1/f noise is seen frequently in nature. It has been carefully measured in the electrical resistance of a variety of solid state systems. For these systems recent measurements are beginning to yield a reasonable physical picture of what is happening. Elegant theory based on deep concepts of scale similarity seems to be essentially irrelevant. As a theorist I find this disappointing, but perhaps we need reminding from time to time that physics is still fundamentally an experimental science.

ACKNOWLEDGMENTS

This work was supported by the United States National Science Foundation through grant number DMR-8108328 and through the Cornell Materials Science Center. I would like to thank Alan Harrison, Walter Lehr, Jonathan Machta, Chris Stanton, and André-Marie Tremblay for their collaboration on work reported here. I would also like to acknowledge stimulating discussions with Joe Mantese, John Scofield, Watt Webb, and Mike Weissman about 1/f noise experiments.

REFERENCES

1. P. Dutta and P. M. Horn, Rev. Mod. Phys. 53, 497 (1981).
2. F. N. Hooge, T. G. M. Kleinpenning, and L. K. J. Vandamme, Rep. Prog. Phys. 44, 479 (1981).
3. J. H. Scofield and W. W. Webb, 7th International Conference on Noise in Physical Systems, Montpellier, France, May, 1983. Proceedings to be published by North Holland.
4. R. D. Black, P. J. Restle, and M. B. Weissman, Phys. Rev. B 28, August 15, 1983.
5. K. Zheng, K.-H. Duh, and A. van der Ziel, Montpellier Conference, (see ref. 3).
6. D. M. Fleetwood, J. T. Masden, and N. Giordano, Phys. Rev. Lett. 50, 450 (1983).
7. R. D. Black, W. M. Snow, and M. B. Weissman, Phys. Rev. B 25, 2955 (1982).
8. Sh. M. Kogan and K. E. Nagaev, Sov. Phys. Solid State 24, 1921 (1982).
9. R. D. Black, P. J. Restle, and M. B. Weissman, preprint, July, 1983.
10. M. Nelkin and A.-M. S. Tremblay, J. Stat. Phys. 25, 253 (1981).
11. P. J. Restle, M. B. Weissman, and R. D. Black, preprint, July, 1983.
12. R. H. Koch, Montpellier Conference (see ref. 3).
13. R. F. Voss and J. Clarke, Phys. Rev. B 13, 556 (1976).
14. J. H. Scofield, D. H. Darling, and W. W. Webb, Phys. Rev. B 24, 7450 (1981).
15. M. B. Ketchen and J. Clarke, Phys. Rev. B 17, 114 (1978).
16. R. F. Voss and J. Clarke, Phys. Rev. Lett. 36, 42 (1976).
17. A. -M. S. Tremblay and M. Nelkin, Phys. Rev. B 24, 2551 (1981).
18. C. Stanton and M. Nelkin, preprint, September, 1983.
19. M. Nelkin and A. K. Harrison, Phys. Rev. B 26, 6696 (1982).
20. K. W. Kehr and J. W. Haus, Physica (Utrecht) 93A, 412 (1978).
21. J. K. E. Tunaley, J. Stat. Phys. 15, 149 (1976).
22. W. Lehr, J. Machta, and M. Nelkin, preprint, September, 1983.
23. J. Machta, J. Stat. Phys. 30, 305 (1983).
24. J. W. Haus, K. W. Kehr, and J. W. Lyklema, Phys. Rev. B 25, 2905 (1982).
25. B. Derrida and R. Orbach, Phys. Rev. B 27, 4694 (1983).
26. B. Derrida, J. Stat. Phys. 31, 433 (1983).
27. E. Marinari, G. Parisi, D. Ruelle, and P. Windey, Phys. Rev. Lett. 50, 1223 (1983).
28. E. Marinari, G. Parisi, D. Ruelle, and P. Windey, Commun. Math. Phys. 89, 1 (1983).
29. P. Manneville, J. Physique 41, 1235 (1980).
30. I. Procaccia and H. Schuster, Phys. Rev. A 28, 1210 (1983).

Index of Contributors